T0201534

COOPERATIVE NETWORKING

COOPERATIVE NETWORKING

Editors

Mohammad S. Obaidat
Monmouth University, USA

Sudip Misra
Indian Institute of Technology, West Bengal, India

A John Wiley & Sons, Ltd., Publication

Library of Congress Cataloging-in-Publication Data

Cooperative networking / [edited by] Mohammad S. Obaidat, Sudip Misra.
      p. cm.
   Includes bibliographical references and index.
   ISBN 978-0-470-74915-9 (cloth)
   1. Internetworking (Telecommunication)   2. Peer-to-peer architecture (Computer networks)   3. Ad hoc networks
(Computer networks)   I. Obaidat, Mohammad S. (Mohammad Salameh), 1952-   II. Misra, Sudip.
   TK5105.5.C675 2011
   004.6 – dc22
                                                                                               2011007852

A catalogue record for this book is available from the British Library.

Print ISBN: 9780470749159
ePDF ISBN: 9781119973591
oBook ISBN: 9781119973584
ePub ISBN: 9781119974277
Mobi ISBN: 9781119974284

Set in 10/12pt Times Roman by Laserwords Private Limited, Chennai, India

This book is dedicated to our parents and families
Mohammad S. Obaidat and Sudip Misra

Contents

About the Editors

Professor Mohammad S. Obaidat (Fellow of IEEE and Fellow of SCS) is an internationally well known academic/researcher/scientist. He received his PhD and MSc degrees in Computer Engineering with a minor in Computer Science from The Ohio State University, Columbus, Ohio, USA. Dr Obaidat is currently a full Professor of Computer Science at Monmouth University, NJ, USA. Among his previous positions are Advisor of the President of Philadelphia University (Jordan), Chair of the Department of Computer Science and Director of the Graduate Program at Monmouth University and a faculty member at the City University of New York. He has received extensive research funding and has published numerous books and numerous refereed technical articles (500 articles as of date of publication of this book) in scholarly international journals and proceedings of international conferences, (about 500 refereed papers as of today), and is currently working on three more books. Professor Obaidat has served as a consultant for several corporations and organizations worldwide. He is the Editor-in-Chief of the Wiley International Journal of Communication Systems, the FTRA Journal of Convergence and the KSIP Journal of Information Processing. He served as an Editor of IEEE Wireless Communications from 2007–2010. Between 1991–2006, he served as a Technical Editor and an Area Editor of Simulation: Transactions of the Society for Modeling and Simulations (SCS) International, TSCS. He also served on the Editorial Advisory Board of Simulation. He is now an editor of the Wiley Security and Communication Networks Journal, Journal of Networks, International Journal of Information Technology, Communications and Convergence, IJITCC, Inderscience. He served on the International Advisory Board of the International Journal of Wireless Networks and Broadband Technologies. Professor Obaidat is an associate editor/editorial board member of seven other refereed scholarly journals including two IEEE Transactions, Elsevier Computer Communications Journal, Springer Journal of Supercomputing, SCS Journal of Defense Modeling and Simulation, Elsevier Journal of Computers and EE, International Journal of Communication Networks and Distributed Systems, The Academy Journal of Communications, International Journal of BioSciences and Technology and International Journal of Information Technology. He has guest edited numerous special issues of scholarly journals such as IEEE Transactions on Systems, Man and Cybernetics, SMC, IEEE Wireless Communications, IEEE Systems Journal, SIMULATION: Transactions of SCS, Elsevier Computer Communications Journal, Journal of C & EE, Wiley Security and Communication Networks, Journal of Networks, and International Journal of Communication Systems, among others. Obaidat has served as the general chair, steering committee chair, advisory Committee Chair and program chair of numerous international conferences. He is the founder of the International Symposium on Performance Evaluation of Computer and Telecommunication Systems, SPECTS and has served as the General Chair of SPECTS since its inception. He is also the founder of the International Conference on Computer, Information and Telecommunication Systems, CITS. Obaidat has received a recognition certificate from IEEE. Between 1994–1997, Obaidat has served as distinguished speaker/visitor of IEEE Computer Society. Since 1995 he has been serving as an ACM distinguished Lecturer. He is also an SCS distinguished Lecturer. Between 1996–1999, Dr. Obaidat served as an IEEE/ACM program evaluator of the Computing Sciences Accreditation Board/Commission, CSAB/CSAC. He has served as the Scientific Advisor for the World Bank/UN Digital Inclusion Workshop- The Role of Information and Communication Technology in Development. Between 1995–2002, he has served as a member of the board of directors of the Society for Computer Simulation International. Between 2002–2004, he has served as Vice President of Conferences of the Society for Modeling and Simulation International (SCS). Between 2004–2006, Professor Obaidat has served as Vice President of Membership of the Society for Modeling and Simulation International SCS. Between 2006–2009, he has served as the Senior Vice President of SCS. Currently, he is the President of the Society for Modeling and Simulation International (SCS). One of his recent co-authored papers has received the best paper award in the IEEE AICCSA 2009 international conference. He also received the best paper award for one of his papers accepted in IEEE GLOBCOM 2009 conference. Dr. Obaidat received very recently the Society for Modeling and Simulation Intentional (SCS) prestigious McLeod Founder's Award in recognition of his outstanding technical and professional contributions to modeling and simulation. He received in Dec 2010, the IEEE ComSoc- GLOBECOM 2010 Outstanding Leadership Award for his outstanding leadership of Communication Software Services and Multimedia Applications Symposium, CSSMA 2010.

He has been invited to lecture and give keynote speeches worldwide. His research interests are: wireless communications and networks, cooperative networking, telecommunications and Networking systems, security of network, information and computer systems, security of e-based systems, performance evaluation of computer systems, algorithms and networks, high performance and parallel computing/computers, applied neural networks and pattern recognition, adaptive learning and speech processing. Recently, Prof. Obaidat has been awarded a Nokia Research Fellowship and the distinguished Fulbright Scholar Award. During the 2004/2005, he was on sabbatical leave as Fulbright Distinguished Professor and Advisor to the President of Philadelphia University in Jordan, Dr. Adnan Badran. The latter became the Prime Minister of Jordan in April 2005 and served earlier as Vice President of UNESCO. Prof. Obaidat is a Fellow of the Society for Modeling and Simulation International SCS, and a Fellow of the Institute of Electrical and Electronics Engineers (IEEE). For more information, see http://bluehawk.monmouth.edu/mobaidat/.

Dr Sudip Misra is an Assistant Professor in the School of Information Technology at the Indian Institute of Technology Kharagpur. Prior to this he was associated with Cornell University (USA), Yale University (USA), Nortel Networks (Canada) and the Government of Ontario (Canada) in different capacities. He received his PhD degree in Computer Science from Carleton University, in Ottawa, Canada, and the masters and bachelor's degrees respectively from the University of New Brunswick, Fredericton, Canada, and the Indian Institute of Technology, Kharagpur, India. Dr Misra has several years of experience working in the academia, government and the private sectors in research, teaching, consulting, project management, architecture, software design and product engineering roles.

His current research interests include algorithm design for emerging communication networks. Dr Misra is the author of over 120 scholarly research papers and book chapters. He has won *six research paper awards* in different conferences. He was also the recipient of several academic awards and fellowships such as the *Young Scientist Award* (National Academy of Sciences, India), *Young Systems Scientist Award* (Systems Society of India), *Young Engineers Award* (Institution of Engineers, India), *(Canadian) Governor General's Academic Gold Medal* at Carleton University, the *University Outstanding Graduate Student Award* in doctoral level at Carleton University and the *National Academy of Sciences, India – Swarna Jayanti Puraskar* (Golden Jubilee Award).

He was also awarded the Canadian Government's prestigious *NSERC Post Doctoral Fellowship* and the *Humboldt Research Fellowship* in Germany. Dr Misra is the *Editor-in-Chief* of two journals – the *International Journal of Communication Networks and Distributed Systems* (IJCNDS) and the *International Journal of Information and Coding Theory* (IJICoT), UK. He has also been serving as the *Associate Editor* of the *Telecommunication Systems Journal* (Springer SBM), *Security and Communication Networks Journal* (Wiley), *International Journal of Communication Systems* (Wiley), and the *EURASIP Journal of Wireless Communications and Networking*. He is also an Editor/Editorial Board Member/Editorial Review Board Member of the *IET Communications Journal, Computers and Electrical Engineering Journal* (Elsevier), the *International Journal of Internet Protocol Technology*, the *International Journal of Theoretical and Applied Computer Science*, the *International Journal of Ad Hoc and Ubiquitous Computing, Journal of Internet Technology*, and the *Applied Intelligence Journal* (Springer).

Dr Misra has edited around six books in the areas of wireless ad hoc networks, wireless sensor networks, wireless mesh networks, communication networks and distributed systems, network reliability and fault tolerance, and information and coding theory, published by reputed publishers such as Springer and World Scientific.

He was invited to chair several international conference/workshop programs and sessions. He has been serving in the program committees of over a dozen international conferences. Dr Misra was also invited to deliver *keynote lectures* in over a dozen international conferences held in USA, Canada, Europe, Asia and Africa.

List of Contributors

Mohamed H. Ahmed
Memorial University of Newfoundland,
St. John's, Newfoundland,
A1B 3X5, Canada
mahmed@mun.ca

J. Barbancho
University of Seville,
EPS. Virgen de África, 7
41011 – Seville, Spain
jbarbancho@us.es

Abderrahim Benslimane
LIA/CERI University of Avignon,
339 Chemin des Meinajaries BP 1228,
84911 Avignon cedex 9, France
abderrahim.benslimane@univ-avignon.fr

Raouf Boutaba
University of Waterloo,
200 University west,
Waterloo, ON, Canada
rboutaba@uwaterloo.ca

Rajkumar Buyya
University of Melbourne,
Parkville, Victoria, 3010,
Australia
raj@csse.unimelb.edu.au

Ramón Agüero
University of Cantabria,
Plaza de la Ciencia s/n,
39005 Santander, Spain
ramon@tlmat.unican.es

R. Canal
Universitat Politècnica de Catalunya,
C/Jordi Girona 1-3, C6-107
08034 Barcelona, Spain
rcanal@ac.upc.edu

D. Cascado
University of Seville,
ETSII. Av. Reina Mercedes s/n
41012 – Seville, Spain
danic@atc.us.es

J. Chang
HP Labs
1501 Page Mill Road,
MS 1183, Palo Alto,
CA 94304, USA
Jichuan.chang@hp.com

Mehrdad Dianati
University of Surrey,
Guildford, GU2 7XH, UK
m.dianati@surrey.ac.uk

N. L. S. da Fonseca
State University of Campinas,
Av Albert Einstein 1251,
13083-852 Campinas SP, Brazil
nfonseca@ic.unicamp.br

Carol Fung
University of Waterloo,
200 University west,
Waterloo, ON, Canada
j22fung@uwaterloo.ca

Saurabh Kumar Garg
University of Melbourne,
111 Barry St. Carlton, Victoria, 3053, Australia
sgarg@csse.unimelb.edu.au

F. Granelli
University of Trento,
Via Sommarive 14, I-38123,
Trento, Italy
granelli@disi.unitn.it

Tarik Guelzim
Res El Hayat B1 BD Emile Zola APT#6,
Belvedere, Casablanca, 20300, Morocco
Tarik.guelzim@gmail.com

E. Herrero
Universitat Politècnica de Catalunya,
C/Jordi Girona 1-3, C6-E208
08034 Barcelona, Spain
eherrero@ac.upc.edu

Salama S. Ikki
University of Waterloo,
Waterloo, Ontario, N2M2C5, Canada
sikki@uwaterloo.ca

C. Khirallah
The University of Edinburgh,
Faraday Building, Mayfield Road,
Edinburgh, EH9 3JL, UK
C.Khirallah@ed.ac.uk

D. Kliazovich
University of Luxembourg,
6 rue Coudenhove Kalergi,
L-1359, Luxembourg
dzmitry.kliazovich@uni.lu

C. León
University of Seville,
EPS. Virgen de África, 7
41011 – Seville, Spain
cleon@us.es

A. Linares
University of Seville,
ETSII. Av. Reina Mercedes s/n
41012 – Seville, Spain
alinares@atc.us.es

J. Liu
Simon Fraser University,
8888 University Drive,
Burnaby, BC, V5A 1S6, Canada
jcliu@cs.sfu.ca

Sudip Misra
Indian Institute of Technology
Kharagpur – 721302,
West Bengal, India
smisra@sit.iitkgp.ernet.in

F.J. Molina
University of Seville,
EPS. Virgen de África, 7
41011 – Seville, Spain
fjmolina@us.es

Hassnaa Moustafa
France Telecom – Orange Labs Networks and Carriers,
38-40 rue General Leclerc,
92794 Issy le Moulineaux Cedex 9, France
hassnaa.moustafa@orange-ftgroup.com

Kshirasagar Naik
University of Waterloo,
Waterloo Ontario, N2L 3G1, Canada
S.Naik@ece.uwaterloo.ca

Mohammad S. Obaidat
Monmouth University,
W. Long Branch, NJ 07764, USA
obaidat@monmouth.edu

H. Rashvand
Advanced Communication Systems Ltd,
University of Warwick,
Coventry, CV4 7AL, UK
h.rashvand@warwick.ac.uk
rashvand.editor@gmail.com

Joel J. P. C. Rodrigues
Instituto de Telecomunicações,
University of Beira Interior,
Rua Marquês D'Ávila e Bolama,
6201-001 Covilhã, Portugal
joeljr@ieee.org

Sidi Mohammed Senouci
University of Bourgogne, 49 rue Mademoiselle Bourgeois,
58000 Nevers, France
Sidi-Mohammed.Senouci@u-bourgogne.fr

J. L. Sevillano
University of Seville,
ETSII. Av. Reina Mercedes s/n
41012 – Seville, Spain
jlsevillano@us.es

Xuemin (Sherman) Shen
University of Waterloo,
Waterloo Ontario, N2L 3G1, Canada
xshen@bbcr.uwaterloo.ca

Vasco N. G. J. Soares
Instituto de Telecomunicações,
University of Beira Interior,
Rua Marquês D'Ávila e Bolama,
6201-001 Covilhã, Portugal

Polytechnic Institute of Castelo Branco,
Av. do Empresário,
6000-767 Castelo Branco, Portugal
vasco.g.soares@ieee.org

G. Sohi
University of Wisconsin-Madison,
1210 West Dayton Street,
Madison, WI 53706-1685 USA
sohi@cs.wisc.edu

V. Stankovic
University of Strathclyde,
Royal College Building,
204 George Street, Glasgow G1 1XW, UK
vladimir.stankovic@eee.strath.ac.uk

L. Stankovic
University of Strathclyde,
Royal College Building,
204 George Street, Glasgow G1 1XW, UK
lina.stankovic@eee.strath.ac.uk

F. Wang
Simon Fraser University,
8888 University Drive,
Burnaby, BC, V5A 1S6, Canada
fwa1@cs.sfu.ca

K. Wu
University of Victoria,
P.O. Box 3055, Station CSc,
Victoria, BC, V8W 3P6, Canada
wkui@cs.uvic.ca

1

Introduction

Mohammad S. Obaidat[1] and Sudip Misra[2]
[1]*Department of Computer Science and Software Engineering, Monmouth University, NJ, USA*
[2]*School of Information Technology, Indian Institute of Technology, West Bengal, India*

Cooperative Networking is an important topic in emerging network technologies characterized by relatively high degrees of autonomy and self-dependent behaviour. Cooperative networking deals with how the different hosts in a resource-constrained communication network cooperate with each other to improve the overall network performance. Different issues are involved in cooperative networking – identifying the bottleneck resource, identifying the peers that when selected would improve the resource utilization, identifying the servers that are loaded and that should be avoided for downloading content at a certain time instant, security issues and so on. The topics that this book covers span these issues.

The issue of cooperation is not new. Successful instances of cooperation exist in biological, chemical, economic, social and telecommunication networks. Instances of cooperation in nature have motivated development of models of cooperation in telecommunication networks. In any telecommunication network, cooperation is important in different degrees to improve the network performance. However, there has been an increased interest in cooperation in the recent years with the growing attention to peer-to-peer networks, and ad-hoc and sensor networks, in which the network throughput largely depends on the degree of cooperation amongst the different nodes. As these technologies are viewed to be very promising for the future, it is expected that cooperative networking will remain an essential subject of interest. Short summaries of the rest of the chapters in this book are provided below. We should emphasize that these summaries provide simplified abstraction of concepts that are described in detail in the later chapters of this book. The summaries in some instances have text, terms, expressions or ideas that are borrowed from the respective chapters.

In Chapter 2, the fundamental issues with cooperation in communication networking are investigated. Today's cooperative networking is one of the leading topics of research around the world. It has huge contributions not only in academic areas such as biology, sociology and economy, but it also has direct applications in communications, robotics and military science. The chapter elaborately discusses the interaction of this field with distributed processing where heterogeneous nodes promise significant enhancement in the capability of the system as well as performance and potentiality. Cooperative communication gives an alternative method to make an advantage of existing network infrastructure by means of spatial diversity. The conviction of user operation consists of the concept of relay channel. The two issues of cooperative quality-of-service (CQoS) and cooperative data caching play a vital role in enhancing the network output, ability and utility.

Chapter 3 discusses the issue of cooperative diversity. Cooperative diversity has revealed an aspiring technique in modern wireless communication systems. According to the authors, the prime concept behind cooperative diversity is that the existing nodes between the pairs of ingress-egress nodes can transmit the signal from the ingress nodes to obtain multiple copies of the same signal at the egress node. This leads to excellent signal quality and amend coverage and acute capacity. Moreover, the authors have represented the main aspects of cooperative diversity including relaying techniques, combining methods and other cooperation schemes (other than cooperative diversity). Efficient algorithms and protocols are necessary to make it easier to accomplish cooperative diversity in order to be able to yield the advantage of cooperative diversity in resource constrained networks.

Chapter 4 reviews the issue of cooperation in Wireless Ad-Hoc Sensor Networks (WAdSN). Commonly, WAdSN are characterized by very small sized nodes having limited radio frequency range, low resources and autonomy. They communicate with one another by transmitting data over multi-hop pathways. However, in this case, collaboration is limited to a certain barter of information. The chapter introduces the new approach taking the network as a whole. It represents cooperation in WAdSN as a collaborative action where network nodes are implicated. Time synchronization, calibration and localization have been emphasized as issues requiring efficient mechanisms of cooperation. Time synchronization is required in wireless sensor networks for the sake of saving of limited energy resident in the nodes. Another issue where cooperation is important is localization, because location information is not only required for monitoring a given area, but can also be exploited to trace a mobile vehicle and animal, or to monitor elderly and disabled people in residencies. In this approach, while a node provides a measure with location information, data fusion techniques can reduce traffic and energy consumption.

Chapter 5 studies cooperation in autonomous vehicular networks. Since chronological advancement of wireless technologies are taking place regularly in all respects, autonomous vehicular networks have become like a new network technology comprising of vehicle-to-infrastructure and vehicle-to-vehicle communication. Cooperation in vehicular networks is categorized into two types: implicit and explicit. Implicit cooperation solicitudes the proficiency of MAC layer protocols for multi-hop communication and for adroit mechanism allowing trusted communication between different vehicles. The behaviours of the drivers focus on explicit communication, and allow the vehicles without having a specific need for a service access to participate in the communication to assist other vehicles that need relay nodes to allow them to access services. It is believed that cooperative techniques can be helpful to amend the enforcement of vehicular networks. Their application ranges from road safety to amusement and commercial.

Chapter 6 investigates the issue of cooperative overlay networking for streaming media content. Currently, media streaming has been recognized as having widespread applications in the networked world. Recently, peer-to-peer content delivery has emanated as one of the inspiring techniques to enable its large-scale deployment. The authors investigate various solutions propounded for peer-to-peer media streaming. The chapter divides the solutions into two categories: tree-based and mesh-based approaches. It states that these two may endure inefficiencies either due to the vulnerability caused by dynamic end-hosts or the efficiency-latency tradeoff. The chapter puts forward a cooperative mesh-tree design, named as *mTreebone*, which clouts both tree and mesh structures. Using simulation analysis and PlanetLab experiments, the authors show that cooperative hybrid solution exhibits superior performance.

Chapter 7 studies the issues of cooperation in Delay Tolerant Network (DTN) based architectures. The DTN architecture consolidates a store-carry-and-forward paradigm by overlaying a protocol layer, called bundle layer, that provides internetworking on heterogeneous networks (regions) operating on different transmission media. DTN is usually exploited in an environment categorized by sparse connectivity, frequent network partitioning, intermittent connectivity, long propagation delays, asymmetric data rates, and high error rates. DTN can be deployed in different kinds of challenged and resource constrained network environments including interplanetary networks, underwater networks, wildlife tracking networks, remote area networks, social networks, military networks, vehicular networks, among others. In this chapter, the authors present a recapitulation of the delay-tolerant networking epitome, including innovative network architecture for vehicular communications, called vehicular delay-tolerant network (VDTN). The chapter also sketches the recent advances related to cooperation on delay tolerant networks. The chapter also highlighted the importance of the nodes' cooperation to revive the delivery ratio, thereby improving the performance of VDTN networks.

The rapid increase in wireless technologies has led to the opening of fast technological areas that have a great impact on our lives. One of the important requirements in wireless technology is cooperation; it improves the network connectivity and also enhances the quality of service of the network. The ambient network architecture includes both scalability and flexibility and it has also the capability of firming up the connection between two networks. Chapter 8 presents the key aspects followed by ambient networks needed to interact with the heterogeneous access networks based upon the cooperation between two functionalities having relevance to 'Generic Link Layer (GLL)' and 'Multi Radio Resource Management (MRMM).' GLL is essential, as it can make comparison among different radio access technologies. On the contrary, MRMM is based on a decision-based scheme. After having gathered information from different sources, the most suitable path for communication is chosen. Additionally, the chapter introduces two ideas that would function in the ensuing wireless communication technologies.

Chapter 9 presents the issue of cooperation in intrusion detection networks. In today's era of advanced technology, we are mostly dependent on Internet-based applications such as email, web-browsing, social networks, remote connections, and online messaging. Concurrently, network intrusions and consequent loss of privacy and security are becoming serious issues for the Internet community. The intrusions are unwanted traffic or computer activities that are generally vicious and troublesome. As stated by the authors, this leads to Denial of Service (DoS), ID theft, spam and phising. Malicious pieces of code are used to succeed in attack goals. This chapter surveyed the cooperation schemes in Intrusion Detection Network. The authors have first classified network intrusions and IDSs according to their behaviour and the techniques they use. Some of the open challenges and future directions in cooperative intrusion detection networks are also discussed.

Chapter 10 reviews the issue of utilizing cooperative diversity in link layer over wireless fading channels. In this chapter, the authors discuss a link level retransmission scheme, named as Node Cooperative Stop and Wait (NCSW). The scheme exploits the inherent cooperative diversity belonging to a multi-user communication system, thereby improving upon the traditional stop-and-wait retransmission. The chapter explains how in conventional retransmission schemes the neighbour nodes remain virtually non-existent to the ongoing transmission between a sender and receiver nodes, whereas in the NCSW scheme, some of the neighbour nodes which may have enough resources may want to cooperate and assist the sender node in retransmission.

Chapter 11 presents a novel concept of cooperative network optimization that is based on inter-layer and inter-node communication. With this concept, protocols from the TCP/IP can be extended to fine tune their configuration parameter values continuously based on the past performance. As stated by the authors, compared to non-cooperative approaches, the results manifest that cooperation between layers of a protocol stack can bring major improvements in data transfer performance. The authors present an analysis of cooperative inter-node and inter-layer networking issues and their respective solutions.

Chapter 12 presents the topic of cooperative network coding (CNC). CNC is a fairly recent methodology which came into existence as a combination of concepts from both network coding and cooperative communications. It has become popular in the last decade or so with the popularity of the future Internet and wireless communications. The authors discuss how the issue of cooperation helps in increasing capacity and minimizing the effect of blackout and how network coding enables more efficient use of the network resources. In addition, this chapter discusses the fundamental issues and definitions underlying the concept of network coding. A summary of currently used cooperative relaying strategies, and different issues of performance is also given.

Chapter 13 reviews the different issues surrounding cooperative caching for chip multiprocessors. In data access patterns, caches are deliberately used to help minimize network activities and storage access latencies. The chapter explores the concept and effectuation of cooperative caching for modern CMPs (i.e., today's multi-core and tomorrow's many-core processors). Cooperative caching helps in supporting various cache sharing behaviours using different techniques relating to cooperative capacity sharing and throttling capabilities.

Finally, Chapter 14 presents a taxonomy and survey of market-oriented resource management and scheduling. Market-oriented computing is currently inevitable for both industry and academia. Grid computing is one of the most important concepts which supports market-oriented computing. Since the last decade, many researchers have investigated issues related to resource management and scheduling in utility grids, but still a lot of work needs to be done. Moreover, the chapter provides a comprehensive taxonomy summarizing the important works on this aspect. A survey of market-oriented resource management systems has been presented as well.

1.1   Major Features of the Book

Below are some important features of this book, which, we believe, would make it a valuable resource for our readers:

- This book is designed, in structure and content, with the aim of making it useful at all learning levels.
- The chapters of this book are authored by prominent academicians/researchers, and practitioners, with solid experience in the subject matter.
- The chapters' authors of the book are distributed across a large number of countries and institutions of worldwide reputation. This gives this book an international flavour.
- The chapter authors have attempted to provide a comprehensive bibliography, which should greatly help the readers interested in exploring the topics in greater detail.
- Throughout the chapters of this book, most of the groundwork research topics of cooperative networking have been covered from both theoretical and practical viewpoints. This makes the book particularly useful for industry practitioners working directly with the practical aspects behind enabling the technologies in the field.
- To make the book useful for pedagogical purpose, all chapters of the book are accompanied by a corresponding set of presentation viewgraphs. The viewgraphs can be obtained as a supplementary resource by contacting the publisher, John Wiley & Sons Ltd, UK.

1.2   Target Audience

The book is written by primarily targeting the student community. This includes the students of both undergraduate and graduate levels – as well as students having an intermediate level of knowledge of the topics, and those having extensive knowledge about many of the topics. To keep up this goal, we have tried to design the overall structure and content of the book in such a manner that makes it useful at all learning levels. The secondary audience for this book is the research community, in academia or in the industry. Finally, we have also taken into consideration the needs of those readers, typically from the industry, who wish to gain an insight into the practical significance of the topics, expecting to discover how the spectrum of knowledge and the ideas are relevant for real-life applications of cooperative networking.

1.3 Supplementary Resources

As previously mentioned, this book comes with sets of viewgraphs for each chapter, which can be used for classroom instruction by instructors who adopt the book as a textbook. Instructors are requested to contact the publisher for access to these supplementary resources.

1.4 Acknowledgements

We are extremely thankful to the authors of the chapters of this book, who have worked very hard to bring this unique resource forward to help students, researchers and community practitioners. We feel it is in context to mention that as the individual chapters of this book are written by different authors, the responsibility of the contents of each of the chapters lies with the concerned authors of each chapter.

We are also very thankful to the John Wiley & Sons publishing and marketing staffs, for taking special interest in publishing this book, considering the current worldwide market needs for such a book. In particular, we would like to thank Sarah Tilley, Anna Smart, Susan Barclay, Jasmine Chang and Gayatri Shanker who efficiently worked with us in the publication process. Special thanks go to our institutions, students and research colleagues who in one way or the other contributed to this book. Finally, we would also like to thank our families, for their patience and for the continuous support and encouragement they have offered during the course of this book project.

2

Fundamentals and Issues with Cooperation in Networking

Mohammad S. Obaidat and Tarik Guelzim
*Department of Computer Science and Software Engineering,
Monmouth University, NJ, USA*

2.1 Introduction

In recent years, wireless point-to-point networks such as ad hoc networks, sensor networks and mesh networks have received a considerable amount of research attention due to their increased applications in both military and civilian domains. A concrete military application would be, for example, a network composed of officers and soldiers who need to share common information, one implementation that takes into consideration the battery power of the mobile units is to centralize the data in the officers and to authorize access to the soldiers. Other research in this field has concentrated on improving cooperative caching in which systems exchange cache data to be reused by all systems in the network thus increasing the overall performance and reducing latency. Cooperation is also improving handover in 3G system architectures given the differences between technologies and mobility principles in 3GPP and non 3GPP networks. This allowed technology to be slowly introduced into the market while aiming at making it interoperable when the technology matures. Cooperative information architecture also applies to security in which many networks cooperate to assure the security of a system. These systems operate in a geographically widely distributed environment with the goal being to manage access to security among stakeholders [1–20]; see Figure 2.1.

2.2 Fundamentals of Cooperating Networks

4G networks can be defined as composite networks made of heterogeneous wireless networks. These networks include, but are not limited to:

- broadcast networks;
- wireless wide area networks (Cellular);
- wireless metropolitan networks (WiMAX);
- short range networks such as WLAN, PAN, and RFID.

Cooperative Networking, First Edition. Edited by Mohammad S. Obaidat and Sudip Misra.
© 2011 John Wiley & Sons, Ltd. Published 2011 by John Wiley & Sons, Ltd.

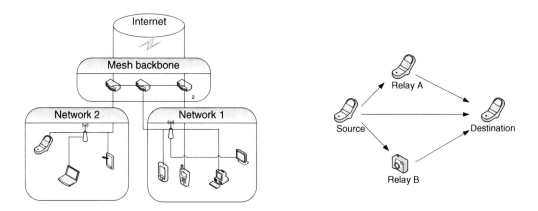

Figure 2.1   Example on cooperative networks.

Convergence between networks, terminals and services will be the main characteristic in 4G networks for both local and wide area setups. With this in mind, cooperation and cognition will become dominant features in the future of wireless networks.

Cooperative communication is a means to enhance the network performance through spatial diversity. Cooperative transmission can be useful for users with single antennas and where there are no dedicated relays. The changing topology and non-centralized nature of cooperative communication is particularly useful for MANET. Relay channel is the basic building block.

Unlike relay channels, in a user-cooperative model each of the cooperating users has data to transmit.

2.2.1   Cooperative Adhoc Network Services

2.2.1.1   Data Caching

Some studies have shown that sharing cache data between nodes can improve significantly the performance in P2P networks; see Figure 2.2. However, this technique is much at the research level leaving many architectural questions without answers. In [1], a new design scheme ensures the best use and sharing of the cache amongst nodes in an ad hoc environment.

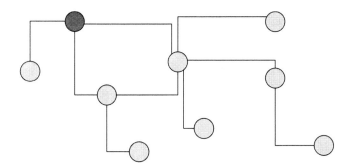

Figure 2.2   Ad hoc cache sharing between nodes.

The presented technique relies on two ideas:

- The cache data requests are transmitted to the cache layers on every node.
- The cache data replies are only transmitted to the cache layer at the intermediate nodes that need to cache the data.

This scheme has in fact helped reduce the overhead between the user space and the kernel space as well as allowing pipelining the data in order to reduce the end-to-end delay. The authors implemented the cooperative cache mechanism as a middleware that sits between the network layer and the application layer. Their implementation consists of:

- A cooperative cache agent (CCA): This is a module that maps application protocol messages to the corresponding cooperative cache layer messages.
- A cooperative cache daemon (CCD): This is a component that implements different cooperative cache mechanisms.
- A cooperative cache supporting library (CCSL): This is the core component that provides primitive operations of the cooperative cache such as checking packets, recording data access history and cache read and writes primitives.

2.2.1.2   Cooperative Quality of Service (CQoS)

Mobile ad hoc networks allow fast and temporary connections among mobile nodes without the help of any infrastructure [2]. In turn, ad hoc routing algorithms are categorized into three groups:

- Table driven protocols: These types of protocols maintain the routing information using tables that are exchanged among all participating nodes. In this case, the delay is significant since not all nodes are aware of all routes at the same instance of time T0. An example of such a protocol is Destination Sequence Distance Vector (DSDV).
- On demand protocols: These types of protocols are reactive in nature since they invoke a route determining process only when a route is needed [2]. Dynamic Source Routing (DSR) is one example of such a scheme.
- Hybrid protocols: These protocols try to find a compromise between the two paradigms above. One example protocol is Zone Routing Protocol (ZRP) which maintains routing information only within a zone and thus there no need to compute routes that are not going to be thoroughly explored.

Quality of Service routing algorithms have been subject to a lot of research in recent years. Core Extraction Distributed Ad hoc Routing (CEDAR) algorithm or the Ticket Based Probing (TBP) and many others like QoS routing based on bandwidth are two examples. Of course the purpose of these routing algorithms is to assure a good quality of service for every node on the same network, but they also suffer from some drawbacks such as the routing overhead and the inefficient use of power since every node has to send a hello packet to the neighbouring nodes. Another breed of protocol that remedies the current drawbacks are the multi-rate aware routing protocols that provide a higher throughput and lower delay by using the smallest path between hops. A version found in IEEE 802.11 networks using multi-rate schemes relies on the amount of timeslots rather than the number of hops. The proposed QoS routing scheme in [2] assumes that:

- The transmission power is the same at every node.
- The modulation scheme is chosen based on the received power strength.

The way this works is that while the distance of one hop decreases, transmitting at a higher bandwidth is most likely to be the chosen option since the required number of timeslots is reduced. It is necessary to transmit the data quickly to the requesting node while we preserve the timeslots needed to transmit more data to other requesting nodes.

2.2.1.3 Cooperative Data Integrity Insurance in Grid Networks

Data integrity has become one of the central concerns of large scale projects of distributed computing systems [3]. In order to maintain this integrity, the system must be resilient to different attacks and tampering methods. In [3] the authors developed a model of trust for grid participants based on the use of the associated feedback mechanism (AFM).

In order for the grid to be successful, the users of that grid must trust the results coming out of the grid computation. One proposed method to accomplish this task is to use system level assurance, in particular all data in and out of the grid. Most of the methods used today such as the Globus System [4] focus entirely on the communication and transport aspect of data integrity by employing authentication and encryption technologies to guarantee the safety of transit. While this method can obtain good results, however, a large scale distributing computing environment does not always protect against result tampering. One such example is the *SETI@Home* project, which experienced data integrity woes due to unknown and non trusted entities tampering with the computation process [5].

There are other models outside the realm of computer science that can be adapted to fit to solve this problem. One such model is the reputation system. There are three types of reputation systems:

- A positive reputation system (PRS): This system rewards good behaviour in order to encourage a desired outcome.
- A negative reputation system (NRS): In contrast to PRS, this model punishes undesirable behaviour.
- A hybrid reputation system (HRS): In this model, both PRS and NRS are accorded points with different weights.

The HRS is the default choice of many reputation systems since it gives the fluctuation of the nodes reputation in both directions. One concrete example is the ebay system. This balanced scheme has buyers and sellers affecting each other's reputations by giving feedback on the interaction experience.

The trustworthiness of a node in grid architecture is determined by speed, accuracy, availability and consistency.

2.2.1.4 Cooperative Intrusion Trace back and Response Architecture (CITRA)

CITRA was originally developed with funding from the DARPA initiative, as an infrastructure for integrating intrusion detection systems with firewalls and routers. This has proven to be very effective and useful in supporting automated security mechanisms that deal with the security of the systems they are protecting. CITRA was developed to help automate intrusion detection and analysis that are usually performed by human administrators. This is critical for two reasons:

- Manual analysis of intrusions can take hours or longer. The cost of resolving the issue rapidly mounts.
- The analysis is a complex task that requires expert administrators and professionals in certain cases. This is not usually available to most organizations.

CITRA uses two levels of organizations. The first is CITRA communities and administrative domain that is controlled by management components named Discovery Coordinator (DC). Second, communities are interconnected with adjacent devices, that is, no third CITRA node is placed between any two nodes. Trace back of intrusion events is done by auditing traffic at the registered devices within the CITRA network. This mechanism is so advanced that it is able to track the malicious source packets. Detection is done as follows:

1. The detector sends a trace back message to each CITRA neighbour.
2. Each boundary Controller (BC) and host along the potential path of an attack uses the network audit trail (AT) to determine if the packets associated with the attack passed through the BC node. If this condition holds, the trace back message is sent to the neighbour node.
3. This loop continues until reaching the source of an attack or an edge of the CITRA system.

Even though this mechanism is very effective in thwarting attacks, it has the disadvantage of causing Denial of Service (DoS) in cases where false alarms rose or if the attacker intentionally launches exploits on a network that has this mechanism deployed.

2.2.1.5 Cooperative Ad hoc Intrusion Analysis (CAIA)

In the previous section, we discussed a first generation intrusion detection mechanism that relies on controller nodes where each has a set of audit trails to watch and analyze for changes in a cooperative manner; however, this type of setup requires a large amount of data that can slow down networks and limit the throughput ratio to be exchanged. This means that this technique is only suitable for wired networks where the bandwidth is in the order of gigabytes. In most of today's wireless networks, this technique cannot be used as we are only dealing with 54 MB/s or so at most installations. In [6], the authors describe a model which relies on a cooperative distributed intrusion detection system where every node in the cooperating network participates in detecting and responding to network intrusions. CAIA takes into consideration two main intrusion issues: cache poisoning and malicious flooding. In the former, an adversary can compromise the information in the routing table through modifying its content or deleting information from it. Malicious flooding on the other hand can flood the whole network or a node with a large amount of data or control packets. Both intrusion attacks lead to DoS via consuming the node's resources along with its battery.

False alarms are considered as one of the main problems that IDSs face and can significantly make them less credible. In order to increase this efficiency, the authors in [6] propose a function that represents both attacks and maps the severity of an intrusion to the corresponding security class.

2.2.1.6 Cooperative Relaying in Ad hoc Networks

The author in [7] proposes a new cooperative relaying technique for delivering high-rate data in mobile ad hoc networks. This approach works as follows. Each node selects only a subset or sub-stream of the data stream to be transmitted. After that, all relay nodes forward their sub-streams simultaneously on the same physical layer. At the receiving end, multiple receive antennas separate the sub-streams and reassemble them based on their spatial characteristics and/or spreading codes, respectively. This approach has many advantages:

- Reducing transmission and reception processing requirements on each relay node.
- Providing significant savings in the overall transmit and receive energy, particularly in the high spectral efficiency regime.
- Attractiveness in situations where each relay can handle only low-rates due to limited resources in terms of energy, bandwidth, hardware and space (size).
- Providing an additional physical layer security mechanism.

Cooperative relaying is gaining attraction amongst network engineers for the benefits mentioned above. This approach employs several nodes as relays in a virtual and distributed antenna battery array to communicate data streams. To date the cooperative relaying techniques have primarily been proposed to achieve diversity gains as the authors in [8–13] suggest. All of these cooperative diversity approaches also known as (C-DIV) allow the network reliability and availability to be improved; however, spectral efficiency or rate is sacrificed to accommodate the diversity gain.

2.2.1.7 Cooperative Sensor Networks

Research in wireless sensor network (WSN) has grown at a very large pace in recent years especially in the field of security. Because of the inherent architecture of WSN, they have been prone to a lot of passive attacks such as stealing, physical damage or active attacks where a group of hackers try to actively exploit certain implantation flaws that are inherent within the system itself. A big effort has been made to create techniques that detect and thwart malicious attacks on the WSN. These are often called intrusion detection systems (IDS) which use multiple approaches and algorithms as we will discuss in the forthcoming sections of this chapter.

In wireless sensor networks, it is primordial that a monitoring scheme be put in place to detect, diagnose and protect attacks on the system. This is no easy task mainly because of two variables: the first is the environment the networks are deployed to operate in and second because of the technical specifications of the wireless sensors themselves that have many restrictions in terms of power consumption, memory requirements and CPUs. Mechanisms to secure such networks are usually different than those in 802.1X networks, because the restrictions mentioned above. As an illustrative example, in wireless local area networks, most of the installations secure their networks using a public key scheme where each node holds a public and private key. Although this is a very secure scheme[1], it cannot be implemented in the WSN architecture because sending multiple keys requires too much energy. The memory requirements to store such a key are also not efficient looking at the small memory that is installed on board of those sensors. In terms of CPU and processing time, calculating such keys every time we need to decrypt and encrypt messages is also unsuitable for such devices. Besides the fact that for the public scheme to function securely and to meet certain requirements such as military grade security, there must be deployment of a third party trusted authority; otherwise we cannot prove that the sensor network is receiving trusted and legitimate encryption keys. In the following section of this chapter we will present the recent advances in this field and how they are used to enhance the security of these networks.

---

[1] As of this writing, there has been no implementation of an attack that showed that the public key scheme is an inherently weak.

2.2.1.8   Characteristics of Wireless Sensor Networks (WSNs)

WSNs are the technology of the future. They are broadly used in a number of crucial and critical applications such as monitoring of seismic activities, performance analysis of manufacturing process and performance analysis, studies of wild life and detection as well as prevention in wild fires. One of the most relevant uses is military application, in which a network of these devices is put on place to track enemy movements as well as for spying purposes. Upon deployment, senor nodes use algorithms to self organize into a mesh of wireless networks and thus be able to collaborate together to extract information and process it together. Each node uses automatic discovery of surrounding nodes using peer-to-peer networking. Usually these are variations of multi hop and cluster-based routing algorithms, which in essence apply technique to dynamically discover resource within that network. Looking at the limitations described above as well as the characteristics of these networks, traditional security techniques are not enough if not inappropriate for such application, which implies that new mechanisms have to be put in place. The following section explains the security issues that are wildly demining these devices.

2.2.2   Cooperative Relaying Network Service

Cooperative communications and networking provides new paradigms of distributed processing and transmission. This gives new capabilities in terms of capacity and diversity gain in wireless networks. For instance, infrastructure networks, 3G networks and wireless sensor networks improve in terms of performance of both area coverage and quality of service (QoS). It also enables a distributed time-space signal processing that can be used in monitoring, localization, distributed measurement, fail over transparency, reduced network complexity and reduced energy consumption per node.

2.2.2.1   Delay Optimization in Cooperative Relaying

In [15], the authors proposed two delay optimization schemes based on the linear approximation of the channel phase and the strongest multi cellular OFDM system with cooperative relays. To achieve greater coverage and capacity, relaying has been widely suggested to be used in the future generation of wireless networks. There are two kinds of relaying techniques in use today. The first is relaying of the signal once after amplification and the second is relaying the signal after decoding. Figure 2.3 shows an example on cooperative relaying cellular networks.

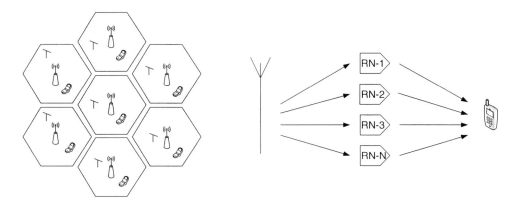

Figure 2.3   Cooperative relaying cellular networks.

Relaying stations can be mobile or stationary. By introducing random cyclic delays at the relay stations along with channel state information about the mobile unit, the best performance is given by the network. However, this technique comes at the expense of network overhead. To optimize this technique, the best segment of the signal is picked from each relay and taken into account to create the full signal, thus, only a fractional feedback is required. In addition, this technique lowers the system complexity and gives high spectrum efficiency.

2.2.2.2 MAC Protocol for a Distributed Cooperative Automatic Repeat reQuest (ARQ) Scheme in Wireless Networks

The ARQ scheme, [19], is made available through the persistent relay carrier sensing multiple access protocol (PRCSMA) and is a novel scheme that allows the execution of a distributed cooperative automatic retransmission request in IEEE 802.11 wireless networks. The underlying idea of this protocol is to enhance the performance of the WLAN MAC layer and to extend its coverage. ARQ works as follows: Once a destination station receives an erroneous packet, it requests its retransmission from any of the relays that 'overheard' the transmission at the first place. Space and time diversity can be used to select the 'best' candidate among the relays that will retransmit the requested packet. With such a scheme, we can improve channel usage and extend the coverage and retransmission of data. When using PRCSMA, every station must listen to all ongoing transmission in order to be able to cooperate on demand. This requires it to keep a copy of any received data packet. These packets are later deleted if they are acknowledged by the destination station. It is worth clarifying that the destination station denotes the next hop of the packet and not the 'final' destination of the transmitted packet. The copy of the transmitted packets are retained by the relays and stored at each station data buffer or in a different dedicated queue. Whenever a data packet is received with errors at the destination station, a cooperative phase can be started or initiated. Cooperation starts by broadcasting a Claim for Cooperation or CFC message in the form of a control packet after sensing the channel idle for a period of time. Cooperative network performance is enhanced by giving priority to cooperation data packets over regular data traffic.

2.2.2.3 Resource Sharing via Planed Relay for HWN

Multi-hop relaying is one way to employ multiple relays to serve a communication channel. It first appeared in the 1940s with an application focused on extending transmission range. Nowadays, it is used to increase network throughput and improve network reliability. In a cooperative relaying model, partnership can take different forms (e.g. multi-hop relaying) with different degrees of complexity [20].

Digital cellular standard GSM networks were first deployed in the 1990s with a 900 MHz band. Due to the uneven nature of the time varying special distribution, network performance metrics are not sufficient for today's wireless networks where more ad hoc features are being introduced. The authors in [20] present an improved version of the adaptive distributed cross layer routing algorithm (ADCR) for hybrid networks with dedicated relay stations (HWN). This technique allows a mobile terminal to borrow radio resources that are available thousands of miles away via secure multi hop relay networks. Each relay network is placed at an ideal location by the network engineers. Figure 2.4 depicts an example on hybrid network with fixed relay stations.

The above scheme has been shown to extend service rang, optimize cell capacity and minimize transmission power. It has also allowed 'shadow' areas to be covered and supported load balancing between networks [21]. In theory, when compared to regular cellular network, relays provide the capability to substitute poor single hop quality of the wireless medium and better link quality

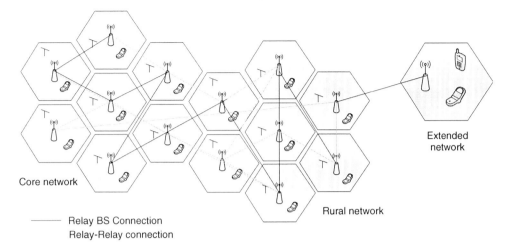

Figure 2.4   Hybrid network with fixed relay stations.

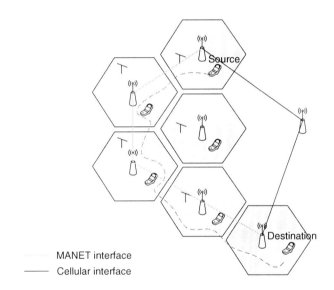

Figure 2.5   Multi hop combined transmission of cellular resource relaying using fixed RNs.

between towers. It also enables a higher end to end data rate by allowing two simultaneously communicating interfaces; see Figure 2.5.

The link between a relay network (RN) and a base station (BS) fosters better antenna gains. Nevertheless, this comes at the price of making network planning more complicated for the receiver design.

2.3   Issues and Security Flaws with Cooperating Networks: Wireless Sensor Networks Case Study

Cooperation has its benefits, but also has its flaws. Since cooperative networking is still in its infancy stage, the issues that are reported are only experimental ones. In order to bring the reader

Table 2.1 Typical Specifications of a
Small Ad hoc Node

Processor Unit	4–8 MHz
RAM	4 KB
Flash memory	128 KB
Radio Frequency	916 MHZ

to some real life security issues in cooperation, we detail in the following sections issues with a dominant cooperative network type: Wireless Sensor Networks (WSNs).

2.3.1 Limitations in Mobile Ad hoc Networks

Hardware Limitations
Security problems of wireless networks start at the physical level of the nodes i.e. their hardware specifications. Table 2.1 summarizes the specs of the nodes.

In addition, wireless sensor networks are deployed in a heterogeneous manner which renders engineering a security solution that works and performs in every one of them a very big challenge. Another issue that arises most often is the physical theft of those devices.

Network Contingents
In addition to the physical limitation of the nodes, often researchers look at the limitations that are imposed by the Ad-HOC network architecture and the insecurities of the wireless media.

Environment Limitations
As mentioned before, WSNs are usually deployed in public locations as well as adverse environments such as in military applications, which make them highly susceptible to theft and damage. Often, to overcome the hurdle of such problems, engineers use materials that can increase the node's resistance, but of course with the expense of extra deployment budget.

2.3.1.1 General Issues and Contingencies

As in any wireless computer network, enhancing operation and access security starts by defining goals as to what security encompasses. The following are the security goals of most forms of cooperative networks [22–23]:

- **Integrity:** There must be a mechanism by which we need to tell whether a message received has been altered by an unauthorized node.
- **Authentication:** This is the most basic form of security in which we need to determine the identity of the source of the information as well as the reliability of the sender node.
- **Confidentiality:** This is also an important security metric through which messages are encrypted using an encryption algorithm to keep the communication among the sensor network nodes private and eavesdrop resistant.
- **Availability:** As in any computer network, having 'healthy' nodes provides and ensures the normal flow of the data across the network without interruption.

The major problems in these networks can be summarized into four categories. Hackers and malicious users often use one of them or a combination of two or more when targeting weak nodes. The first to mention is the interruption technique. In this kind of attack, the hacker cuts off the communication link in the sensor network thus rendering the node unavailable. Usually this can be accomplished using methods such as code injection in which additional code is added maliciously to the device to be able to accomplish operations of interest to the hacker. The second mechanism that threatens the wireless network is the interception technique in which the hacker compromises the sensor network by capturing data and after analysis using its content to gain unauthorized access to the entire network; an example of such an attack is the node capture attack. The third method of gaining unauthorized access to the WSN is through modification of transmitted packets by tampering with their contents. This might lead to problems such as Denial of Service (DoS) attack and network flooding in which a large amount of unsupported data is consuming the network bandwidth. The fourth type of attack is the fabrication attack. Here, the malicious user inserts false information and data into the network and thus forces the sensor nodes to think that the network cannot be trusted anymore, that is, compromised.

The following section defines specific attack on wireless sensor networks:

Information Gathering
In WSNs, this is a type of passive attack that is usually implemented by a hacker who proves to have very powerful resources that enable him to extract and collect information on the fly and mine it for useful data that can allow access to the network such as keys, paraphrases and so on.

Node Insurrection
In this kind of attack, the malicious user leads a cracking attack on the entire network that leads to obtaining encryption keys as well as other cryptographic data and thus compromising the whole sensor network.

Sensor Node Injection
In this type of attack, the hacker injects a similar node to the network and lures peer nodes that it is legitimate. Hence, attack such as packet forwarding and data analysis can be implemented. Although this sounds like an easy attack to conduct, however, it is very hard in certain setups such as in military installations, for example. In addition, in order to lure other sensor nodes that the current 'injected' node is legitimate, this latter must obtain encryption keys to enter the network in the first place. This is not a trivial task to do, and often is retrained to the same contingents that WSNs suffer from and that is power limitations where the false node is also limited in terms of power and thus in terms of computational strength.

Faulty Sensor Node
In this kind of attack, the attacker targets the lead sensor node and injects code in it in order to force it to generate inaccurate information. Aggregate nodes such as the leader node are very essential to the integrity of the network as a whole. This is because they provide information about the performance of the network to other peer aggregate nodes. Any false information can force other nodes to restrain from sending information for example or switching operation states to match the current presented 'false' condition. This can bring the entire network to a halt state.

Node Outage

This is an extension to the above attack in which the leader sensor node ceases to function. Current protocols fail to provide immunity against such an attack because they do not provide a robust mechanism to detect it nor correct it. If a network becomes compromised, a human intervention is required in order to replace the compromised hardware for the network to continue functioning properly.

Traffic Mining

This is another attack in which the attacker employs tools to analyze the traffic between the nodes as well as extract information in terms of keys, encryptions algorithm parameters used and so on. This occurs in the early stage of conducting a successful attack of a sensor network.

Major Developed Attacks

As mentioned above attacks in wireless sensor networks are based on the inherent limitations of the existing infrastructure of the nodes as well as the existing protocols that govern them. The following are the known attacks that were conducted successfully on WSNs.

Routing Information Spoofing

Spoofing routing information is a very widely used attack against wireless sensor networks. To intrude such a network, the attacker manipulates the routing tables either in software by injecting loop records in the lead nodes, or hardware based, by positioning devices that are able to attract or repel network traffic. Both of these attack methods are aimed at shortening or extending the routing graph of the network and/or falsifying error messages. It is worth mentioning that implanting malicious hardware, or zombies, in the network is very hard to implement considering the possible security mechanism that are put in place. Moreover, a hacker must first obtain the keys necessary for the illegitimate device to be authorized to use the sensor network.

Selective Filtering

For this attack to be implemented, an attacker must first hijack a sensor node, either physically or remotely. After a successful compromise of the node, this latter is altered to control the information that is to be sent or forwarded. A malicious node can for example refuse to pass along the packets that it receives. This can bring the entire sensor network to a halt. Nevertheless, modern intrusion detection software as well as hardware can detect such zombie nodes and reroute the information around them. Once the IDS system flags the malicious nodes, it then informs all other nodes to retrain from either receiving or sending data packets from it.

Sybil Attack

The Sybil attack is one where the malicious node impersonates and acts as multiple, distinct nodes. We can view the wireless sensor network as a peer to peer (P2P) network also in which the reputation of a node determines its role within it. In P2P, the mapping between machine and identities is many-to-one. This implies that one user can declare multiple identities for the purpose of redundancy. By doing so, the malicious node can gain a substantial 'trust' inside the network and thus, use its new role to introduce more damage to the entire WSN. There are many counter-measures to this attack; one is to introduce validation mechanisms to the network. Another method is to enforce a one to one relationship between the sensor node and the identity of the node. This can be considered as a variant to wormhole attacks in which the malicious user attacks the network by tunneling messages through over low-latency links in order to confuse the routing mechanisms of the sensor networks. This is usually done by convincing the sensor nodes that are many hops down the line that they are currently closer than they think.

Hello Flood Attack

As mentioned earlier, leader nodes keep the integrity of the entire network. The Hello flood attack is an attack in which the malicious node sends a packet at high transmission power in order to appear as if it came from the lead node. Following the protocol, every node thinks that the sender node is the closest node, thus, they send all of their 'Hello' messages to that node, where in reality it is the farthest one. This presumably leads to wasting the node's energy and deteriorating the whole network.

2.4 Conclusions

Cooperative communication offers an alternative method to leverage existing network infrastructure by means of spatial diversity. Cooperation helps users to get the most from available resources. The philosophy of user cooperation is based on the theory of the relay channel. In this chapter, we have discussed the fundamentals of cooperative networks and the services that they present. We scratched the surface on some services that have been experimented in and are in use today such as CQoS and cooperative data caching. These two play an important role in enhancing network throughput and capacity as well as availability. Cooperation today is focused on layers 1, 2 and 3 of the OSI model allowing existing application (layers 6 and 7) to benefit from it transparently or with minimal change.

On the other hand, there are some security issues with cooperation that are inherent in wireless ad hoc networks. Some flaws that attackers can use to either provoke a DoS on the network or to leverage the capacity of their nodes to the detriment of others, but hogging cooperative resources or launching WAN level attacks.

References

[1] Jing Zhao, Ping Zhang and Guohong Cao, 'On *Cooperative Caching in Wireless P2P Networks*,' The 28th International Conference on Distributed Computing Systems, ICDCS 2008.

[2] Zae-Kwun Lee, Hwangjun Song, and Gyeong Cheol Lee, '*Cooperative QoS-aware Routing Algorithm based on IEEE 802.11 Multi-rateover Mobile Ad Hoc Networks*,' International Conference on Intelligent Information Hiding and Multimedia Signal Processing, IIH-MSP, pp. 1009–1012, 2008.

[3] Austin Gilbert, Ajith Abraham and Marcin Paprzycki, '*A System for Ensuring Data Integrity in Grid Environments*,' Proceedings of the International Conference on Information Technology: Coding and Computing (ITCC'04), Vol. 1, p. 435, 2004.

[4] V. Welch, F. Siebenlist, I. Foster, J. Bresnahan, K. Czajkowski, J Gawor, C. Kesselman, S. Meder, L. Pearlman, S. Tuecke, '*Security for Grid Services*,' ANL/MCS-P1024-0203, February 2003. In Proceedings, 12th IEEE International Symposium on High Performance Distributed Computing (HPDC'03), pp. 48–57, Seattle, Washington, USA, June 2003.

[5] D. Molnar, '*The Seti@Home Problem*', ACM Crossroads, September 2000, available: http://www.acm.org/crossroads/columns/onpatrol/september2000.html

[6] Hadi Otrok, Mourad Debbabi, Chadi Assi and Prabir Bhattacharya, '*A Cooperative Approach for Analyzing Intrusions in Mobile Ad hoc Networks*,' 27th International Conference on Distributed Computing Systems Workshops (ICDCSW'07), p. 86, 2007.

[7] Sang Wu Kim, '*Cooperative Spatial Multiplexing in Mobile Ad Hoc Networks*,' 2005 Mobile Adhoc and Sensor Systems Conference, pp. 387–395, Washington, DC, 2005.

[8] R. Pabst, et al., '*Relay-based deployment concepts for wireless and mobile broadband radio*,' IEEE Communications Magazine, pp. 80–89, Vol. 42, No. 9, Sept. 2004.

[9] A. Nosratinia, T. E. Hunter, and A. Hedayat, et al., '*Cooperative communication in wireless networks*,' IEEE Communications Magazine, pp. 74–80, Vol. 42, No. 10, Oct. 2004.

[10] A. Sendonaris, E. Erkip, and B. Azhang, '*Increasing uplink capacity via user cooperation diversity,*' Proc. of the 1998 IEEE International Symposium on Information Theory, ISIT 1998, Cambridge, MA, p. 156, Aug. 1998.

[11] R.U. Nabar, H. Bolcskei, and F.W. Kneubuhler, '*Fading relay channels: performance limits and space-time signal design,*' IEEE Journal on Selected Areas in Communications, pp. 1099–1109, Vol. 22, No. 26, Aug. 2004.

[12] J.N. Laneman and G.W. Wornell, '*Distributed space-time coded protocols for exploiting cooperative diversity in wireless networks,*' IEEE Transactions on Information theory, pp. 2415–2425, Vol. 49, No. 10, Oct. 2003.

[13] S. Barbarossa, G. Scutari, '*Cooperative diversity through virtual arrays in multihop networks,*' Proc. IEEE ICASSP, pp. IV-209-IV-212, April 2003.

[14] Huanyu Zhao and Xiaolin Li, '*H-Trust: A Robust and Lightweight Group Reputation System for Peer-to-Peer Desktop Grid,*' The 28th International Conference on Distributed Computing Systems Workshops, pp. 235–240, 2008.

[15] Slimane Ben Slimane, Bo Zhou, Xuesong Li, '*Delay Optimization in Cooperative Relaying with Cyclic Delay Diversity,*' EURASIP Journal on Advances in Signal Processing, Vol. 2008, 9 pages.

[16] M. Schwarz, W. R. Bennett, and S. Stein, '*Communications Systems and Techniques,*' IEEE Press, Piscataway, NJ, USA, 1996.

[17] J. G. Proakis, '*Digital Communications,*' McGraw-Hill, New York, NY, USA, 4th edition, 2001.

[18] J. H. Winters, '*Smart antennas for wireless systems,*' IEEE Personal Communications, Vol. 5, No. 1, pp. 23–27, 1998.

[19] J. Alonso-Z'arate, E. Kartsakli, Ch. Verikoukis, and L. Alonso, '*Persistent RCSMA: MAC Protocol for a Distributed Cooperative ARQ Scheme in Wireless Networks,*' EURASIP Journal on Advances in Signal Processing, 2008.

[20] Chong Shen, Susan Rea, and Dirk Pesch, '*Resource Sharing via Planed Relay for HWN,*' EURASIP Journal on Advances in Signal Processing, 2008.

[21] C. Shen, S. Rea, and D. Pesch, '*Adaptive cross-layer routing for HWN with dedicated relay station,*' in Proceedings of the International Conference on Wireless Communications, Networking and Mobile Computing (WiCOM '06), pp. 1–5, Wuhan, China, September 2006.

[22] M. S. Obaidat and N. Bourdriga, 'Security of e-Systems and Computer Networks,' Cambridge University Press, 2007.

[23] P. Nicopolitidis, M. S. Obaidat, G. Papadimitriou and A. S. Pomportsis, 'Wireless Networks,' Wiley, 2003.

3

To Cooperate or Not to Cooperate? That is the Question!

Mohamed H. Ahmed[1] and Salama S. Ikki[2]

[1]*Faculty Engineering, Memorial University, St. John's, NL, Canada*
[2]*Department of Electrical and Computer Engineering, University Waterloo, Ontario, Canada*

3.1 Introduction

The last two decades have witnessed substantial efforts in developing new wireless technologies that can provide higher data rates and better quality of service. Among the proposed technologies, there are two promising ones, namely *multiple-input multiple-output space-time processing* (MIMO-STP) and multi-hop relaying.

In MIMO-STP, multiple antenna elements are used at the transmitter and/or the receiver so that signals can be processed using time and space dimensions. MIMO-STP systems can provide many advantages including diversity gain, antenna array gain, directivity gain and multiplexing gain. However, the implementation of MIMO-STP systems requires the installation of multiple antennas, which is not feasible in some cases such as with mobile terminals and in wireless sensor networks.

Multi-hop relaying, on the other hand, provides nodes with better links (smaller path-loss and less shadowing), so that higher signal-to-noise ratio (SNR) can be obtained, which can be translated into higher data rates and/or better signal quality.

The synergy of these two technologies (MIMO-STP systems and multi-hop relaying) provides a new paradigm where most of the advantages of MIMO antennas can be obtained by node cooperation using signal relaying without using collocated multiple antennas. This new paradigm, called cooperative diversity, provides substantial improvement in the performance but it also brings some drawbacks, challenges and complexity issues as discussed below. In this chapter, we analyze the benefits, challenges and complexity issues of cooperative-diversity systems in order to gain a deeper insight into the performance and feasibility of cooperative-diversity systems in practice.

The remainder of this chapter is organized as follows. In Section 3.2, we present an overview of cooperative-diversity systems. Then, we discuss their main benefits and advantages in Section 3.3. In Section 3.4, the major challenges and complexity issues of cooperative-diversity systems are given. Finally, we conclude with a brief discussion answering the raised question *whether we should cooperate or not cooperate*.

Cooperative Networking, First Edition. Edited by Mohammad S. Obaidat and Sudip Misra.
© 2011 John Wiley & Sons, Ltd. Published 2011 by John Wiley & Sons, Ltd.

3.2 Overview of Cooperative-Diversity Systems

In this section, we discuss the main aspects of cooperative diversity including *relaying techniques, combining methods* and *other cooperation schemes* (other than cooperative diversity). Figure 3.1 shows a simplified model of a cooperative-diversity system, where the source transmits the signal to the destination directly and indirectly (through M cooperating relays). Then, the destination combines the multiple signal replicas to produce a more robust signal. Although Figure 3.1 shows a two-hop cooperative communication model, this model can be generalized to multi-hop scenarios (with three or more hops). A larger number of hops can lead to shorter links and better signal quality but this might increase the required amount of resources and the end-to-end transmission delay.

3.2.1 Relaying Techniques

There are several relaying techniques that can be employed by the cooperating relays. These techniques vary in the required signal processing, performance and implementation complexity as briefly discussed below.

3.2.1.1 Amplify-and-Forward

In *amplify-and-forward* (AF), relays simply amplify the received signal (including the desired signal and added noise as well.) Although the AF concept is simple, it is difficult to implement in TDMA systems because it requires the storage of a large amount of analog data. Also, AF suffers from the noise amplification problem which can degrade the signal quality, particularly at low SNR. The relaying gain (G) in AF can be a constant or a variable. The variable gain can be chosen to be inversely proportional to amplitude of the source-relay channel coefficient ($G = \frac{1}{|h_{SR}|}$) to eliminate the source-relay fading effect [1]. The main problem with this option is that the gain can take very large values, especially during deep fading conditions, which might lead to signal saturation and clipping distortion. Hence, the following variable gain is usually instead

$$G = \sqrt{\frac{1}{N + |h_{SR}|^2}},\qquad(3.1)$$

where N is the noise power. This gain is more practical and avoids the saturation and clipping distortion problem.

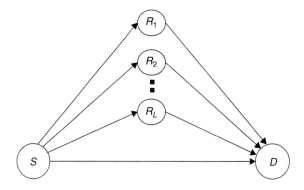

Figure 3.1 A multi branch dual-hop cooperative-diversity system.

3.2.1.2   Decode-and-Forward

In order to avoid the noise amplification problem, the *decode-and-forward* (DF) technique removes the noise by detecting and decoding the received signals and then regenerating and re-encoding the signal to be forwarded to the destination. However, DF suffers from the error propagation problem which may occur if the relay incorrectly detects/decodes a message and forwards this incorrect information to the destination. Thus, adaptive DF has been proposed, where the relay forwards the signal to the destination only if it is able to decode the signal correctly. The correct decoding can be checked using some error detection check or SNR threshold. A special case of the decode-and-forward technique is that when the relay detects the signal but does not decode it. In this case the scheme is called detect-and-forward although it is sometimes referred to in the literature as decode-and-forward.

3.2.1.3   Other Relaying Techniques

There are other relaying techniques that have been proposed in the literature. For instance, in *hybrid AF and DF*, the relay switches between AF and DF depending on the channel conditions [2]. In *demodulate-and-forward*, the relay demodulates (without detecting) the received signal and then forwards the signal to the destination [3].

In order to improve the efficiency of resource utilization in cooperative systems, some techniques such as *Incremental Relaying* [1] and *Best-Relay Selection* [4] are proposed in the literature.

The *Incremental relaying* protocols limit the cooperation to necessary conditions only by exploiting a feedback signal from the destination terminal indicating the success or failure of the direct transmission [1]. In the case of unsuccessful detection, one or more relays forward the signal to the destination. Otherwise, the relays do nothing and the source can send another signal. Therefore, the additional resource needed for relaying will be used only if the direct transmission is not adequate.

When multiple relays are available, the *Best-relay selection* is used to improve the resource utilization. In this case, the *best* relay only is selected to forward the signal to the destination [4]. Hence, two channels only are required in this case (regardless of the number of relays). However, it has been shown that a full diversity order can still be achieved [4].

Such techniques like incremental relaying and best-relay selection can be used with AF, DF or any other relaying method.

3.2.2   Combining Techniques

Cooperative-diversity systems use traditional diversity combining techniques such as *Maximum Ratio Combining* (*MRC*), *Equal Gain Combining* (*EGC*) and *Selection Combining* (*SC*).

In MRC, the received signals are weighted by the optimal weights ($w_i = h_i^*/N_i, \forall i$); where h_i^* and N_i are the complex conjugate of the effective channel coefficient[1] (between the source and the destination) and the average noise power of the ith signal, respectively. Then, the weighted signals are added coherently. In classical diversity techniques (e.g., space diversity), N_i can be assumed to be constant for all signals, and as a result N_i can be dropped from the optimal weight expression which reduces to h_i^*. However, in cooperative-diversity systems, N_i cannot be dropped because the average noise power is different for each signal. In fact there are several factors that affect the value of N_i in the indirect links such as the relaying technique (AF or DF), the relaying gain (for AF) and the relay-destination channel coefficient. It is known that MRC

---

[1] The effective channel coefficient for the indirect link is equal to '$G\, h_{SR}\, h_{RD}$' for AF and 'h_{RD}' for DF, where h_{RD} is the channel coefficient between the relay and the destination.

is optimal (maximizing the total SNR) in noise-limited networks. However, the main drawback of the MRC technique is that it requires full knowledge of the channel state information (CSI).

EGC can be used as a simplified suboptimal combining technique, where the destination combines the received copies of the signal by adding them coherently. Therefore, the required channel information at the receiver is reduced to the phase information only. EGC can be implemented in a differential mode so that the phase information is not required for detection. However, differential encoding must be employed in this case [5]. The combiner can be simplified more using SC, and the combiner simply selects the signal with the largest SNR. Although SC removes the overhead of the CSI estimation, its performance is inferior to MRC and EGC.

There are other combining techniques such as *Optimum Combining* (OC) that maximizes SNR in interference-limited networks [6] and *Generalized Selection Combining* (GSC), which is a generalization of SC technique, where N of M relayed signals are selected based on some signal-quality criterion [7]. However, to the best of authors' knowledge, both techniques (OC and GSC) are not analyzed in cooperative-diversity systems yet despite the importance of the two schemes.

3.2.3   *Other Cooperating Techniques*

In addition to the cooperative-diversity technique described above, there are other techniques for cooperation such as *cooperative space-time coding, superposition transmission* and *cooperative beamforming*.

3.2.3.1   **Cooperative Space-Time Coding**

In *cooperative space-time coding* (STC), relays cooperate with the source to form a space-time code (e.g., Alamouti scheme). There are various ways to implement cooperative STC in cooperative-diversity systems, but the main idea is that relays are used to replace the collocated multiple antenna elements required for STC. For instance, in [8] the source sends two symbols (S_1 and S_2) to two relays. Then, the two relays use these two symbols to form Alamouti code $\begin{pmatrix} S_1 & -S_2^* \\ S_2 & S_1^* \end{pmatrix}$ and send it to the destination as shown in Figure 3.2(a). This 2×1 Alamouti

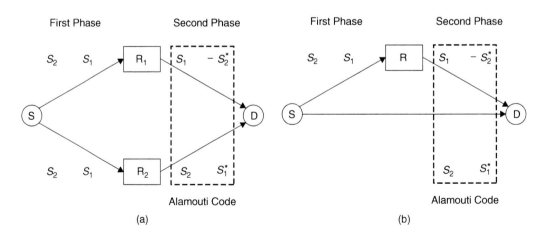

Figure 3.2 Implementation of space-time coding (Alamouti code) in cooperative wireless systems using (a) two relays (b) one relay only.

scheme can even be implemented using one relay only as shown in Figure 3.2(b). After the source sends the two symbols, the source and the relay can form together Alamouti code and send it to the destination. The same concept is extended to emulate 2×2 Alamouti using two relays and a source (base station) with 2 antennas [9]. In general, $M \times N$ STC code can be employed using M relays and a source with N antennas or $M + N-1$ relays when the source has a single antenna.

3.2.3.2   Superposition Transmission

In [10], the authors proposed a cooperation strategy between a pair of users. Each node sends its own signal to the BS. While each node sends its own signal, it listens to the transmitted signal of its partner. Of course, this implies a full duplex transmission, which is implemented using orthogonal channels using CDMA codes. Then, each node sends its own signal plus the previously detected signal of its partner. Results show a substantial increase in the data rate and a considerable reduction to the outage and error rates. However, the realization of this protocol is not straightforward due to the full-duplex transmission assumption and the need to acquire CSI (at least the phase value) at the source.

3.2.3.3   Cooperative Beamforming

In this scheme each cooperating node multiplies the signal (to be forwarded) by a specific weight so that the cooperating nodes can form a beam directed towards the destination. These weights are calculated under the assumptions that cooperating nodes know the exact location (or at least the direction) of the destination node and that each cooperating node can accurately estimate the channel between the source and itself. However, these two assumptions, especially the first one, might be hard to achieve in reality.

3.3   Benefits of Cooperative-Diversity Systems

Compared with classical direct transmission, cooperative diversity can offer several potential benefits. These benefits include signal-quality improvement, reduced power consumption, better coverage and capacity enhancement.

3.3.1   Signal-Quality Improvement

The signal quality is improved in cooperative diversity using three main mechanisms: *Diversity Gain*, *Link Improvement* and *Virtual Antenna Gain*.

3.3.1.1   Diversity Gain

Cooperative-diversity systems, as the name implies, offer diversity gain due to the independent fading experienced by the different links (direct and indirect ones). Actually, it is highly unlikely that all links have deep fading simultaneously. This feature improves the SNR significantly and reduces the bit error rate (BER) and the outage probability.

3.3.1.2   Better Links

Even if different links do not have independent fading, we still can achieve a better signal quality in cooperative diversity. This is due to the fact that the source-relay and relay-destination links can have a better quality than the source-destination link since the relays can always be selected and strategically-located so that the source-relay and relay-destination link have smaller path-loss and less shadowing.

3.3.1.3 Virtual Array Gain

In cooperative diversity, the destination receives multiple copies of the signal (from the source and cooperating relays), to be combined coherently. Therefore, the received signal power at the destination is the sum of the power of the received signal copies from the source and all cooperating relays.

The improvement in the signal quality due to the three aforementioned mechanisms can be seen by analyzing total SNR at the output of the MRC combiner which is given by

$$\gamma_{total} = \gamma_{SD} + \frac{\gamma_{SR}\gamma_{RD}}{\gamma_{SR} + \gamma_{RD} + 1}, \quad (3.2)$$

when AF relaying with variable gain is employed, where γ_{SD}, γ_{SR} and γ_{RD} are the SNRs of the source-destination, source-relay and relay-destination links, respectively. The second term in the right-hand side represents the effective SNR of the indirect link. The diversity gain can be shown from this equation as follows. If one of the two links, for example, the direct link, experiences deep fading, while the indirect link does not have deep fading, γ_{total} will not degrade because the second term representing the SNR of the indirect link is still high. The second mechanism, the better links, takes place because of the higher values γ_{SR} and γ_{RD} should have (compared with γ_{SD}) due to the smaller path-loss and less shadowing, which makes γ_{total} considerably high. The virtual antenna gain is represented by the fact that γ_{total} is the sum of the SNRs of the direct and indirect links, which improves γ_{total} regardless of the fading independence or better links.

In all results presented in this chapter, we use a single relay (M = 1) except in Figure 3.3(b), where we use three relays (M = 3) to show the performance of the best-relay selection technique. Also, in all results (Figures 3.3–5) we use AF relaying.[2] However, in Figure 3.3(b) we also show the results of adaptive DF for comparison. Direct transmission results are shown as a reference. These results are obtained analytically and verified by simulations. The reader is referred to [5], [11], [12] for the derivations and the details of the analytical solutions and computer simulations.

Figure 3.3 shows the BER performance of cooperative diversity (with different relaying and combining schemes). It is evident that cooperative diversity reduces the BER by several orders of magnitude, particularly at high SNR. Furthermore, it is apparent that a full diversity order (M + 1) can be achieved regardless of the relaying or the combining technique except for incremental relaying at high SNR. In fact, incremental relaying does not give diversity gain at high SNR because at high SNR values, the relays rarely forward the signal to the destination. However, at low to medium values of SNR, incremental relaying achieves almost the same performance as fixed relaying. Also, Figure 3.3(b) shows that AF slightly outperforms adaptive DF, particularly at high SNR. Furthermore, it is evident in Figure 3.3 that the BER reduction is not only because of the diversity gain but also due to the SNR enhancement from the link improvement and the virtual antenna gain. This can be easily noticed at low SNR (particularly Figure 3.3(a)) when the cooperative diversity does not have high diversity gain but it still significantly outperforms the direct transmission.

---

[2] In all figures, fixed relaying refers to regular AF to differentiate it from AF with incremental relaying and AF with best-relay selection.

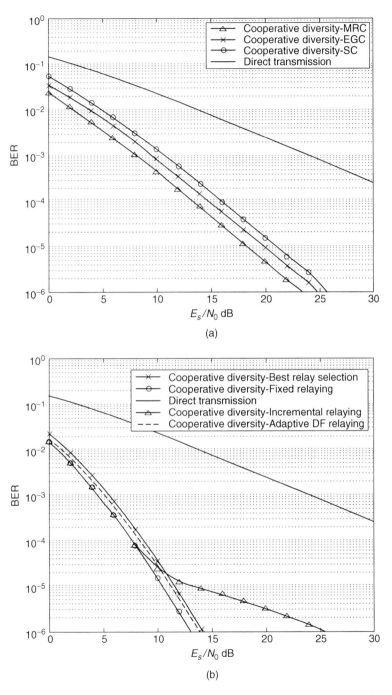

Figure 3.3 Bit error rate performance of BPSK using direct transmission and cooperative diversity and (a) Different combining schemes (M = 1) (b) Different relaying schemes using 3 relays (M = 3). All cooperative diversity cases use AF relaying except the adaptive DF curve in Figure 3.3(b).

3.3.2 Reduced Power

If we trade the signal-quality improvement (due to the diversity gain, better links and virtual antenna gain) with the transmission power, the total transmission power of cooperative diversity can be reduced to be significantly less than the transmission power of the traditional direct transmission for the same end-to-end SNR or received power levels. This power reduction is a very desirable feature in battery-operated devices as in mobile terminals and wireless sensor nodes.

3.3.3 Better Coverage

Another advantage of cooperative-diversity systems is the extension of the signal coverage and communication range to remote users experiencing large path-loss by utilizing the signal-quality improvement (in terms of higher SNR and received power). Also, if the relay locations are carefully chosen, cooperative diversity can overcome large shadowing that may exist in the direct link due to blockage by large objects.

3.3.4 Capacity Gain

Cooperative diversity can improve the capacity in terms of the achievable throughput. This throughput gain can be realized by trading the enhancement in SNR and the received power with the transmission rate using, for example, adaptive modulation in TDMA/OFDMA systems or variable spreading gain in CDMA networks. The Shannon capacity of cooperative-diversity systems with M relays can be expressed as

$$C = \frac{1}{M+1} \int_0^\infty \log(1+\gamma) f_{\gamma_{total}}(\gamma) d\gamma \qquad (3.3)$$

where $f_{\gamma_{total}}(\gamma)$ is the probability density function (pdf) of the total SNR at the output of the MRC combiner.

Figure 3.4 shows the Shannon capacity and the average throughput of the cooperative-diversity system as well as the average throughput of the traditional direct transmission. Adaptive modulation is employed in the cooperative-diversity system and the direct transmission to translate the high SNR to throughput gain. It is evident that the average throughput of the cooperative diversity is substantially higher than that of direct transmission despite the fact that the average throughput of the cooperative diversity system is reduced by 50% due to the use of two time slots to send one message. At high SNR values, the throughput of cooperative diversity does not continue to outperform the throughput of direct transmission because at high SNR the throughput gain in the cooperative-diversity system (due to the SNR improvement) becomes less than the 50% loss.

3.4 Major Challenges of Cooperative-Diversity Systems

The benefits of cooperative diversity discussed in the previous section do not come without a price. There are several challenges and complexity issues that need further investigation to facilitate the realization of cooperative-diversity systems in practice.

3.4.1 Resources Over-Utilization

As mentioned above, cooperative-diversity systems need $M + 1$ channels to send one message from the source to the destination using M relays. Efficient relaying techniques such as

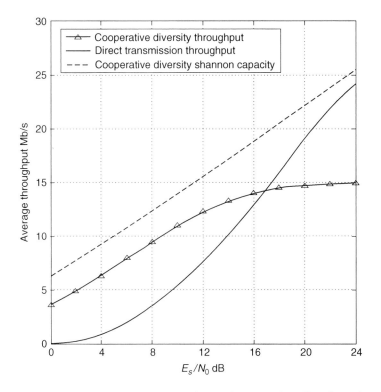

Figure 3.4 Average throughput of direct transmission and cooperative diversity using a single relay (M = 1), AF relaying, and a bandwidth of 5 MHz and adaptive modulation (BPSK and 4, 8, 16, 32, 64, 128 and 256 QAM) with a target BER = 10^{-3}.

Best-Relay Selection and *Incremental Relaying* can reduce the amount of required resources without causing significant degradation to the BER performance as shown in Figure 3.3(b). Meanwhile, best-relay selection uses two channels only regardless of the number of relays. On the other hand, the average number of channels in incremental relaying is even less and ranges between one and two[3] depending on the value of the SNR as shown in Figure 3.5. Actually, at high SNR values, the average number of channels approaches one since the relay rarely needs to forward the signal to the destination. The resource over-utilization problem can also be improved using superposition modulation [13]. In superposition modulation, the source sends, in odd slots, its symbol(s) with high power and superimposed symbol(s) (for the relay) with lower power. In even slots, the relay sends its symbol(s) with higher power and superimposed symbol(s) (for the source) with lower power. The implicit assumption here is that both the source and the relay have traffic to send (not only the source). In addition, the demodulation here becomes more complicated due to the need to iterative detection or interference canceling.

3.4.2 Additional Delay

When cooperative-diversity systems (with M relays) use TDMA as the access scheme, M + 1 time slots are needed to send the signal from the source to the destination. Even if other multiple access techniques (e.g., FDMA, CDMA or OFDMA) are used, we still need at least two time

---

[3] It is assumed here that one relay only will forward the signal if the SNR is low.

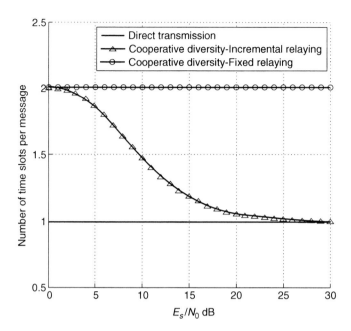

Figure 3.5 Number of time slots per message for direct transmission and Cooperative diversity with AF relaying and a single relay (M = 1).

slots; one for the source transmission (to the destination and relays) and one for the relays transmission (to the destination). This is due to half duplex transmission since full duplex transmission is not always feasible for several practical reasons. Therefore, it is obvious that the signal detection at the destination will experience extra M time-slot delay if TDMA is used or one additional time-slot delay if other multiple access techniques are employed. Full-rate schemes are proposed in the literature [14], where the source keeps sending while the relay is forwarding the signal to the destination. However, these schemes are not always feasible as they require directive antenna beams at the destination and at least two cooperating relays.

Simultaneous transmission by the cooperating relays can alleviate the two problems of *resource over-utilization* and *additional delay*. In this case, the source sends the signal in the first time slot while the second time slot is given for all cooperating relays to send simultaneously. However, the coherent combining of the relayed signals at the destination becomes more challenging because of the phase mismatching of the relayed signals as a result of the different values of the propagation delay between the relays and the destination [15].

3.4.3 Complexity

There are several complexity issues in cooperative diversity. These issues may require more processing, novel algorithms, new software and additional processing capability. The main complexity issues are as follows.

3.4.3.1 Signal Combining and Detection

When cooperative diversity is employed, the signal detection at the receiver becomes more complex compared with traditional direct transmission. The reception of multiple replicas of the signal requires some signal processing to produce the combined signal for detection. The MRC technique, for example, requires M + 1 signal multiplications and M additions operations. In [16], it is shown that the design of an optimal detector that minimizes the bit error rate is quite complex because it requires complicated mathematical calculations and some long term statistics.

3.4.3.2 Channel Estimation

Channel estimation is essential in cooperative diversity since some combining and relaying techniques need CSI. For example, the MRC combining technique needs full CSI (amplitude and phase) of all links to calculate the optimal weights. Also, AF relaying with dynamic gain uses CSI to adjust the amplification gain based on the source-relay channel coefficient as mentioned above.

Channel estimation in cooperative diversity uses training sequences and feedback signaling, which causes signaling overhead and wastes some of the scarce radio resources. Furthermore, the channel estimation errors can degrade the performance and reduce the achievable gains of cooperative diversity.

3.4.3.3 Relaying Protocols

The implementation of the relaying protocols adds some complexity to the design of the wireless nodes. Most of the signal functionalities to be performed by the relay (e.g., amplification, detection, etc.) do not require any additional software or hardware because they are already available in the node for its own traffic signal transmission and reception. However, the implementation of signaling and control functions of the relaying protocols requires new (or at least) modified medium access control (MAC). A new cross-layer PHY-MAC protocol is proposed in [17] for cooperation in IEEE 802.11. This protocol allows basic functions of cooperative diversity such as signal forwarding and combining. However, several functions still need to be added in MAC protocols. For instance, MAC protocols should be able to determine the best nodes that can cooperate with the source and assign to them the relaying task.

3.4.3.4 Synchronization

Perfect synchronization at the frame, symbol, chip[4] and carrier level is imperative for effective results of cooperative diversity systems. Synchronization can be achieved using various techniques with different degrees of complexity and reliability (e.g., GPS, pre-ample/post-ample synchronization bits, pilot carrier transmission and blind estimation.) More analysis is needed to address the performance and complexity of such schemes in practical wireless systems particularly in distributed infrastructure-less networks. Even if all nodes are synchronized at all levels, coherent combining might be hard to achieve due to different propagation delays of different paths so that some timing advance technique is proposed for solving the problem of different delay values [15].

3.4.3.5 Resource Management

Resource management in cooperative diversity is obviously more complex than that in traditional direct transmission. For example, the admission control problem must be revisited since admitting more users in cooperative-diversity systems, unlike traditional direct transmission networks, can improve the system capacity as shown in [17]. Furthermore, finding the optimal power allocation and channel assignment in cooperative diversity is very challenging taking into

---

[4] Chip synchronization is needed if synchronous CDMA is used.

account the existence of direct and indirect links as well as the different cooperation techniques. Traffic scheduling also becomes more complex since cooperating nodes must handle their own traffic and other nodes' traffic, which complicates the scheduling scheme taking into account the various quality of service levels, fairness requirements and priority levels of different users and traffic streams.

3.4.4 Unavailability of Cooperating Nodes

Cooperation schemes are always discussed under the assumption that relay nodes are available for cooperation. However, there is no guarantee that other nodes will be always available or willing to cooperate. In fact, some users might still be reluctant to cooperate in order to save their terminal power or for security concerns. Thus, cooperative diversity might face the problem of the unavailability of cooperating relays unless some incentives (e.g., monetary reward or credit time) are used to motivate users to cooperate. Fixed relays are proposed in [18] as a solution of this problem. However, this will increase the cost of the network infrastructure and adds more complexity to the network planning.

3.4.5 Security Threats

Like other multi-hop wireless networks, cooperative-diversity systems face a serious challenge in securing users' data simply because different users share their data with each other. In addition, security techniques are vital to protect nodes from malicious attacks that might lead to resource depletion, or unreliable cooperation. For example, the attacking node might try to overwhelm cooperating relays by a vast amount of data packets (for forwarding) to deprive the attacked relays of resources. Attacks can also take the form of exchanging false information by forwarding to the destination incorrect or distorted versions of the signals. Therefore, robust encryption, authorization and authentication algorithms must be employed in cooperative-diversity systems to protect the data privacy and integrity.

3.5 Discussion and Conclusion

In the previous two sections we have discussed the main benefits and challenges of cooperative diversity. It has been shown that cooperative diversity brings several benefits including *better signal quality, reduced transmission power, better coverage* and *higher capacity*. Meanwhile, cooperative diversity has some drawbacks and challenges including *resource over-utilization, additional delay, implementation complexity, unavailability of cooperating nodes* and *security threats*. Nevertheless, it is evident that overall the benefits of cooperative diversity overweigh the drawbacks. For instance, although cooperative diversity causes additional delay and uses extra resources, the throughput of cooperative diversity still outperforms its counterpart in traditional direct transmission, particularly at low and medium SNR.

On the other hand, it should be emphasized that the implementation of cooperative diversity is shown to be challenging due to the various complexity issues discussed above. Therefore, reduced-complexity algorithms and protocols are needed to simplify the implementation of cooperative diversity in order to be able to harvest the benefits of cooperative diversity. Moreover, effective measures and schemes are required to increase users' motivation for cooperation and to enhance information security and service authenticity in cooperative diversity. Therefore, the final conclusion is that cooperative diversity is very promising and should be considered in existing and future wireless networks. However, more efforts and novel schemes and protocols are needed to facilitate the implementation and smooth integration of cooperative-diversity systems.

References

[1] J. Laneman, D. Tse, and G. Wornell, 'Cooperative diversity in wireless networks: efficient protocols and outage behavior,' *IEEE Trans. on Inform. Theory*, vol. 50, no. 11, Dec. 2004, pp. 3062–3080.

[2] J. Fricke, M. Butt and P. Hoeher, 'Quality-Oriented Adaptive Forwarding for Wireless Relaying,' *IEEE Communications Letters*, vol. 12, no. 3, March 2008, pp. 200–202.

[3] D. Chen and J. N. Laneman, 'Modulation and demodulation for cooperative diversity in Wireless Systems,' *IEEE Trans. Wireless Comm.*, vol. 5, July 2006, pp. 1785–1794.

[4] A. Bletsas, A. Khisti, D. Reed, and A. Lippman, 'A simple cooperative diversity method based on network path selection' *IEEE Journal on Selected Areas in Commun.*, vol. 24, no. 3, March 2006, pp. 659–672.

[5] S. Ikki and M. H. Ahmed, 'Performance of Cooperative Diversity Using Equal Gain Combining (EGC) over Nakagami-*m* Fading Channels,' *IEEE Trans. Wireless Commun*, vol. 8, no. 2, Feb. 2009, pp. 557–562.

[6] J. Winters, 'Optimum Combining in Digital Mobile Radio with Channel Interference,' *IEEE J. Select. Area. Commun. (JSAC)*, vol. SAC-2, July 1984, pp. 528–539.

[7] L. Xiao and X. Dong; 'Unified analysis of generalized selection combining with normalized threshold test per branch,' *IEEE Transactions on Wireless Communications*, vol. 5, no. 8, Aug. 2006, pp. 2153–2163.

[8] P. Anghel, G. Leus, and M. Kaveh, 'Multi-user space-time coding in cooperative networks,' *Proc. IEEE International Conference on Acoustics, Speech, & Signal Processing 2003 (ICASSP'03)*, pp. 73–76.

[9] M. Dohler, E. Lefranc, and H. Aghvami, 'Virtual Antenna Arrays for Future wireless Mobile Communication Systems,' *Proc. IEEE ICT 2002*, Beijing, China, Jun. 2002.

[10] A. Sendonaris, E. Erkip, and B. Aazhang, 'User cooperation diversity – Part I: system description,' *IEEE Trans. on Commun.*, vol. 51, no. 11, Nov. 2003, pp. 1927–1938.

[11] S. Ikki, *Performance Analysis of Cooperative-Diversity Networks with Different Relaying and Combining Techniques*, Ph.D. Thesis, Memorial University of Newfoundland, 2008.

[12] S. Ikki and M. H. Ahmed, 'Performance of Multiple-Relay Cooperative Diversity Systems with Best-Relay Selection over Rayleigh Fading Channels,' *EURASIP Journal on Wireless Communications and Networking, the Special Issue on Wireless Cooperative Networks*, 2008.

[13] K. Ishii, 'Cooperative Transmit Diversity Utilizing Superposition Modulation,' *IEEE Radio and Wireless Symposium*, Jan. 2007, pp. 337–340.

[14] P. Rost and G. Fettweis, 'A Cooperative Relaying Scheme Without The Need For Modulation With Increased Spectral Efficiency,' *Proc. IEEE Vehicular Technology Conference (VTC-F'06)*, Montreal, Canada, Sept. 2006, pp. 1–5.

[15] J. Mietzner and P. Hoeher, 'Distributed Space-Time Codes for Cooperative Wireless Networks in the Presence of Different Propagation Delays and Path Losses,' *Proc. IEEE Sensor Array and Multichannel Signal Processing Workshop*, 2004, pp. 264–268.

[16] A. Sendonaris, E. Erkip, and B. Aazhang, 'User cooperation diversity – Part II: Implementation Aspects and Performance Analysis,' *IEEE Trans. on Commun.*, vol. 51, no. 11, Nov. 2003, pp. 1939–1948.

[17] P. Liu, Z. Tao, Z. Lin, E. Erkip, and S. Panwar, 'Cooperative wireless communications: a cross-layer approach,' *IEEE Wireless Communications*, vol. 13, no. 4, Aug. 2006, pp. 84–92.

[18] A. Adinoyi and H. Yanikomeroglu 'Cooperative relaying in multi-antenna fixed relay networks,' *IEEE Transaction on Wireless Communications*, vol. 6, no. 2, Feb. 2007, pp. 533–544.

4

Cooperation in Wireless Ad Hoc and Sensor Networks

J. Barbancho[1], D. Cascado[2], J. L. Sevillano[2], C. León[1], A. Linares[2] and F. J. Molina[1]

[1]*Department of Electronic Technology, University of Seville, Spain*
[2]*Department of Computer Architecture, University of Seville, Spain*

4.1 Introduction

Cooperation is a key issue in Ad Hoc and sensor networks (Wireless Ad Hoc Sensor Networks, WAdSN) for several reasons. Usually ad-hoc and sensor networks are two concepts related to networks consisting of wireless nodes. Our approach in this chapter considers that these nodes have common features such as tiny size, low computational and storage resources and low autonomy (understood as battery lifetime). Frequently, these nodes can acquire information from the environment via different sensors and can interact with it through actuators.

The term ad hoc refers to a type of network configuration associated with broad networks, such as WiFi. However, this type of network faces different challenges that are not covered in this chapter.

The rapid development of low-cost Micro-Electro-Mechanical-Systems has led to a sharp increase in the use of these devices in several fields such as home automation, industry, health, biology, and so on. The number of potential applications is huge. However, the design of these applications is subject to important constraints: power consumption, limited storage, limited computing capacity, low data rate, and so on.

A typical scenario for application could be summarized as follows: Nodes are spread out during deployment and each device has a limited radio range. Nodes have different roles assigned to them: data source, data sink and gateway.

Two different paradigms exit when trying to identify the main objective of such a network. The classical approach assumes that the main purpose of the network is to transport data from the sources to the sinks (which we could name data transport paradigm). Data often needs more than one hop to reach the sink node. As there is no network manager (ad hoc approach), nodes have to collaborate to form routes through which data can be transmitted. Once the data reaches the sinks, a base station (usually a personal computer) processes the collected information. In this paradigm, cooperation is limited to an exchange of information. In contrast, the new approach

Cooperative Networking, First Edition. Edited by Mohammad S. Obaidat and Sudip Misra.
© 2011 John Wiley & Sons, Ltd. Published 2011 by John Wiley & Sons, Ltd.

considers the network as a whole. In this way, although nodes may have low resources they can store, process and send small pieces of information. So, cooperation is understood as a distributed way of achieving an objective, like in a colony of ants. The main characteristics are: only local computations are involved, no global knowledge is assumed (except for the common objective), no centralized control is needed, and scalability is relatively easy to achieve [1]. In this second approach, cooperation becomes a much more important issue (cooperative paradigm).

Collaboration can be established in such a network in pursuit of different goals, as detailed in in the following text.

Time synchronization, calibration and localization are topics that can be studied under the new approach of cooperation in wireless sensor networks. Every node has its own clock that sets the rhythm the microprocessor has to follow. Usually these clocks have limited precision, and microprocessors tolerate changes in the time measurement to work correctly. If there is no external time reference the network has to create the rules to define a global clock, so that nodes can synchronize their clocks estimating and compensating clock skew and offset. These rules could be designed in a cooperative way, enabling all network nodes to exchange synchronization messages. Calibration could be approached in the same way. Given that a physical magnitude is measured by a network of sensors, they can cooperate themselves to reach a given calibration accuracy. In this sense, localization also uses measured data where the nodes need to match cooperatively. Synchronization, localization and calibration are discussed in Section 4.2.1.

Having to manage data routes in the whole network is a problem that can also be solved with cooperative methods. Routing can be understood from the point of view of power consumption. Nodes have to cooperate to achieve a common trade-off: data delivery versus energy expenditure. Radio transmission is the process that uses most energy in the network. Reducing the number of packets that are transmitted is one of the strategies that we can follow to extend network lifetime. The selection of the path that the packets have to follow is directly related to the number of packets that are used in the task of delivering information. In Section 4.2.2 we study an example that shows how power consumption can be reduced by managing the quality of the radio links involved in data transmissions. The quality of the radio links is estimated by the nodes using Artificial Neural Networks. This information is exchanged in a cooperative way to dynamically decide packet flow. The effect that this behaviour produces is similar to that of the blood flow in a human body. There is no central control; the decision about where the blood has to flow is taken by all the arteries and veins.

Other issues that are studied under the paradigm of cooperative networking are data aggregation and data fusion. Several examples of applications based on these topics are described in Section 4.2.3 to illustrate the usefulness of cooperative networking.

Having discussed why cooperation is needed in WAdSN, we shall then look at how cooperation can be implemented. Wireless sensor nodes can cooperate among themselves in many different ways in pursuit of the common goal. Researchers have proposed a wide array of mechanisms for implementing cooperation, such as multi-agent intelligent systems (MAS), neural networks, middleware systems, and so on. These techniques are covered in Section 4.3.

Finally, Section 4.4 presents a summary of the chapter, focusing on future lines of research into cooperation in ad hoc and sensor networks.

4.2 Why Could Cooperation in WAdSN be Useful?

4.2.1 Time Synchronization, Localization and Calibration

Observation of a physical phenomenon requires the specification of the time and location of occurrence, with a measure of how confident we can be about the observation. Therefore, if a

Wireless Ad Hoc Sensor Network (WAdSN) is used for this observation, nodes must know their own position and the time, as well as being able to refer their output to a well-defined scale. This is the main argument for providing WAdSNs with localization, synchronization and calibration techniques. However, there are many other reasons why these techniques may be useful.

For instance, location information may not only be used to stamp a given measure; it may also be a requirement in many sensor network applications: tracking of mobile vehicles or animals; monitoring of elderly and disabled people in residencies; support for navigation, logistics and inventory management, and so on. In these cases, localization can be seen as a service: some nodes provide location information and others require this service to determine their locations or in turn to provide location based services or perform location-aware functions. On the other hand, when a node provides a measure with location information, data fusion techniques can reduce traffic and energy consumption. For instance, data aggregation can be used to combine the information coming from a number of sensors in the same area (i.e. temperature) to provide more meaningful information (i.e. forest fire alarm). Data aggregation requires the nodes to be positioned with a level of accuracy that depends on the application.

Time synchronization is particularly important in wireless sensor networks. One of the most important reasons is energy saving. Data communication is the major source of energy expenditure (much higher than sensing or data processing [2]), and in data communication there are four major sources of energy waste [3]:

- **Collision:** Colliding packets must be discarded and therefore waste energy. Retransmissions also increase energy consumption.
- **Overhearing:** When a node picks up packets that are destined to other nodes.
- **Control packet overhead:** As sending and receiving control packets also consumes energy.
- **Idle listening:** i.e., listening to receive possible traffic that is not sent.

The energy consumed by idle listening may be of the same order of magnitude as the energy required for receiving [3], [4]. As WAdSNs are characterized by low data rates, according to the node features we have assumed in this chapter, nodes can be in idle mode for most of the time and therefore idle listening becomes the major source of energy waste in WAdSNs. Nodes should be kept awake the minimum time required to exchange data to save energy. There is often only specific time periods of interest. Nodes should be synchronized to perform the observation within these time periods, and then switch off the radio during the inactive periods.

Collisions also waste energy, and this is one of the drawbacks of contention-based MAC (Medium Access Control) mechanisms. Usually, protocols based on reservation and scheduling are preferred when energy consumption is the main concern. For instance, a TDMA (Time Division Multiple Access)-based MAC allows energy saving because it avoids collisions, reduces overhead and eases the implementation of the duty cycle of the radio. Of course, TDMA protocols require tight time synchronization.

It is true that some WAdSNs may use contention-based MACs provided they transmit with low power and low data rate, so that the probability of collision is very low. For instance, this is the case of the beaconless mode of the IEEE 802.15.4 protocol (the basis of the ZigBee standard) which is essentially an unslotted CSMA/CA mode. However, this simple solution presents several drawbacks. For instance, when any node is mobile (i.e. not fixed), the beaconless mode is not suitable because there is no periodical communication with the coordinator node, so the mobile node may assume its association although it may have lost the link with the coordinator [5]. On the other hand, the 802.15.4 protocol includes a beacon-enabled mode where a so-called ZigBee Coordinator (ZC) periodically transmits beacon frames that establish a superframe structure. In this superframe, devices may transmit using a slotted CSMA/CA medium access, with an optional

Contention-Free Period (CFP). Even if these CFPs are not used, the periodic transmission of beacons provides a way of synchronizing all nodes. It is also possible to change the duty cycle to achieve low power consumptions: the device, upon receiving the first beacon, gets current superframe structure and thus knows when to activate its receiver for the next beacon (just a bit earlier to save power). The device periodically enables its radio to receive the beacon frame, performs the required transmissions and then switches off the radio during the inactive period. With mobile devices in the beacon-enabled mode, if a node does not receive a predetermined number of beacon frames then it considers itself as an 'orphan' node, beginning a realignment procedure to be associated again with a coordinator. Therefore, generally speaking, when there are energy and/or timing requirements the right choice is the beacon-enabled mode which means that nodes are time synchronized.

A distributed alternative to the 802.15.4 beacon enabled mode is S-MAC [3], where instead of a central coordinator, neighbouring nodes coordinate their sleep schedules. The so-called 'coordinated sleeping' requires each node to maintain a schedule table that stores the schedules of its neighbours. This way they know their neighbours' scheduled listen time and they exchange their schedules within these periods. This is a good example of how collaborative methods can be used not only to reduce energy waste but also control overhead and latency.

There are many other uses of time synchronization. For instance, at lower layers tight time synchronization may also be fundamental for increasing data rates (short bit times) or enhancing noise immunity (as in the frequency hopping mechanism). At higher layers, data fusion requires synchronization for two tasks: time scheduling and time stamping. The first is needed when the nodes coordinate to perform cooperative communications. The second is commonly used when data is fused taking into account the collecting instant; for example, to perform event detection, tracking, reconstruction of system's state for control algorithms, off line analysis, and so on. These issues will be further discussed in the following sections.

There is a close relationship between calibration, time synchronization and localization in wireless sensor networks. Calibration can be broadly defined as the mapping of the output of a sensor to a well-defined scale, which as pointed out in [6] includes time synchronization as a special case simply considering the hardware clock as a sensor whose output has to be mapped to a timescale. Localization is also closely connected to synchronization [7] and by extension to calibration. Localization is usually performed measuring angle-of-arrival (AoA), time-difference-of-arrival (TDOA) or Received Signal Strength (RSS), and in any case it also entails mapping one physical magnitude to a well-defined scale to infer the node position. Of course, RSS based localization systems require careful calibration of the receivers. Furthermore, some of these localization methods are based on the measurement of time of flight or difference of arrival time of a signal so they require synchronized time. For instance, in [8] a location system is described where the position is obtained by measuring the time of flight of ultrasonic signals sent by the 'beacons'. These beacons must be synchronized so that the ultrasonic signals are sent by the beacons at the same time. In a sense, the opposite is also true: synchronization accuracy depends on the relative location of the nodes. If nodes are located too far away from each other, multiple hops are needed to connect them so synchronization messages suffer from higher delays and these delays decrease synchronization accuracy [9].

Differences in time synchronization, localization and calibration (SLC for short in what follows) do not stop similar techniques and algorithms being transferred among the three. For instance, unlike other physical variables time is not bounded, so the time difference between two sensors will also grow unbounded unless periodic time synchronization is performed. However, it should be noted that this is also the case with calibration and localization in many wireless sensor networks. Obviously, if nodes are mobile or simply their position changes because of the effect of environmental changes (wind, rain, etc.), periodic location information has to be exchanged among the nodes to perform localization. Similarly, although calibration of conventional sensors is usually performed only once before operation, periodical calibration may also be required. The reason is that changes in environmental parameters during the lifetime of typically low-cost sensors may affect their operation [6]. This may be especially important in the case of mobile sensors or in safety critical applications. It is for all these reasons that we present in this chapter a unified discussion of synchronization/localization/calibration (SLC). The three share the same classes identified by [6], for instance:

- **External vs. internal:** If SLC information is provided by an external node, then we say that the SLC is external. For instance, a GPS (Global Positioning System) receiver obtains its position or absolute time from external sources. Calibration is external if the output of all sensors map to a given scale [6]. For internal calibration, the only thing that is needed is for all nodes to output the same value (at least within a given interval) if they are exposed to the same stimulus, but not necessarily equal to what an external sensor would output. Similarly, internal localization or synchronization only tries to maintain consistency among the nodes, but without conforming to an absolute position or time. In some applications, this internal SLC could be limited to a subset of the nodes (for instance, those observing a given phenomenon), or could have different precision requirements in different subsets.
- **Continuous vs. on demand:** Continuous SLC means that nodes try to maintain synchronization, localization and calibration at all times. This approach is not very efficient in WAdSN especially in terms of energy consumption. On demand SLC is much more appropriate in this case, particularly because the system does not waste energy in SLC during the long idle periods. On demand SLC may be:
  - **Event-triggered:** when the event has occurred the node is synchronized/located/calibrated.
  - **Time triggered:** periodically or when an external node sends the order to perform the SLC. Due to the uncertainty of delays in a WAdSN, especially with multi-hops, this order may be anticipated: nodes receive the order to take the sample at some future time (of course, this requires nodes to be synchronized).
- **Master-slave vs. peer to peer:** Usually, two kinds of nodes are considered; 'masters' that know their time and/or location (which we call 'reference nodes') and 'slaves' that must estimate their location and/or time by some algorithm ('common nodes'). Reference nodes (sometimes called 'anchors' [10] or 'beacons' [11]) can obtain their coordinates from an external system (such as GPS) or simply act as a time or location reference for the other nodes. This would be the case of an internal SLC, where only relative time or location information is required. Some algorithm must be used for the common nodes to estimate their SLC information based on the known information from the reference nodes. However, note that sometimes reference nodes are not required, especially with internal SLC. In these cases (most of which are time synchronization protocols), peer-to-peer solutions offer more flexibility but are more difficult to control.

A recent survey of localization techniques is [10]. When all common nodes are at one-hop distance from a sufficient number of reference nodes, then localization techniques do not require nodes to cooperate. They can simply use some signal measuring system to infer the distance between the reference node(s) and the common node (or the element to be located). As mentioned before, they simply use methods based on measuring Angle Of Arrival (AoA), Time Difference Of Arrival (TDOA) or Received Signal Strength (RSS). Once the distance is obtained, multilateration is used to obtain the coordinates of the element to be located. Redundant measurements are needed to cope with difficulties such as non-line of sight errors [11]. Less accurate methods exist that are not based on these measurement techniques. For instance, in [12] a node estimates its position as the centroid of the coordinates of the reference nodes that are within its communications range. Of course, this simple solution requires high node density and regular topologies, and in any case it can only provide a coarse estimation. More complex algorithms are described in [13] and [10].

However, cooperative algorithms are more useful for common nodes that are not necessarily at one-hop distance. For instance, in [14] all reference nodes flood messages with their location, and these messages, which are propagated hop-by-hop, include a hop-count so that each receiver can record their hop-count distance to the reference nodes. When another reference node receives this message, it computes the mean one-hop distance using the known location of both reference nodes and the hop-count. This one-hop estimation is sent back to the network so that common nodes can estimate their actual distance to reference nodes from their own hop-count.

As for time synchronization, cooperation is more important as the clock skew (differences in the frequencies of the nodes' clocks) and drift (differences in the rate at which frequencies change) increase over time even in static WAdSNs. Many different techniques for clock synchronization exist, and they include all the classes discussed above: external vs. internal synchronization, continuous vs. on demand, Master-slave vs. peer-to-peer, and so on. [15], [6]. Regarding cooperation, it is also interesting to distinguish between methods where there is a sender transmitting the current clock values to the receiver as timestamps (sender-to-receiver) and methods where synchronization messages are broadcast so that all receivers receive the message at approximately the same time (receiver-to-receiver). According to Maróti et al., the most significant delays when transmitting messages over a wireless link are those from the send, receive and access processes [16]. Sender-to-receiver protocols like TPSN (Timing-sync Protocol for Sensor Networks) or the Flooding Time Synchronization Protocol (FTSP) avoid the indeterminism of these delays working at the MAC (Medium Access Control) layer to precisely timestamp messages, which is not always possible when using standard or complex protocols. In receiver-to-receiver methods, the use of reference broadcast messages eliminates the time uncertainty introduced by the send and access processes and sets a temporal reference shared by all the nodes.

One of the best known receiver-to-receiver synchronization methods is MBS [17]. Elson and Estrin propose a post-facto synchronization method; that is, a method that performs synchronization only when it is needed [18]. The scattering method they propose does not give a common time reference to the broadcast sender; it only synchronizes receivers. However, several authors point out the need for setting a network time to propagate the synchronism over a multi-hop network using broadcasts [17], [19]. Thus, nodes in broadcast domain need to cooperate by sharing timing information among them to determine the global network time (GNT). This makes them non-usable in large networks because the amount of data exchange required is huge [20]. On the other hand, Multi-hop Broadcast Synchronization protocol [9] is a receiver-to-receiver synchronization scheme that nonetheless obtains a GNT working at the application layer. In this protocol each reference broadcast informs about the timestamp of the previous one. This way message exchanging is minimized, being similar to other sender-to-receiver methods, which makes MBS suitable for large multi-hop networks.

Finally, works on calibration in WAdSNs are very scarce. Römer et al. [6] discuss external and internal calibration methods. Collaboration among nodes is important mainly in internal methods such as Collaborative In-Place Calibration (CIC) [21]. It is interesting to note that in this method location, synchronization and calibration are again closely related. For instance, nodes need to be synchronized. Node location is also important, as CIC begins with collocated pairs of nodes that are calibrated against each other.

A final comment is that SLC in cooperative wireless sensor networks can be seen from two different points of view: as we have shown, cooperation is needed to perform SLC. Many of the previously discussed algorithms are based on collaboration among nodes. On the other hand, SLC is often needed to perform cooperation. For instance, in dense WAdSNs nodes that are close to each other may be redundant from a routing point of view. The network can then be partitioned into a grid, with some active nodes assuming the role of grid 'leaders' while others are kept asleep except for periodical wake-ups to communicate with the active nodes. This way nodes in sleep mode save energy. Active and asleep nodes cooperate by alternating their roles to prevent active nodes from running out of energy. Some of these algorithms need nodes to know their position to select which neighbouring nodes will take the role of grid leaders [22].

4.2.2 Routing

The need for connectivity among nodes introduces the routing problem. There are several reasons that make routing in wireless sensor networks a challenge:

- First, a global addressing scheme, as in Internet Protocol, cannot be maintained. This is because the network is formed by a huge number of nodes. Furthermore, sensors are deployed in such a way (ad-hoc) that self-configuration techniques are required.
- Second, the data usually flow from several sources to a few number of sinks.
- Third, the nodes provide highly correlated information. So, the implementation of data aggregation and fusion techniques is required. We'll raise this problem in Section 4.2.3.

Due to these features, routing in WAdSN has traditionally been studied as a problem that could be solved by solutions based on distributed systems. In other words, collaboration among nodes is needed to route data.

In the following we shall focus our approach to the routing problem on the need for a multi-hop scheme to let data travel from a source to a destiny. The paths the packets have to follow can be established based on a specific criterion. Possible criteria can be a minimum number of hops, minimum latency, maximum data rate, minimum error rate, etc. For example, imagine that all the nodes wish to have a path to route data to the base station. In this situation, the problem could be solved by a technique called network backbone formation. In general, sensor networks do not use node addresses. In contrast, sensor locations could be used to solve the routing problem, as seen in Section 4.2.1. Another approach to routing is to query all sensor nodes matching a certain criterion instead of utilizing individual node addresses; this is discussed in Section 4.3.1.

An example to enhance this solution is based on the introduction of artificial intelligence techniques in the WAdSNs: expert systems, artificial neural networks, fuzzy logic and genetic algorithms. Although there are many authors who have proposed the introduction of different AI techniques in several applications over WAdSNs [23], [24], [25], only a few (e.g. [26]) have considered the possibility of implementing an AI technique inside a sensor node. Due to the processing constraints we have to consider in a sensor node, the best suited of all these techniques, is the self-organizing-map (SOM). This kind of artificial neural network is based on the self organization concept.

The network backbone formation has been studied in mathematics as a particular discipline called *Graph Theory*, which studies the properties of graphs. A directed graph G is an ordered pair $G := (V, A)$ with V, a set of vertices or nodes, v_i, and A, a set of ordered pairs of vertices, called directed edges, arcs, or arrows. An edge $v_{xy} = (x, y)$ is considered to be directed from x to y; where y is called the head and x is called the tail of the edge.

In 1959, E. Dijkstra proposed an algorithm that solves the single-source shortest path problem for a directed graph with nonnegative edge weights. In our wireless sensor network we assume that all the links are symmetrical, in the sense that if a node A can reach a node B, then the node

B can reach the node A. With these kinds of links, we can model our network as an undirected graph $G := (V, E)$.

Barbancho et al. [27] propose a modification of Dijkstra's algorithm to form the network backbone, with the minimum cost paths from the base station or root, r, to every node in the network. The proposed algorithm is named Sensor Intelligence Routing, SIR. In Dijkstra's algorithm the graph has arrows and in SIR the graph has edges. Every edge between nodes v_i and v_j has a weight, w_{ij}, and it is easy to prove that $w_{ij} = w_{ji}$. The distance from the base station to a node v_i is named $d(v_i)$. The set of nodes which are successors or predecessors of a node v_i is denoted by $\Gamma(v_i)$, and is defined in this way: $\Gamma(v_i) = \{v_j \in V | (v_i, v_j) \in E\}$. If we denote a path from the root node to a node v_k by p, we can define $\Gamma_p(v_j)$, if $v_j \in p$, as the subset of nodes which are predecessors or successors of node v_j. It is also assumed that $V = \{r, v_i\}_i$ and that there is a subset of V, T, defined as $T := V - \{r\}$. With this terminology, the SIR algorithm can be described as detailed in Table 4.1.

In the first step, every node is assigned an initial cost to reach the sink. In the following steps this cost is updated depending on the neighbourhood. The algorithm ends when there are no more possible updates.

Once the backbone formation algorithm has been designed, a way of measuring the edge weight parameter, w_{ij}, must be defined. On first approach it could be assumed that w_{ij} can be modelled with the number of hops. According to this assumption, $w_{ij} = 1 \, \forall i, j \in \mathbb{R}, i \neq j$. However, let us imagine that we have another scenario in which the node v_j is located in a noisy environment. The collisions over v_j can introduce link failures increasing power consumption and decreasing reliability in this area. In this case, the optimal path from node v_k to the root node may be p', instead of p. Modification of w_{ij} is required to solve this problem. The evaluation of the quality of service (QoS) in a specific area can be used to modify this parameter.

The traditional view of QoS in communication networks is concerned with end-to-end delay, packet loss, delay variation and throughput. Several authors have proposed architectures and integrated frameworks to achieve guaranteed levels of network performance [28], [29]. However, other performance-related features, such as network reliability, availability, communication security and robustness are often neglected in QoS research. The definition of QoS requires some enlargement if we want to use it as a criterion to support the goal of controlling the network.

Table 4.1 Network Backbone Formation Algorithm

	Set up phase:
	$d(r) = 0$
Step 1:	$d(v_i) = \begin{cases} w_{ri} & \text{if } v_i \in \Gamma(r) \\ \infty & \text{if } v_i \notin \Gamma(r) \end{cases}$
	$\Gamma_p(v_i) = \begin{cases} r & \text{if } v_i \in \Gamma(r) \\ 0 & \text{if } v_i \notin \Gamma(r) \end{cases}$
Step 2:	Find a $v_j \in T$ such as $d(v_j) = \min\{d(v_i) | v_i \in T\}$ Do $T = T - \{v_j\}$
Step 3:	$\forall v_i \in T \cap \Gamma(v_i)$ calculate $t_i := d(v_j) + w_{ji}$ If $t_i < d(v_i)$ do $d(v_i) = t_i$
Step 4:	If $|T| > 0$ go to step 2 If $|T| = 0$ stop

This way, sensors participate equally in the network, conserving energy and maintaining the required application performance. What is sensor network QoS? Iyer and Kleinrock proposed in [30] a definition of sensor network QoS based on sensor network resolution. They define resolution as the optimum number of sensors sending information toward information-collecting sinks, typically base stations. Kay and Frolik defined sensor network QoS in terms of how many of the deployed sensors are active [31]. The same idea is discussed in [32] by Perillo et al., and in [33] by Rakocevic et al.

Barbancho et al. [27] uses a definition based on three types of QoS parameters: timeliness, precision and accuracy. Due to the distributed feature of sensor networks, our approach measures the QoS level in a spread way, instead of an end-to-end paradigm. Each node tests every neighbour link quality with the transmissions of a specific packet named ping. With these transmissions every node obtains mean values of latency, error rate, duty cycle and throughput. These are the four metrics we have defined to measure the related QoS parameters.

Once a node has tested a neighbour link QoS, it calculates the distance to the root using the obtained QoS value. Expression 4.1 represents the way a node v_i calculates the distance to the root through node v_j, where qos is a variable whose value is obtained as an output of a neural network.

$$d(v_i) = d(v_j) \cdot qos \qquad (4.1)$$

According to this strategy, data from source nodes travels through dynamic paths, avoiding the regions with the worst quality of service levels.

4.2.3 Data Aggregation and Fusion

Data fusion is an analogous procedure to the human cognitive process. This procedure is like a set of multidisciplinary techniques with the objective of integrating data from several sensors. The main goal of this integration is to produce a relevant output that identifies the state of the phenomenon being studied. In this sense data fusion could be understood as a connection between a sensory layer and a reaction layer.

The Joint Directors of Laboratories (JDL), an organism that belongs to the Department of Defense of the United States of America has defined a process model to describe the data fusion process. The model has five levels [34]:

- **Level 0:** Source preprocessing. Data is standardized on this level. This process is executed on the data acquisition hardware associated with the sensor. This could be called pre-fusion. On this level every sensor is provided with independent data from other sensors.
- **Level 1:** Object refinement. The data is identified with an entity. This entity is understood as an application which is being carried out. For example: a medical diagnosis.
- **Level 2:** Situation refinement. The results obtained with the last level are evaluated according to the speed of data processing. If the speed is too low, data aggregation algorithms have to be carried out to increase the speed.
- **Level 3:** Impact assessment. The phenomenon is studied with the acquired data. This study intends to predict the performance of the phenomenon in the future.
- **Level 4:** Process refinement. General refinement in real time.
- **Level 5:** Cognitive refinement. On this last level, the human interprets the results obtained by the data fusion processing. This way possible error can be detected.

Data fusion has been considered in wireless sensor networks as a mechanism for reducing the number of packets transmitted among sensors. This reduction consequently entails lower

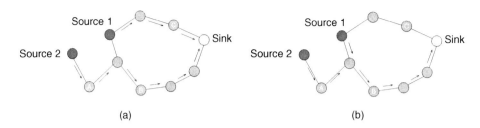

Figure 4.1   Different routing approaches: (a) address-centric routing and (b) data-centric routing.

average network power consumption. In such a network, the information obtained by a node and the nodes that belong to its neighbourhood is often redundant and highly correlated. In this scenario, it makes sense to process data locally instead of sending it to a central station for processing.

Routing is a problem that needs to be borne in mind when considering data delivery to sinks. Conventional address-centric routing finds the shortest routes from sources to sinks. However this approach does not provide an optimal solution. Better energy and bandwidth efficiency are obtained with the data-centric routing approach. It considers data aggregation along a route with multiple sources and one sink. This approach is particularly efficient in scenarios where the sources are located close to one another and far away from the sink [35]. This comparison is depicted in Figure 4.1. In the address-centric routing approach, source 1 chooses the path through node C to deliver data to the sink with the minimum numbers of hops. At the same time, source 2 chooses the path through node B following the same criterion. The total amount of hops and the number of packets travelling through the network are 9. In the data-centric routing approach, source 1 chooses the path through node B. In this approach the node between A and B relays and aggregates data from sources 1 and 2. Consequently only seven hops are required for data delivery. Research into data aggregation and fusion can be split into three categories:

- **Research based on the fusion function.** The main goals are to suppress redundancy and estimate a system parameter [36].
- **Research based the system architecture.** The main goal is to determine how many data aggregators must be used in the wireless sensor network and where.
- **Research based on trade-off and resources.** This research focuses on several trade-off, such as energy vs. estimation accuracy, energy vs. aggregation latency and bandwidth vs. aggregation [37].

The following is an example of an application in which fusion can bring these three categories into play:

The South of Spain is home to one of the largest wildlife reserves in Europe: the National Park of Doñana which covers a huge area of about 520 km^2. It has a variety of ecosystems: Mediterranean forest, marshland, wetlands, dunes, beaches, and so on. Doñana is a wildlife reserve protected by the Spanish Government; it is an area that has seen little human interference through history. Nowadays no-one can enter Doñana without special permission. As a result, it is a privileged scenario for studying populations of wild animals and plants.

The international scientific community has been conducting research in Doñana for over 50 years. Biologists have traditionally deployed different kinds of sensors to gather information about natural phenomena such as vegetation stress, water levels of the aquiferous system, growth of animal populations, etc. These sensors are usually connected to a data-logger that stores data for a period of time. Nowadays the Park has a WiFi infrastructure that provides connections to an intranet from every part of the Park. This way, it is possible to access an environmental variable in real time. For example, anyone can observe online the ultraviolet irradiation at Santa Olalla's lake through the information provided by the Doñana Biological Station, EBD [38].

The deployment of wireless sensor networks in this Park has increased the number of information sources. Consequently data fusion and aggregation techniques are needed to manage such a huge amount of data. The solution that is being implemented is based on the integration of computational intelligence in the wireless sensor nodes. Some approaches to this integration are described in Section 4.3.3.

In our approach every node executes an artificial neuronal network to estimate the water level in flood zones. This task is performed once a day. The neural network is a Self-Organized Map (SOM) trained with historical data obtained by the EBD over several years. The SOM receives data from environmental sensors installed in the node (level 0, source preprocessing according to the JDL definition of fusion) and produces an output that identifies the state of the water level with a neuron at the output layer of the SOM (level 1, object refinement). This procedure constitutes a data fusion technique that suppresses the redundancy at the estimation of the system parameter, that is, the water level in the flood zone under study (level 3, impact assessment, and level 4, process refinement).

A node processes this information in the following way. If the state is different to the state estimated the day before, the node notifies this event to a base station. If the state is equal to the state obtained the day before, the node sends no information. This way the nodes waste neither energy nor bandwidth. In this situation we have a case of research based on trade-off and resources: energy vs. estimation accuracy.

Hence the nodes are deployed over a flood zone, and every node produces its own estimation; it is clear that there will be several estimations spread over the sensor network. Some nodes will change their daily estimation with a higher frequency than others. Consequently the former will produce more traffic than the latter. As the information is sent to a base station using multihop protocols based on IEEE 802.15.4 standard, this information needs to be aggregated. At this point a second fusion technique is implemented to determine how many data aggregators must be used in the wireless sensor network and where. The use of aggregators enhances the speed of data processing (level 2, situations refinement).

Once the base station receives the aggregated data, it sends it to the EBD through the intranet network (WiFi and Internet). At the EBD the biologists interpret the results obtained by the sensor networks and evaluate the accuracy (level 5, cognitive refinement).

This application example illustrates the cooperative activities performed by all the nodes in the network towards a common goal, that is, the estimation of the water level of a flood zone. From the individual tasks developed at each node to the communications between them, all the activities constitute a cooperative networking scheme.

4.3   Research Directions for Cooperation in WAdSN

A set of network devices is a system that could be quite hard to manage if only the physical and MAC layers are implemented. As we mentioned above, aspects such as time synchronization, calibration and location are fundamental for implementation in a WAdSN. In this sense, routing, data aggregation and fusion can also be understood as fundamental. Obviously the upper layers of the network stack are the proper place for implementation. Therefore cooperation networking, as we have seen in the previous section, must be implemented in these upper layers. There are many different ways of doing this and we will discuss some of them in the following subsections. In particular, we will look at some of the current approaches that may facilitate the implementation of cooperation in WAdSN.

4.3.1 Middleware for WAdSN

All the nodes in a wireless sensor network have to run a program written in a specific programming language. Compatibility problems may arise if the executable code of this program runs in different hardware platforms to the network nodes. The use of operating systems solves part of these problems, bringing a hardware abstraction that makes it slightly easier to write programs (the program can be written in the same language for all network nodes but, if different hardware and/or operating systems are used, it would be diffcult to create a source code without incompatibilities between the different versions of hardware, compilers and operating systems). Moreover, additional problems arise when different networking technologies are used to communicate the nodes of the system (imagine for instance that one set of tiny nodes coexists with another set of more powerful nodes, all running the same program), or when one client/server architecture has to be developed in a simple way.

These problems, specific to distributed systems, can be solved by adding an additional layer between applications and operating system: the middleware (Figure 4.2). This layer provides a common interface for communicating clients and services, methods for service discovering (how to know if there are services in the network), service description (how to know what it does), service control (how to use it), and even a common language for developing applications, services and clients. Sharing of services is a good way of performing collaboration among nodes in a WAdSN. Furthermore, the unified view that the middleware provides allows the system to be considered as a single system and not just as the sum of a (high) number of sensor nodes. There are several middleware architectures that meet these requirements:

- **Distributed Tuples:** Offers abstraction for distributed tuples. Examples: L2IMBO, LINDA.
- **Remote Procedure Call:** Abstraction of external procedure invocation: Java Remote Method Invocation, Modula-3, XML-RPC, .NET Remoting.
- **Message-Oriented Middleware:** Message queue abstraction for messages between distributed applications. Examples: Advanced Message Queuing Protocol, Java Message Service.

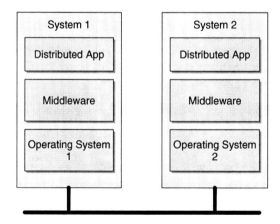

Figure 4.2 Middleware in distributed systems.

- **Object Request Broker:** Remote object management abstraction allows using remote objects as if they were in the user's own space. Examples: DCOM, CORBA, COM.

But, why should we have to do these things with a WAdSN? In a way, a WAdSN can be seen as a distributed system. It has a relatively large set of computational nodes that work together to perform a specific task: for instance gathering environmental information and sending it to a central node for processing. But this vision can be interpreted in another way: each network sensor performs a service that offers actions such as reading the environmental values obtained in the sensors, reset, calibration, and an event subscription that advises clients if a sensor reading changes. At the same time, nodes like the central node mentioned above can be clients of these services and conduct operations in the service of the sensors/actuators.

The use of middleware has until now been restricted to large systems where resource restrictions have not been a problem and the networking technologies used have been very fast, homogeneous and reliable. However, in WAdSN, the resources (memory and computing power) are very limited, the small batteries of sensor nodes impose limits on data communications and operating systems and hardware platforms are very heterogeneous. This heterogeneity may simply be due to the fact that the system is made up of relatively simple sensor nodes plus a number of central nodes with larger resources and better capabilities. In addition, this new layer imposes the use of more memory for programs, more computation power requirements, and more energy wasted in communications due to the protocols used to maintain client-server architecture (for example, the protocol for service discovering requires more messages to be sent between nodes).

Thus, the design principles of a middleware oriented towards WAdSN have to be quite different to a 'standard' middleware for these reasons. We must also consider the nature of applications implemented in this kind of system [39]:

- Distributed by nature.
- Limited by low machine resources.
- Focused on energy saving.
- Dynamic availability and quality of data collected from the environment.
- Constrained applications quality of services.

So, the design principles of WAdSN middleware could be summarized as follows [40]:

- Data-centric: focused on supporting distributed data requests.
- Distributed and collaborative algorithms.
- Support of data aggregation.
- Lightweight.
- Cluster based architecture: The whole system is a set of clusters, each of which consists of several sensor nodes and one head node responsible for cluster coordination.
- Virtual-machine abstraction and (preferably) a common language to specify the applications in the whole system, regardless of the operating system and hardware platform.

Nowadays, some implementations of middleware can be accommodated in WAdSN, and they follow the above-mentioned principles.

- **Cougar [41] and SINA [42]:** Both provide a distributed database interface to query WAdSN data in a SQL-like style. Energy saving policies are implemented to make the queries at the lowest power consumption. SINA has a cluster-based architecture.

- **AutoSec [43]:** Provides access to sensor network's resources for control and data. This implementation takes into account the quality of service of each query launched to the system, adapting the data collecting strategies to the sensor network.
- **DSWare [44]:** Very similar to AutoSec, is capable of maintaining clusters of nodes and obtaining information from them transparently to provide reliable measurements regardless of network failures.
- **IMPALA [44]:** Proposes asynchronous event-based methods for WAdSN to implement a multi-agent system. Agents can move through the network depending on the system's requirements.
- **MILAN [39]:** Designed to help in the development of portable applications, its applications are capable of adapting to changes in network topology and energy constraints. It supports a cluster-based architecture where the system itself identifies the clusters to meet the application's requirements, the energy consumption and the desired bandwidth. MILAN is not only an intermediate layer between applications and network/operating system layer: it extends into the network layer to control certain policies related to obtaining the optimum performance at minimum consumption.
- **Mires [45]:** Implemented under TinyOS/NESC, this middleware offers a message oriented architecture (with publish/subscription to data services) that complies with the traditional middleware for large and distributed systems. Adaptations for WAdSN are made, such as routing features and data aggregation.
- **COCA [46]:** Cross-Layer Cooperative Caching is a cluster-based middleware that provides data caching and management services to the applications in Ad-Hoc networks. Although not specifically oriented towards WAdSN, it shows how the very important issue of data caching could be implemented in a cooperative way, and many of their results may also be useful for WAdSNs.

As we can see, there are several solutions for implementing a middleware in a WAdSN. However, there is still work to be done [47], [40]. Middleware solutions have no solid definition of functionality. As we saw above, the extension of the middleware layer and the functionalities of each one are different and it is not possible to extract a common set of functionalities from all of them. Support for inclusion of time and location of measurement extracted from the environment is not present in these implementations and this is a very important feature within this type of network. Proactivity has been widely considered and marks a trend to follow for future implementations. The whole system must take actions before an important event occurs. It must look for changes in the network topology, QoS, and other aspects to adapt the whole system before things go wrong.

However, one of the most interesting advances in the middleware systems of the future will be the increased capabilities of future sensor nodes. Moore's Law guarantees growths in speed, resources and cost. In the future it may be possible to apply middleware solutions currently used in large systems to these nodes, making the absolute integration of wireless networks within Internet a near reality, an idea that has been pursued for several years, since the researchers at MIT defined the term The Internet of Things.

4.3.2   Multi-Agent Systems in WAdSN

In the previous section we looked at the advantages of using a middleware framework: Easy to program in several hardware platforms at a time; abstraction of network and operating system services; easy to discover and use resources scattered in the network and (desirably for WAdSN middleware) control and automatic adaptation of the whole system to network constraints. Middleware therefore, empowers the capabilities of systems that use WAdSN together with other hardware platforms, integrating WAdSN nodes in a unique distributed system and making it possible for programs (or services) operating within the system to run, communicate and use all the system's resources regardless of which hardware is being used and which network technology has to be used to communicate with them.

The only diffculty with these frameworks is to obtain an implementation lightweight enough to be fitted into the very constrained hardware of a WAdSN node. If we may overcome this, the following step is to include certain degree of intelligence in WAdSN.

Knowing that an agent is an entity located in one environment, capable of performing flexible and autonomous actions to achieve its goals [48], how can this concept be applied to WAdSN? Initially, the application of agents was a technology designed to implement a kind of 'life' in a system to solve complex algorithms; like in a machine with lots of small programs coexisting with each other. The rules for interacting, surviving and evolving in this 'world' led to the solution of the problem in a collaborative fashion. The following evolution in this technology was to scatter agents between various machines, creating a system consisting of several machines, each containing one or more agents. In both cases, we are talking about 'multi-agent system' (MAS) but the more recent evolution allows the migration of agents between machines to meet the system's computational requirements (Figure 4.3).

If these machines all have the same hardware and software architecture, the only remaining problem for the implementation of multi-agents is the issue of establishing the migration and communication mechanisms between machines. But, when heterogeneity in hardware and software exists, the middleware framework becomes a necessary solution for multi-agent systems technology. Several middleware solutions provide a hardware abstraction and a common language for defining and describing applications regardless of hardware implementation so, with one of them, the creation of a multi-agent system is solved. However, implementing mobility in agents requires another type of capability that the majority of middleware solutions (except for IMPALA) do not have. Every agent system requires a platform providing an identification service, mobility, communication, and so on.

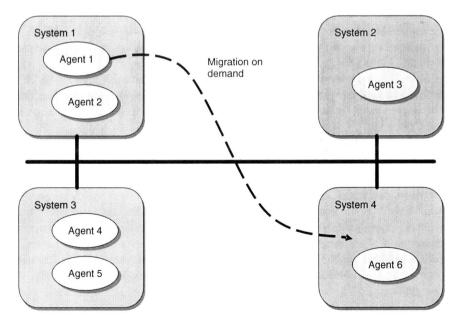

Figure 4.3   The concept of multi-agent system.

This approach is implemented in a WAdSN through a set of agents that reside in the nodes of the WAdSN [49]. These agents can interact with other agents and gather information about the environment (for example: try to classify the information) or another important aspect of the system (for example: they can be aware of routing tasks in the WAdSN to minimize energy consumption). They can take actions based on the information gathered, communicate with other agents in the system, and move from one sensor to another if required. These agents and their actions are supported by a middleware architecture that guarantees [50], [51]:

• Local services: Agents can access local resources through the platform.
• Agent discovery and maintenance: Detection of agent in the neighbourhood. Also, agents must be initiated and stopped when required.
  – Agent = code + data + state + meta-information.
• Common language: For defining and describing agents.
• Reactivity: Capability to detect and react to changes in the environment.
• Asynchronous operation: Agents must react to changes in the environment. This operation mode can save energy in WAdSN.
• Autonomy: Agents can control their own actions.
• Communications: With other agents, regardless of where they are.
• Collaboration and cooperation: Agents must be capable of processing information provided by other agents.
• Mobility: Capability to migrate between nodes.

This set of features seems quite large for a WAdSN node. There are implementations for multi-agent systems that operate in medium-large machines (as in middleware, the set of protocols and functionalities required is large and so, therefore, is the footprint).

FIPA [52] defined a reference architecture for MAS but it is too large to be included in WAdSN. However, this architecture must be taken into account because it is the base of some 'standardized' frameworks of MAS. LEAP is an example of one of these frameworks. It is a lightweight version of FIPA architecture but not light enough for WAdSN because it is implemented in Java.

Another implementation is the above-mentioned IMPALA [44], which enables application modularity, adaptability to network failures and reparability in WAdSN under an asynchronous operation mode. Agilla [53] is a framework that supports code mobility of agents written in a high-level code. Data communications are made based on a shared tuple space. MASIF [54] is a set of interfaces for promoting interoperability and mobility of agents in distributed systems but it cannot be implemented in WAdSN because of its large footprint and resource consumption (it is based on CORBA).

4.3.3 Artificial Neural Networks in WAdSN

Nowadays, neural networks cover a wide spectrum of applications: they can be used when no algorithm exists to solve a specific task, but the solution can be expressed on the basis of a rich set of input examples. Another application area of artificial neural networks (ANN) is the classification of information and pattern recognition (as in weather forecasting, etc.) or when algorithms are incapable of solving a problem. There are several reasons for using ANN: high robustness against noise, malfunctions or even failure; easy to parallelize and distribute; and lightweight.

Basically, the concept of artificial neuron is an entity whose output is defined as a function of a set of weighted inputs (from each of the inputs, see Figure 4.4). The weights of each input define the interaction between neurons so, with a low weighted input coming from a neuron,

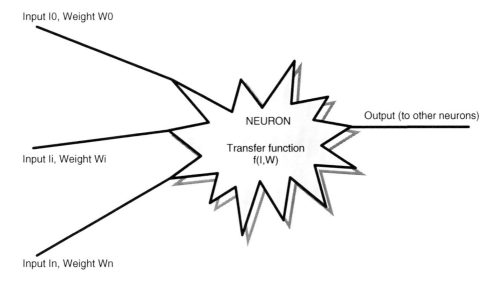

Figure 4.4   Model of an artificial neuron.

it seems that there is almost no interaction with it. This output can be codified as a number within a range (for example: $0-1$) or as a series of spikes distributed in time. In the latter case, we are talking about Spiking Neurons [55], [56]. The frequency of the spike train or the space between spikes (these are two methods of coding outgoing information in this type of neural network) defines the neuron's excitation level, or state. Of course, the same coding system is also used in the neuron's inputs. Therefore, the output function is defined as a weighted function of neuron's incoming pulse trains, allowing the building of layers of interconnected neurons to work together. Spiking neural networks is a more realistic way of implementing a neural system because the way this type of neuron runs is closer to the natural entities that they try to imitate.

Neurons are organized in layers and usually, one neuron from layer i has an input to all neurons in layer $i - 1$. In this case, we can talk about feed-forward ANN (Figure 4.5), where information is spread from input to outputs. In the event of any layer i spreading information to layer $i - k$; $k > 0$, the ANN can be classified as a recurrent ANN (Figure 4.6). In an ANN, there must be an input layer and a (sometimes is the same) output layer. Multiple layers can

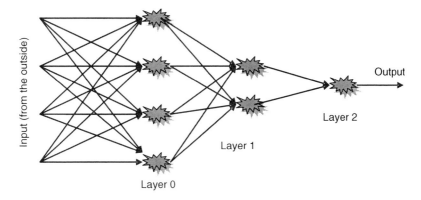

Figure 4.5   Example of a 3-layer feed-forward ANN.

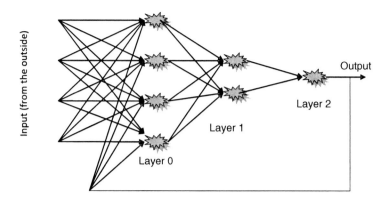

Figure 4.6   Example of a 3-layer recurrent ANN.

be placed between them, forming a multilayer ANN. The stage prior to operating with an ANN is the learning procedure. During this stage, the weight of each neuron takes its final value through a process in which a known input is presented to the ANN and the weights of neurons are modified to achieve the desired output at ANN's output layer. There are several classes of learning procedures: supervised, not supervised and reinforcement. When the ANN has finally been configured, each ANN neuron executes its output function only when there is any change in input values. This algorithm is usually simple and fast.

How can an ANN be implemented in a WAdSN? Bearing in mind the similarity between neurons and sensor nodes, one may think that the only way to implement an ANN is to assign one neuron to each node in the network. This way, the neurons implemented in the WAdSN will be the neurons of the input layer. Further layers could be implemented in a more powerful network, where the information processing can be finished without exhausting the resources of the WAdSN. The ANN built in this way would be valid for pattern recognition, data classification or predictions from the input data. An example of this kind of implementation can be seen in [57].

Another type of implementation of ANN in WAdSN is to keep a complete ANN within each network node. This way, each network node has a trained ANN to solve autonomously a specific problem related to some aspect of the WAdSN. For example, [27] use this type of implementation to solve the problem of routing optimization. In [26] ANN and genetic algorithms are used to solve two important problems in WAdSN: sensor self-calibration and the insertion of new nodes in runtime. Basically, each node has a complete ANN that can be trained by a genetic algorithm once it is inserted in the network. The genetic algorithm is used to contrast the output of the ANN (that classifies the outputs of the sensor) with an ideal value extracted from a lookup table. This way, a node can be inserted at anytime in the network and the genetic algorithm will adjust the ANN of this node based on its inputs. Finally, spiking neural networks are the subject of intense research efforts to implement bio-inspired devices where information flows in spike trains transported by AER buses (Address Event Representation). One example of these kinds of devices can be found in [58]. In this project, the chain of bio-inspired devices is connected by wire and instead of focusing on low power consumption the system has been designed as a prototype to demonstrate the suitability of these devices for certain tasks. In [59], [60] neuro-inspired devices and wireless sensor networks are combined to offer a fast and low power detection device (each one implemented by spiking neural networks) and a persistent and low-cost form of communication (each WAdSN node is responsible for its sensor communicating with a central node that collects and processes all the information provided by the sensors) that enables the system to work for long periods of time without maintenance.

4.4 Final Remarks

Throughout this chapter cooperation in wireless ad-hoc and sensor networks (WAdSN) has been shown as a collaborative action which network nodes are involved in. According to the Open System Interconnection model, OSI, proposed by the International Organization for Standardization, ISO, the development of cooperation could be understood as being located between the application and low level layers (physical and link layer).

Time synchronization, localization and calibration (SLC) are services that cooperation must provide. Consequently, there needs to be some mechanism to allow the provision of these services. As explained in Section 4.3.1, this mechanism works as a container of cooperation services, allowing their implementation in a consistent manner. These services may include not only SLC, but also data caching, data management, etc. With middleware, cooperation services can be implemented and accessed by the application regardless of the hardware or network technology being used.

The use of Multi-Agent Systems (MAS) is another approach to network cooperation. Several implementations of MAS over WAdSN are being carried out by different research groups, as explained in Section 4.3.2. In some ways MAS could also be understood as a middleware.

Middleware and multi-agent systems are concepts that have been created bearing in mind other kinds of networks. For this reason current implementations of them are complex and hardly feasible over ad hoc and sensor networks. However, the adaptation of MAS and middleware to these special networks will change the current conception of cooperation in WAdSN. Conceptually it is possible to adapt them. So, we are faced with the challenge of adapting current solutions to the case of resource and energy limited nodes that communicate in an ad-hoc fashion, providing at the same time scalability, reliability and dynamic adaptation to the environment.

4.5 Acknowledgements

This work was supported by contracts TIN2006.15617.C03.03, **AmbienNet**: Ambient Intelligence Supporting Navigation for People with Disabilities; P06TIC-2298, **SemiWheelNav**: External sensing based semiautonomous wheelchair navigation; P07-TIC-02476, **ARTICA**: Wireless Sensor Networks and Computational Intelligence in Environmental Applications, as well as by the Telefonica Chair 'Intelligence in Networks' of the University of Sevilla (Spain).

References

[1] T. Vercauteren, D. Guo and X. Wang. Joint multiple target tracking and classification in collaborative sensor networks. *IEEE Journal on Selected Areas in Communications*, 23(4), April 2005.

[2] I. Akyildiz, Y. Su, W. Sankarasubramaniam and E. Çayirci. *Wireless sensor networks: A survey. Computer Networks*, Elsevier, 38: 393–422, December 2002.

[3] W. Ye, J. Heidemann and D. Estrin. Medium access control with coordinated adaptive sleeping for wireless sensor networks. *IEEE/ACM Transactions on Networking*, 12(3), 2004.

[4] M. Kohvakka, M. Kuorilehto, M. Hännikäinen and T. D. Hämäläinen. Performance analysis of IEEE 802.15.4 and Zigbee for large-scale wireless sensor network applications. In Proc. Third ACM International Workshop on Performance Evaluation of Wireless Ad Hoc, Sensor, and Ubiquitous Networks, pp. 48–57, Málaga, Spain, 2006.

[5] K. Zen, D. Habibi, A. Rassau and I. Ahmad. Performance evaluation of IEEE 802.15.4 for mobile sensor networks. In 5th IFIP Int. Conf. on Wireless and Optical Communications Networks, WOCN'08, pp. 1–5, May 2008.

[6] K. Römer, P. Blum and L. Meier. *Handbook of sensor networks: algorithms and architectures*. John Wiley & Sons Ltd, 2005.

[7] G. Pottie and W. Kaiser. *Principles of Embedded Networked Systems Design*. Cambridge University Press, 2005.

[8] A. Marco, R. Casas, J. Falco, H. Gracia, J. Artigas and A. Roy. Location-based services for elderly and disabled people. *Computer Communications*, 31(4): 1055–1066, 2008.

[9] A. Marco, R. Casas, J. Sevillano, V. Coarasa, J. Falcón and M. Obaidat. Multi-hop synchronization at the application layer of wireless and satellite networks. In Proc. IEEE Global Communications Conference (IEEE GLOBECOM 2008), pp. 1–5, New Orleans, LA, USA, December 2008.

[10] M. Mao, B. Fidan and B. Anderson. Wireless sensor network localization techniques. *Computer Networks*, 51(4): 2529–2553, 2007.

[11] R. Casas, A. Marco, J. Guerrero and J. Falcón. Robust estimator for nonline-of-sight error mitigation in indoor localization. In *EURASIP Journal on Applied Signal Processing*, pp. 1–8, 2006.

[12] N. Bulusu, J. Heidemann and D. Estrin. GPS-less low cost outdoor localization for very small devices. *IEEE Personal Communications Magazine*, 7(5): 28–34, 2000.

[13] P. Baronti, P. Pillai, V. Chook, S. Chessa, A. Gotta, and Y. Fun Hu. Wireless sensor networks: A survey on the state of the art and the 802.15.4 and Zigbee standards. *Computer Communications*, 30(3): 1655–1695, 2007.

[14] D. S. Niculescu and B. Nath. Dv based positioning in ad hoc networks. *Telecommunication Systems*, 22(1-4): 267–280, 2003.

[15] B. Sundararaman, U. Buy and A. Kshemkalyani. Clock synchronization for wireless sensor networks: A survey. *Ad Hoc Networks*, 3: 281–323, 2005.

[16] M. Maróti, B. Kusy, G. Simon and A. Lédeczi. The flooding time synchronization protocol. In Proceedings of the 2nd International Conference on Embedded Networked Sensor Systems, pp. 39–49, Baltimore, MD, USA, 2004.

[17] J. Elson, L. Girod and D. Estrin. Fine-grained time synchronization using reference broadcasts. In *Proc. 5th Symp. Op. Sys. Design and Implementation*, pp. 147–163, Boston, MA, USA, 2002.

[18] J. Elson and D. Estrin. Time synchronization for wireless sensor networks. In *International Parallel and Distrib. Processing Symp.*, San Francisco, CA, USA, 2001.

[19] S. PalChaudhuri, A. Saha and D. Johnson. Adaptive clock synchronization in sensor networks. In Proceedings of the 3rd International Symposium on Information Sensor Networks (IPSN), pp. 340–348, Berkeley, CA, USA, April 2004.

[20] Y. Hong and A. Scaglione. A scalable synchronization protocol for large scale sensor networks and its applications. *IEEE Journal on Selected Areas in Communications* (JSAC), 23(5): 1085–1099, 2005.

[21] V. Bychkovskiy, S. Megerian, D. Estrin and M. Potkonjak. A collaborative approach to in-place sensor calibration. In Proceedings of the 2nd International Workshop on Information Processing in Sensor Networks (IPSN'03), Berkeley, CA, USA, April 2004.

[22] Y. Xu, J. Heidemann and D. Estrin. Geography informed energy conservation for ad hoc routing. In Proc. of the 7th Int. Conf. on Mobile Computing and Networking (MobiCom 2001), pp. 70–84, Rome, Italy, July 2001.

[23] P. Sheu, S. Chien, C. Hu and Y. Li. An efficient genetic algorithm for the power-based QoS many-to-one routing problem for wireless sensor networks. In International Conference on Information Networking, ICOIN 2005, pp. 275–282, Jeju, Korea, February 2005.

[24] S. Jin, M. Zhou and A. S. Wu. Sensor network optimization using a genetic algorithm. In Proceedings of the 7th World Multiconference on Systemics, Cybernetics, and Informatics, Orlando, FL, USA, July 2003.

[25] A. Talekder, R. Bhatt, S. Chandramouli, L. Ali, R. Pidva and S. Monacos. Autonomous resource management and control algorithms for distributed wireless sensor networks. In The 3rd ACS/IEEE International Conference on Computer Systems and Applications, 2005, pp. 19–26, Cairo, Egypt, Enero 2005.

[26] R. Abielmona, V. Groza and W. Pretiu. Evolutionary neural network network-based sensor self-calibration scheme using IEEE 1451 and wireless sensor networks. In International Symposium on Computational Intelligence for Measurement Systems and Applications, CIMSA 2003, pages 38–43, Lugano, Switzerland, July 2003.

[27] J. Barbancho, C. León, F. Molina and A. Barbancho. Using artificial intelligence in routing schemes for wireless networks. *Computer Communications*, 3842: 2802–2811, June 2007.

[28] C. Aurrecoechea, A. Campbell and L. Hauw. *A Survey of QoS Architectures. Multimedia Systems*, Springer-Verlag, 6: 138–151, 1998.

[29] B. Sabata, S. Chatterjee, M. Davis, J. Sydir and T. Lawrence. Taxonomy for QoS specifications. In Proceedings of the Third International Workshop on Object-Oriented Real-Time Dependable Systems, pp. 100–107. IEEE Press, 1997.

[30] R. Iyer and L. Kleinrock. QoS control for sensor networks. In IEEE International Conference on Communications, ICC'03, volume 1, pp. 517–521. IEEE Press, May 2003.

[31] J. Kay and J. Frolik. Quality of service analysis and control for wireless sensor networks. In 2004 IEEE International Conference on Mobile Ad-hoc and Sensor Systems, pp. 359–368. IEEE Press, October 2004.

[32] M. Perillo and W. Heinzelman. Sensor management policies to provide application QoS. *Ad Hoc Networks*, 1: 235–246, 2003.

[33] V. Rakocevic, M. Rajarajan, K. McCalla and C. Boumitri. QoS constraints in bluetooth-based wireless sensor networks. *Lecture Notes in Computer Science*, Springer Verlag, 3266: 214–223, 2004.

[34] D. Hall and S. McMullen. *Mathematical Techniques in Multisensor Data Fusion*. Artech House Publishers, 2004.

[35] B. Krishnamachari, D. Estrin and S. Wicker. Modelling data-centric routing in wireless sensor networks. In Proceedings of INFOCOM 2002, New York, USA, June 2002.

[36] M. M. Rabbat and R. Nowak. Distributed optimization in sensor networks. In Proceedings of the 3rd International Symposium on Information Sensor Networks (IPSN), Berkeley, CA, USA, April 2004.

[37] A. Scaglione and S. Servetto. On interdependence of routing and data compression in multi-hop sensor networks. In Proceedings of the 8th Annual ACM/IEEE International Conference on Mobile Computing and Networking (MobiCom'02), Atlanta, Georgia, 2002.

[38] Doñana Biological Stations. http://www.ebd.csic.es/website1/Principal.aspx.

[39] Heinzelman, W.B. Middleware to support sensor network applications. *IEEE Network*, 18(1): 6–14, 2004.

[40] Y. Yu, B. Krishnamachari and V. Prasanna. Issues in designing middleware for wireless sensor networks. *IEEE Network Magazine*, 12(3), January 2004.

[41] P. Bonnet, J. Gehrke and P. Seshadri. Querying the physical world. *IEEE Personal Communication*, 7(10), October 2000.

[42] C. Jaikaeo, C. Srisathapornphat and C. C. Shen. Querying and tasking in sensor networks. In International Symposium on Aerospace/Defense Sensing, Simulation, and Control (Digitization of the Battlespace V). In SPIE's 14th Annual, pp. 24–28, Orlando, FL, USA, April 2000.

[43] Q. Han and N. Venkatasubramanian. Autosec: An integrated middleware framework for dynamic service brokering. *IEEE Distributed Systems Online*, 2(7), 2001.

[44] S. Li, S. Son and J. Stankovic. Event detection services using data service middleware in distributed sensor networks. In Proceedings of the 2nd International Workshop on Information Processing in Sensor Networks, April 2003.

[45] E. e. a. Souto. Mires: A publish/subscribe middleware for sensor networks. *Personal Ubiquitous Computing*, 10(1), 2005.

[46] M. Denko, J. Tian, T. Nkwe and M. Obaidat. Cluster-based cross-layer design for cooperative caching in mobile ad hoc networks. *IEEE Systems Journal*, 3(4): 499–508, 2009.

[47] K. K. Römer, O. Kasten and F. Mattern. Middleware challenges for wireless sensor networks. In *Proceedings of the ACM SIGMOBILE Mobile Computing and Communication Review (MC2R)*, 6(4), pp. 59–61, October 2002.

[48] N. Jennings, K. Sycara and M. Wooldridge. A roadmap of agent research and development. *Autonomous Agents and Multi-Agent Systems*. Kluwer Academic Publishers, 1: 7–38, 1998.

[49] H. Qi, Y. Xu, and X. Wang. Mobile-agent-based collaborative signal and information processing in sensor networks. In Proceedings of the IEEE, 8(91), August 2003.

[50] J. Martínez, A. García, A. Sanz, L. López, V. Hernández and A. Dasilva. An approach for applying multi-agent technology into wireless sensor networks. In Proceedings of the 2007 Euro American Conference on Telematics and information Systems, Faro, Portugal, Mayo 2007. EATIS, ACM.

[51] IEEE/FIPA WG Mobile Agents. Ulrich Pinsdorf. Presentation of WG Mobile Agents Group, in FIPA. 2005.

[52] Foundation for Intelligent Physical Agents. http://www.fipa.org.

[53] C. Fok, G.-C. Roman, and C. Lu. Rapid development and flexible deployment of adaptive wireless sensor network applications. Technical Report WUCSE-04-59, Washington University, Department of Computer Science and Engineering, St. Louis, USA, 2004.

[54] Object Management Group. http://www.omg.org.

[55] W. Mass and C. M. Bishop. *Pulsed Neural Networks*. The MIT Press, 2001.

[56] W. Gerstner and W. Kistler. *Spiking Neurons Models. Single Neurons, Populations, Plasticity*. Cambridge University Press, 2002.

[57] Q. Huang, T. Xing and H. T. Liu. Vehicle classification in wireless sensor networks based on rough neural network. *Lecture Notes in Computer Science*. Springer-Verlag: Berlin Heidelberg, pp. 58–65, 2006.

[58] R. Serrano-Gotarredona, M. Oster and P. Lichtsteiner. Caviar: A 45kneuron, 5m-synapse, 12g-connects/sec AER hardware sensory-processing-learning-actuating system for high speed visual object recognition and tracking. *IEEE Transactions on Neural Networks*, 20(9): 1417–1438, September 2009.

[59] F. Zhengming, E. Culurciello, P. Lichtsteiner and T. Delbruck. Fall detection using an address-event temporal contrast vision sensor. In IEEE International Symposium on Circuits and Systems, 2008. ISCAS 2008, pp. 424–427, May 2008.

[60] http://pantheon.yale.edu/~dk6/TDsensor.html.

5

Cooperation in Autonomous Vehicular Networks

Sidi Mohammed Senouci[1], Abderrahim Benslimane[2] and Hassnaa Moustafa[3]
[1]*DRIVE Laboratory, University of Bourgogne, France*
[2]*LIA/CERI, University of Avignon, France*
[3]*France Telecom – Orange Labs Networks and Carriers, France*

5.1 Introduction

Vehicular networks are considered to be a novel class of wireless networks that have emerged thanks to the advances in wireless technologies and automotive industry. Vehicular networks are spontaneously formed between moving vehicles equipped with wireless interfaces that could be of homogeneous or heterogeneous technologies. These networks, also known as VANETs (Vehicular Ad hoc Networks), are considered to be one of the ad hoc networks real-life applications enabling communications among nearby vehicles as well as between vehicles and nearby fixed equipments, usually described as roadside equipments. Vehicles can be either private, belonging to individuals or private companies, or public transportation means (e.g., buses and public services vehicles such as police cars). Fixed equipments can belong to the government, or private network operators or service providers.

Vehicular networks applications range from road safety applications oriented to the vehicle or to the driver, to entertainment and commercial applications for passengers, making use of a plethora of cooperating technologies. This new computing paradigm is promising by allowing drivers to detect hazardous situations to avoid accidents, and to enjoy the plethora of value-added services. The increased number of vehicles on the road magnifies significantly the unpredictable events outside vehicles. In fact, accidents arrive rarely from vehicles themselves and mainly originate from on-road dynamics. This means that cooperation using vehicular networks must be introduced into transportation networks to improve overall safety and network efficiency, and to reduce the environmental impact of road transport.

As an example, let's take the Cooperative Collision Avoidance (CCA) application. There are two different ways to achieve cooperative collision warning: a passive approach and an active approach. In a passive approach, a vehicle broadcasts frequently its location, speed, direction, and so on, and it is the responsibility of the receipt vehicle to take the decision on the eminent danger

Cooperative Networking, First Edition. Edited by Mohammad S. Obaidat and Sudip Misra.
© 2011 John Wiley & Sons, Ltd. Published 2011 by John Wiley & Sons, Ltd.

if it judges its existence. In an active approach, a vehicle causing an abnormal situation broadcasts an alarm message containing its location in order to warn vehicles in its neighbourhood.

In this chapter we are exploring cooperation issues in large-scale vehicular networks, where vehicles communicate with each other and with the infrastructure via wireless links. High-level services are built following a cooperative model that depends exclusively on the participation of contributing vehicles. Hence, we will focus on the major technical challenges that are currently being resolved from cooperation perspectives for various OSI layers, such as physical and medium access control layers, network and application layers, authentication and security, and so on.

The remainder of this chapter is organized as follows. Section 5.2 gives an overview on vehicular networks. Section 5.3 highlights some existing contributions in the field of cooperative vehicular networks. Finally, Section 5.4 summarizes and concludes the chapter.

5.2   Overview on Vehicular Networks

Vehicular networks can be deployed by network operators, service providers or through integration between operators, providers and a governmental authority. The recent advances in wireless technologies and the current and advancing trends in ad hoc networks scenarios allow a number of deployment architectures for vehicular networks, in highways, rural and city environments. Such architectures should allow the communication among nearby vehicles and between vehicles and nearby fixed roadside equipments. Three alternatives include: i) a pure wireless Vehicle-to-Vehicle ad hoc network (V2V) allowing standalone vehicular communication with no infrastructure support, ii) an Infrastructure-to-Vehicle or Vehicle-to-Infrastructure (I2V, V2I) architecture with wired backbone and wireless last hops, iii) and a hybrid architecture that does not rely on a fixed infrastructure in a constant manner, but can exploit it for improved performance and service access when it is available. In this latter case, vehicles can communicate with the infrastructure either in a single hop or multi-hop fashion according to the vehicles' positions with respect to the point of attachment with the infrastructure.

Vehicular networks applications ranges from road safety applications oriented to the vehicle or to the driver, to entertainment and commercial applications for passengers, making use of a plethora of cooperating technologies. The primary vision of vehicular networks includes real-time and safety applications for drivers and passengers, allowing for the safety of these latter and giving essential tools to decide the best path along the way. These applications thus aim to minimize accidents and improve traffic conditions through providing drivers and passengers with useful information including collision warnings, road sign alarms and in-place traffic view. Nowadays, vehicular networks are promising in a number of useful drivers and passengers oriented services, which include Internet connections facility exploiting an available infrastructure in an 'on-demand' fashion, electronic tolling system, and a variety of multimedia services.

However, to bring its potency to fruition, vehicular networks have to cope with some challenging characteristics [1] that include:

- *Potentially large scale*: As stated in the last section, most ad hoc networks studied in the literature usually assume a limited network size. However, vehicular networks can in principle extend over the entire road network and include so many participants.
- *High mobility*: The environment in which vehicular networks operate is extremely dynamic, and includes extreme configurations: in highways, a relative speed of up to 260 kms/h may occur, while density of nodes may be one to two vehicles per kilometer on low busy roads. On the other hand, in the city, relative speed can reach up to 100 kms/h and nodes' density can be high, especially in rush hours.

- *Network Partitioning*: Vehicular networks will be frequently partitioned. The dynamic nature of traffic and a low penetration of the technology may result in large inter-vehicle gaps in sparsely populated scenarios, and hence in several isolated clusters of nodes.
- *Network topology and connectivity*: Vehicular networks scenarios are very different from classical ad hoc networks ones. Since vehicles are moving and changing their position constantly, scenarios are very dynamic. Therefore the network topology changes frequently as the links between nodes connect and disconnect very often. Indeed, the degree to which the network is connected is highly dependent on two factors: the range of wireless links and the fraction of participant vehicles, where only a fraction of vehicles on the road could be equipped with wireless interfaces.
- *Security*: Security is a crucial aspect in vehicular networks in order to become a reliable and accepted system bringing safety onto public roads. Vehicular communication and its services will only be a success and accepted by customers if a high level of reliability and security can be provided. This includes authenticity, message integrity and source authentication, privacy and robustness,
- *Applications distribution*: From a general view, we can notice that building distributed applications involving passengers in different vehicles requires new distributed algorithms. As a consequence, a distributed algorithmic layer is required for managing the group of participants, and ensuring data sharing among distributed programs. Such algorithms could assimilate the neighbourhood instability to a kind of fault. However, the lack of communication reliability necessitates employing fault tolerant techniques.

Several technical challenges are not yet resolved in vehicular networks. Consequently, research works and contributions are needed to investigate such challenges aiming to resolve them. We will focus on some of these technical challenges that are being resolved from cooperation perspectives. Some of our related research contributions will be also presented in the following sections.

5.3   Cooperation at Different OSI Layers

We are interested in designing vehicular networks protocols, for which we would like to quantify performance gains due to relaying and cooperation. In this section, we will concentrate on cooperation at the various OSI layers.

5.3.1   Cooperation at Lower Layers

Cooperation from MAC layer viewpoint is classified into two classes: the homogenous MAC cooperation, where one distinct MAC layer is present in the system; and the heterogeneous MAC, where MAC protocols from different systems are used for cooperation [2]. Efficient MAC protocols [3], [4] need to be in place, while adapting to the high dynamic environment of vehicular networks, and considering messages priority of some applications (ex, accidents warnings). In spite of the dynamic topology and the high mobility, fast association and low communication latency should be satisfied between communicating vehicles in order to guarantee: i) service's reliability for safety-related applications while taking into consideration the time-sensitivity during messages' transfer, and ii) the quality and continuity of services for non-safety applications.

Many MAC protocols for vehicular ad hoc networks have been introduced in the literature. But, they do not involve any cooperation between vehicles except if we consider the competition to access a given channel (as in IEEE 802.11p or DSRC) is a kind of cooperation which is not realistic. So, we proposed in a recent work a cooperative collision avoidance system which consists of two fold contributions, that is, the cluster-based and risk-conscious approaches [3]. Our adopted strategy is referred to as the Cluster-based Risk-Aware CCA (CRACCA) scheme. First, we have presented a cluster-based organization of the target vehicles. The cluster is based upon several criteria, which define the movement of the vehicles, namely the directional bearing and relative velocity of each vehicle, and also the inter-vehicular distance. Second, we have designed a cooperative risk-aware Media Access Control (MAC) protocol in order to increase the responsiveness of the proposed CCA scheme. According to the order of each vehicle in its corresponding cluster, an emergency level is associated with the vehicle that signifies the risk to encounter a potential emergency scenario. In order to swiftly circulate the emergency notifications to collocated vehicles for mitigating the risk of chain collisions, the medium access delay of each vehicle is set as a function of its emergency level.

5.3.2 Cooperation at Network Layer

Cooperation from a network viewpoint concerns the cooperation mechanisms between network elements for traffic forwarding. More specifically, it is about the design of an efficient routing protocol that enables effective network resource management [5], [6]. We note that it is important to study the node behaviour in the case of infrastructure-less vehicular networks. In fact, in such networks, where no centralized entity exists, a malicious or self-interested user can misbehave and does not cooperate. A malicious user could inject false routing messages into the network in order to break the cooperative paradigm. The basic vehicular network functions subject to selfishness are dissemination and routing. For our propositions dealing with cooperative routing protocols and presented afterward, we considered that all vehicles are not selfish and cooperate to route data for the others.

Furthermore, vehicular networks face a number of new challenges like scalability and high mobility. An effective solution is also to define a robust self-healing and self-organizing architecture that facilitates the cooperation between vehicles. Depending on the application, this cooperation will be based on either proactive or reactive self-organization architecture [7], [8]. The two architectures are cross layer and structure intelligently the vehicular network in a permanent manner by portioning roads into adjacent segments seen as geographic fix clusters.

In the following, we give details of some of these network protocols and quantify performance gains due to relaying and cooperation.

5.3.2.1 Cooperative Routing in Vehicular Networks

In vehicular networks consisting of distributed vehicles, the information is routed from the source node to the destination node using intermediate nodes in a multi-hop fashion. These intermediate nodes cooperate with each other in transmitting the information, and through this cooperation effectively enhance the end-to-end delay. It is important to use the best intermediate nodes when multiple nodes exist in the transmission [9], [10]. Several questions arise in this context: What level of coordination among the cooperating nodes is needed? And how must the route selection be done to minimize the end-to-end delay? Here, we are interested by these problems. Hence, we developed a formulation that captures the benefit of cooperative transmission and developed a routing algorithm for selecting the optimal route under this setting.

Topology-based and position-based routing are two strategies of data forwarding commonly adopted for vehicular networks. The increasing availability of GPS equipped vehicles makes position-based routing a convenient routing strategy for these networks. Several variants of position-based concept have been proposed for data forwarding in vehicular networks [11]–[17]. Three classes of forwarding strategies can be identified for position-based routing protocols:

1) restricted directional flooding, 2) hierarchical forwarding, and 3) greedy forwarding [5]. Most of these protocols do not take into account the vehicular traffic, which means that such algorithms may fail in case they try to forward a packet along streets where no vehicles are moving. Such streets should be considered as 'broken links' in the topology. Moreover, a packet can be received by a node that has no neighbours nearer to the receiver than the node itself. In this case, the problem of a packet having reached a local maximum arises. These problems can be overcome to some extent knowing the real topology, by opting to use only streets where vehicular traffic exists. In addition, in [16], forwarding a packet between two successive intersections is done on the basis of a simple greedy forwarding mechanism. This classic greedy approach works well since it is independent of topological changes but it suffers from inaccurate neighbour tables since it does not consider the vehicle direction and velocity. Thus, it may be possible to lose some good candidate nodes to forward the packets. Our objective was to conceive a routing protocol that overcomes the above limitations.

We proposed GyTAR (improved Greedy Traffic Aware Routing protocol) [5], [6]; an intersection-based geographical routing protocol, capable of finding robust and optimal routes within urban environments. GyTAR scheme is organized into three mechanisms: i) a mechanism for the dynamic selection of the intersections through which packets are forwarded to reach their destination, and ii) an improved greedy forwarding mechanism between two intersections. Using GyTAR, packets will move successively closer towards the destination along the streets where there are enough vehicles providing connectivity. We do not impose any restriction to the communication model, and GyTAR is applicable to both completely ad hoc and infrastructure-based routing.

For the first mechanism 'Intersection selection', GyTAR adopts an anchor-based routing approach with street awareness. Thus, data packets are routed between vehicles, following the street map topology. However, unlike GSR [15] and A-STAR [16], where the sending node statically computes a sequence of intersections, the packet has to traverse in order to reach the destination, intermediate intersections in GyTAR are chosen dynamically and in sequence, considering both the variation in the vehicular traffic and distance to destination. Partial successive computation of the path has a threefold advantage: i) the size of packet header is fixed; ii) the computation of subsequent anchors is done exploiting more updated information about vehicular traffic distribution; iii) subsequent anchors can be computed exploiting updated information about the current position of the destination. When selecting the next destination intersection, a node (the sending vehicle or an intermediate vehicle in an intersection) looks for the position of the neighbouring intersections using the map. A score is attributed to each intersection considering the traffic density and the curvemetric distance to the destination. The best destination intersection (that is, the intersection with the highest score) is the geographically closest intersection to the destination vehicle having the highest vehicular traffic. After determining the destination intersection, the second mechanism 'improved greedy strategy' is used to forward packets towards the intersection. For that, all data packets are marked by the location of the next intersection. Each vehicle maintains a neighbour table in which the velocity vector information of each neighbour vehicle is recorded. Thus, when a data packet is received, the forwarding vehicle predicts the position of each neighbour using the corresponding recorded information (velocity, direction and the latest known position), and then selects the next hop neighbour (the closest to the destination intersection). Note that most of the existing greedy-based routing protocols do not use the prediction and consequently, they might lose some good candidates to forward data packets. Despite the improved greedy routing strategy, the risk remains that a packet gets stuck in a local optimum (the forwarding vehicle might be the closest to the next intersection). Hence, a recovery strategy is required. The recovery strategy adopted by GyTAR is based on the idea of 'carry- and-forward' [18]: the forwarding vehicle of the packet in a recovery mode will carry the packet until the next intersection or until another vehicle, closer to the destination intersection, enters/reaches its transmission range.

GyTAR efficiently utilizes the unique characteristics of cooperative vehicular environments like the highly dynamic vehicular traffic, road traffic density as well as the road topology in making routing and forwarding decisions. The selection of intermediate intersections among road segments is performed dynamically and in-sequence based on the scores attributed to each intersection. The scores are determined based on the dynamic traffic density information and the curvemetric distance to the destination. Simulation results showed that GyTAR performs better in terms of throughput, delay and routing overhead compared to other protocols (LAR and GSR) proposed for vehicular networks. The robust intersection selection and the improved greedy carry-and-forward scheme with recovery, suggests that GyTAR should be able to provide stable communication while maintaining high throughput and low delays for vehicular routing in urban environments.

Cooperation between vehicles can also help a given vehicle to access the Internet. When a node in a vehicular ad hoc network wants Internet access, it needs to obtain information about the available gateways and it should select the most appropriate of them. Exchanging information messages between vehicles and gateways is important for V2I. We can distinguish three different approaches to discover gateways: i) proactive gateway discovery, ii) reactive gateway discovery, and iii) hybrid gateway discovery. To connect vehicles to the Internet, our objectives are to reduce the overhead during the gateway discovery process, create a relatively robust network, and make the handovers seamless. We suggested a hybrid gateway discovery process that restricts broadcasts to a pre-defined geographical zone, while letting only some relays re-broadcast the advertised messages [19]. Stability metrics (for example, speed, direction and location) of vehicles can help us to predict the future location of vehicles, and the period that they stay in the transmission range of each other. We applied this information to estimate the link lifetime, and recursively the lifetime of routes from vehicles to gateways. Vehicles select the most stable route to gateways, and extend the lifetime of their connection. The most stable route is not necessarily the shortest one, it is the path with the longest lifetime. Here we are more interested in the lifetime of the connection rather than the number of hops to the destination. Having a list of routes to different gateways, a vehicle can hand-over the connection to the next available gateway before the current connection fails. If a vehicle does not receive advertisement messages, it should start sending out solicitation messages to find a new gateway. Internet access is provided by gateways implemented in roadside infrastructure units, and vehicles initially need to find these gateways to communicate with them. Gateway discovery is the process through which vehicles get updated about the neighbouring gateways. Gateways periodically broadcast gateway advertisement messages in a geographically restricted area using geocast capabilities. Gateway discovery aims at propagating the advertisement messages in VANET through multiple hops in this area. We call this area the broadcast zone of a gateway: a message that originated from that gateway should not be broadcast outside this zone. This area can be a rectangle or a circle, and is defined according to the distance between gateways, transmission range of the gateways, and density of the vehicles (whether it is a highway or city, traffic congestion, etc). For instance, suppose that the broadcast zone is selected to be a circle. Gateways send their location (xg, yg), as the centre of this circle, and a predefined radius along with the advertisement messages. Upon receiving the message, vehicles extract this information and can perceive if they are located inside or outside of the broadcast zone of a gateway.

To accomplish the task of proactive gateway discovery, we consider Optimized Dissemination of Alarm Messages (ODAM) [20], which is based on geographical multicast, and consists of determining the multicast group according to the driving direction and the positioning of the vehicles in a geographically restricted area using geocasting capabilities. These messages

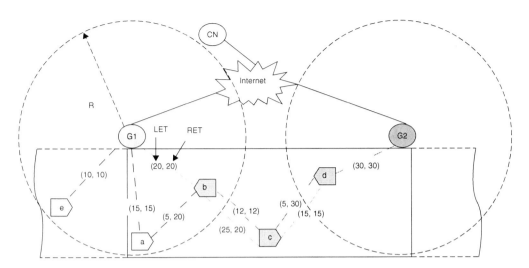

Figure 5.1 A gateway broadcasting an advertisement message. Transmission range R and broadcast zone of gateway G1 is shown. On the links, a couple of values (LET, RET) means (link expiration time, route expiration time). CN means the correspondent node (as in mobile IP).

are then re-broadcast in the network by some particular nodes called relays. Figure 5.1 shows a simple scenario, in which gateway 1 starts to broadcast advertisement messages to neighbouring vehicles. The broadcast zone is considered as a rectangle here, and has an intersection with the transmission range zone of gateway 2. In this case, messages from gateway 1 will be broadcast through multi hops to some of the nodes which are connected to gateway 2. Each advertised message contains the gateway address, relay address, message sequence number, broadcast zone, and the stability parameters. Stability parameters (sender position, sender speed, sender direction, and the estimated route expiration time) are used by each vehicle receiving the message to predict the link lifetime.

5.3.2.2   Cooperative Dissemination in Vehicular Networks within City Environment

Many of the vehicular network applications rely on disseminating data, for example, on the current traffic situation, weather information, road works, hazard warning, and so on. Typically, such applications are based on some form of proactive information dissemination in an ad hoc manner. Proactive information dissemination is, however, a difficult task due to the highly dynamic nature of vehicular networks. Indeed, vehicular networks are characterized by their frequent fragmentation into disconnected clusters that merge and disintegrate dynamically. One of the largely accepted solutions towards efficient data dissemination in vehicular networks is by exploiting a combination of fixed roadside infrastructures and mobile in-vehicle technologies. There are some recent examples of broadcasting protocols specifically designed for vehicular networks with infrastructure support [21], [22]. While such infrastructure-based approaches may work well, they may prove costly as they require the installation of new infrastructures on the road network, especially if the area to be covered is large.

In this context, our contribution was to propose a self-organizing mechanism to emulate a geo-localized virtual infrastructure (GVI) by a bounded-size subset of cooperating vehicles populating the concerned geographic region [23]. This is realized in an attempt to both approaching the performance of a real infrastructure while avoiding the cost of installing it. As we are dealing with the city environment, an intersection sounds suitable as a geographic region because of its better line-of-sight and also because it is a high traffic density area. Hence, the proposed GVI mechanism can periodically disseminate the data within a signalized (traffic lights) intersection area, controlled in fixed-time and operated in a range of conditions extending from under-saturated to highly saturated. Thus, it can be used to keep information alive around specific geographical areas [24] (nearby accident warnings, traffic congestion, road works, advertisements and announcements, and so on). It can also be used as a solution for the infrastructure dependence problem of some existing dissemination protocols like ODAM [20].

The geo-localized virtual infrastructure mechanism consists on electing vehicles that will perpetuate information broadcasting within an intersection area. To do so, the GVI is composed of two phases: i) selecting the vehicles that are able to reach the broadcast area (that is, a small area around the intersection centre, where an elected vehicle could perform a local broadcast); then, ii) among the selected vehicles, electing the local broadcaster which will perform a local single-hop broadcast once it reaches the broadcast area (that is, at the intersection centre).

In the first phase and as shown in the next Figure 5.2, among the vehicles which are around the intersection, only those within the notification region A_i (a cell centred on C_i and delimited by a ray of R/2 where R is radio range) could participate in the local broadcast. They are selected as candidates if they are able to reach the intersection centre C_i. In the second phase, a waiting time is assigned to each candidate vehicle. This waiting time considers the geographical location, direction and speed of the vehicle and also the desirable broadcast cycle time T of GVI.

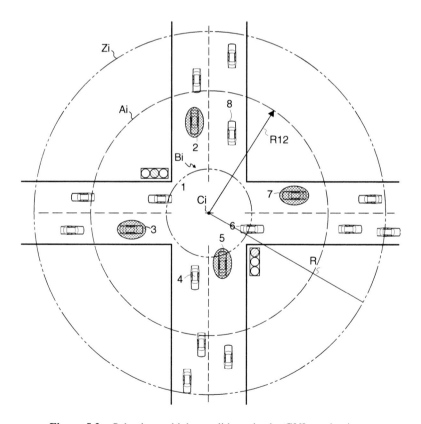

Figure 5.2 Selecting vehicle candidates in the GVI mechanism.

The candidate vehicle with the shortest waiting time will broadcast a short informative message telling other candidate vehicles that it has been elected as the local broadcaster.

Analytical and simulation results show that the proposed GVI mechanism can periodically disseminate data within an intersection area, efficiently utilize the limited bandwidth and ensure high delivery ratio. More precisely, with varying the broadcast cycle time T, we can have a kind of compromise between two metrics, namely the number of copies of the same message (which corresponds to a measure of the cost to provide the service) and the probability of informing a vehicle (which corresponds to a measure of quality of service). Indeed, if we want all vehicles to receive the message, we should decrease the broadcast cycle time value which will generate an overhead. However, we can minimize the number of copies of the same broadcast message received by a vehicle as long as we tolerate the fact that certain vehicles fail to receive the message. Analytical models showed that GVI fails only when the traffic density is extremely low with no sufficient cooperative vehicles within the intersection.

5.3.2.3   Cooperative Dissemination in Vehicular Networks within a Highway

Cooperative Collision Warning (CCW) is an important class of safety applications that target the prevention of vehicular collisions using vehicle-to-vehicle (V2V) communications. In CCW, vehicles periodically broadcast short messages for the purposes of driver situational awareness and warning. However, a classical broadcast cannot be used since it causes a protocol overhead and high number of message collisions which can be harmful for the safety of drivers. To overcome this limitation, we introduced an Optimized Dissemination of Alarm Messages (ODAM) [20] while restricting re-broadcast to only special nodes, called relays, and in restricted regions, called risk zones.

ODAM works as follow. When a crash occurs, a damaged vehicle or any other vehicle which detects this problem must broadcast an alarm message to inform the other vehicles about the danger. Several methods were used for the crash detection. For example, when an accident occurs, the activation of the airbag can initiate the alarm message broadcast. Among all the neighbours of this vehicle, only those which are in the risk zones take into account the message (Figure 5.3). Vehicles in these risk zones constitute a dynamic multicast group. Among all neighbours in the same zone, only one vehicle, called relay, must react to ensure the rebroadcast of the alarm message to inform vehicles which have not received the message yet. The relay is completely selected in a distributed way. Each vehicle can know, from the transported information in the alarm message which it receives, if it will become relay or not. Moreover, the relay must be

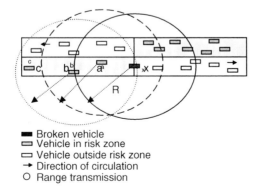

■ Broken vehicle
▭ Vehicle in risk zone
▢ Vehicle outside risk zone
➝ Direction of circulation
O Range transmission

Figure 5.3   Relevant areas and relay selection in ODAM.

selected in order to ensure the coverage of the greatest zone not yet covered by the sender. Consequently, the relay must be the furthest neighbour away from the sender.

In Figure 5.3, a damaged vehicle (x) broadcasts an alarm message. We remark that if the vehicle (a) were taken as relay, then (c) cannot be reached because it was out of the transmission range of (a). Against, if (b) were selected as relay, then (c) would have been reached and informed. A relay is designated as the vehicle having the minimum value of computed defertime. The vehicle which receives an alarm message should not rebroadcast it immediately but must wait during a defertime. The defertime value is inversely proportional to the distance from the sender to the receiver. At the expiration of this time, if a node does not receive another alarm of the same message, coming from another node, then it rebroadcasts the message. In this way it is chosen as relay.

To favour the furthest vehicle from the sender to becoming relay, defertime of each vehicle must be inversely proportional to the distance which separates it from the sender. As the distance between these two vehicles is large so defertime is small. The value of defertime (x), computed by a vehicle (x) receiving a message and which is a candidate to retransmit it, is given by the following formula:

$$defertime(x) = \max\_defer\_time \cdot \frac{(R^\varepsilon - D_{sx}^\varepsilon)}{R^\varepsilon}$$

where ε is a positive integer.

If we suppose that the distribution of the vehicles is uniform, the choice for $\varepsilon = 2$ will give a uniform distribution of the various values of defertime in [0, max_defer_time]. Dsx is the distance between the sender (s) and the receiver (x). The value of max_defer_time is equal to twice the average of communication delay.

For each received message, the vehicle must determinate its location in report with the broken vehicle. Indeed, we presented in [20] a technique to compute a vehicle location with GPS. Also, it allows the direction of circulation and position to be determined in conjunction with the broken vehicle. Thus, this technique allows the broadcast to be restricted to relevant zones only.

A study of broadcast enhancement techniques for CCW applications over Dedicated Short Range Communication (DSRC) reveals interesting trade-offs inherent to the latency perceived by periodic broadcast safety applications [25]. A broadcast-based packet forwarding mechanism is proposed in [26] for intra-platoon cooperative collision avoidance using DSRC MAC protocol. An implicit acknowledgement mechanism was introduced to reduce the amount of broadcast traffic for enhanced packet delivery rate. Due to a high frequency of link breaks, a standard approach cannot cope with high mobility. A recent approach based on virtual routers has been proposed to address this problem. In this new environment, virtual routers are used for forwarding data. The functionality of each virtual router is provided by the mobile devices currently within its spatial proximity. Since these routers do not move, the communication links are much more robust compared to those of conventional techniques. To enforce collaboration among mobile devices in supporting the virtual router functionality, some techniques are investigated in [27]. These techniques are Connectionless approaches for Street (CLA-S). According to application requirements, authors in [28] design a vehicle-to-vehicle communication protocol for cooperative collision warning. It comprises congestion control policies, service differentiation mechanisms and methods for emergency warning dissemination.

5.3.2.4   Self-Organizing Cooperative Vehicular Networks

To overcome some of the challenges that face a vehicular network, a self-organizing architecture has to be set up to simplify the network management task and to permit the deployment of a lot of services. The term 'Self-organization' was introduced in the 1960s in cybernetics and in the 1970s in the physics of complex systems. It is described as a mechanism through which individual elements in a group cooperate locally in order to give the group a macroscopic property, often described as an organization or a structure. This architecture should take advantage of vehicle properties to issue a global virtual structure enabling the network self-organization. It should be sufficiently autonomous and dynamic to deal with any local change.

Most research suggests virtual backbone [29] and clustering [30] as the most efficient structures to self-organize mobile ad hoc networks and achieve scalability and effectiveness in broadcasting. The idea of defining a virtual backbone structure is brought from the wired networks. The principle of this solution is to constitute a dorsal of best interconnected nodes. The other nodes will be associated with the dorsal nodes. The constraint is the judicious choice of backbone members to avoid the rapid loss of interconnection between them. The second self-organizing structure is clustering where 'vehicles-cooperation' is used to group the nodes into homogeneous groups named clusters. Each cluster has at least one cluster head and many members. Cluster-based solutions represent a viable approach in propagating messages among vehicles. Thus, the clustering structure is usually used as a support of backbone structure.

We proposed CSP (Cluster-based Self-organizing Protocol) [7], [8]; a vehicular network proactive self-organizing architecture that is based on geographical clustering to ensure a permanent self-organization of the whole network. The key idea is to divide each road stump into segments seen as fixed clusters and electing a cluster head for each segment to act as a backbone member. CSP adapts itself to vehicular network characteristics and permits the improvement of inter-vehicles or vehicle-to-infrastructure connectivity without producing a great overhead.

We demonstrated that CSP facilitates the network management task and permits a wide panoply of services to be deployed. For example, it allows telecommunication/service providers to better exploit/extend the existing infrastructure by overcoming its limitations using cooperative vehicles. We demonstrate via simulations that CSP is optimal when using an advertisement diffusion application on the top of it. In addition CSP does not generate a great routing overhead since it relies on fixed segments to organize the network.

5.3.3 Security and Authentication versus Cooperation

Cooperation between nodes in vehicular networks should be guaranteed in order to assure the correct service provision. Although cooperation in vehicular networks is important and beneficial to allow service access in a multi-hop distributed fashion, it could penalize the service access and the whole communication if malicious nodes were involved in the communication. To assure secure and hence reliable cooperation, it should be ensured that only authorized users are granted network's access.

Two main types of attacks could exist in vehicular networks and could allow non-cooperative behaviour in such an environment: i) external attacks, where the attackers do not participate in the network, however they could carry out some attacks and malicious acts impacting the communication and the network and services performance, and ii) internal attacks, where the attackers participate in the network and have legitimate service access, however they penalize the network performance through malicious and non cooperative acts. Consequently, efficient counter measures against these attacks need to be employed in order to ensure secure and reliable cooperation in vehicular networks. These counter-measures include authentication and access control that are important counter-attack measures in vehicular networks deployments, allowing only authorized users to have connectivity. Although authentication and access control can reinforce cooperation through prevention against external attackers, internal attackers could always exist even in the presence of effective authentication and access control mechanisms. Internal attackers are nodes that are authenticated and authorized to participate in the network;

however, they can be harmful nodes causing network and service performance degradation mainly through non cooperative behaviours (selfishness, greediness and Denial-of-Services or DoS). Hence, there is a need for complementary mechanisms to authenticate and access control.

Prevention Against External Attacks. Indeed, authentication and access control are important counterattack measures in vehicular networks deployments, allowing only authorized clients to be connected and preventing external attackers from sneaking into the network disrupting the normal cooperative operation or service provisioning. A simple solution to carry out authentication in vehicular networks is to employ an authentication key shared by all nodes in the network. Although this mechanism is considered as a plug and play solution and does not require communication with centralized network entities, it is limited to closed scenarios of a small number of participants in limited environments and belonging to the same provider. In addition, this shared secret authentication has two main pitfalls. Firstly, an attacker only needs to compromise one node to break the security of the system. Secondly, mobile nodes do not usually belong to the same community, which leads to difficulty in installing/pre-configuring the shared keys. A challenge for wide scale services deployment in vehicular networks is to design authentication mechanisms for the more vulnerable yet more resource-constrained environment of vehicular networks having multi-hop ad hoc communication. In most commercial deployments of WLANs, authentication and access control is mostly provided through employing IEEE 802.11i (IEEE 802.11i, 2004) authentication in which a centralized server is in place. In the context of vehicular networks, the challenge for applying the 802.11i approach mainly concerns the multihop characteristics and the hybrid infrastructure based/less architecture. Hence, the 802.11i authentication model should be adapted to such an environment through mainly considering two issues: i) introducing distributed authentication mechanisms, and ii) ensuring cooperation between nodes to support the hybrid architecture.

A possible approach for distributed authentication is the continuous discovery and mutual authentication between neighbours, whether they are mobile clients or fixed APs/BSs. Nevertheless, if mobile nodes move back to the range of previous authenticated neighbours or fixed nodes, it is necessary to perform re-authentication in order to prevent an adversary from taking advantage of the gap between the last security association and the current security association with the old neighbour. An approach adapting the 802.11i authentication model to multihop communication environments is presented in [31], proposing an extended forwarding capability to 802.11i and allowing mobile node authentication with the authentication server in a multihop fashion. The notion of friend nodes is introduced allowing each mobile node to initiate the authentication process through a selected node in its proximity, which plays the role of an auxiliary authenticator and forwards securely the authentication requests to the authentication server. Friend nodes are chosen to be trusted and cooperating nodes. This approach is suitable for the hybrid infrastructure-based/less architecture in vehicular networks, allowing mobile nodes beyond the APs/BSs coverage zone to get authenticated in a cooperative manner, through communicating with the authentication server at the infrastructure while passing by cooperative nodes (friend nodes). In addition, this approach allows authentication keys storage among intermediate (friend) nodes which optimizes the re-authentication process in the case of roaming.

In addition, [32] presents a distributed authentication and services' access control solution for services' commercialization in ad hoc networks with a possible application to vehicular networks environments. This work extends the Kerberos authentication model to provide each mobile node with a number of keys that are encapsulated in the Kerberos authentication ticket and are based on the sliding interval principle, where each key is only valid for a certain interval of time. Consequently, each pair of communicating nodes could authenticate and setup a secure link if they share the key that corresponds to the interval of communication and hence could cooperate when relaying each others' packets during services access. The number of keys obtained by each node reflects the node's duration for services' access. In addition, the Kerberos services' tickets are used by each node to authorize access to the corresponding services.

Another way to facilitate multihop authentication is to employ a Protocol for carrying Authentication and Network Access or PANA [33]. PANA allows the encapsulation of the used authentication protocol messages and their routing to the authentication server. The advantage of PANA mainly lies in its independence of the wireless media, and thus it is suitable for future vehicular networks allowing cooperation between heterogeneous deployments and operator co-existence. However, PANA necessitates the existence of a routing infrastructure, which is a technical challenge in cooperative vehicular networks as previously outlined.

Prevention Against Internal Attacks. Although authentication and access control can reinforce cooperation through prevention against external attackers, internal attackers could always exist even in the presence of effective authentication and access control mechanisms. Internal attackers are nodes that are authenticated and authorized to participate in the network; however, they can be harmful nodes causing network and service performance degradation mainly through non cooperative behaviours (selfishness, greediness and Denial-of-Services or DoS). Hence, there is a need for complementary mechanisms to authentication and access control. Nodes may behave selfishly by not forwarding packets for others in order to save power, bandwidth or just because of security and privacy concerns. Watchdog [34], CONFIDANT [35] and Catch [36] are three approaches developed to detect selfishness and enforce distributed cooperation and are suitable for vehicular networks multihop environment. Watchdog is based on monitoring neighbours to identify a misbehaving node that does not cooperate during data transmission. However, CONFIDANT and Catch incorporate an additional punishment mechanism making misbehaviour unattractive through isolating misbehaving nodes. On the other hand, nodes may behave greedily in consuming channel and bandwidth for their own benefits at the expense of the other users. The DOMINO mechanism [37] solves the greedy sender problem in 802.11 WLANs with a possible extension to multihop wireless networks and hence vehicular networks. Internal attackers may also cause DoS through either faked messages injection or messages replay. DoS is a challenging problem greatly impacting cooperation, however it could be partially resolved through effective authentication of messages and messages' sources.

5.3.4 Cooperation at Upper Layers

Several cooperative applications are based on cooperation between vehicles and the infrastructure belonging to the government, or private network operators or service providers. Based on the CVIS project [38], these services fall under three main categories: *urban, inter-urban and freight and fleet management.*

- *CURB – Cooperative Urban Applications*: Aims to improve the efficient use of the urban road network at both local junction and network level, and enhance individual mobility. The main innovation is the cooperative exchange of data between individual vehicles and the roadside equipment, and provision of dedicated, targeted services to individual vehicles from the road-side. This will create a cooperative system for detailed travel data collection, personalized travel information, greatly improved management of traffic at all urban levels and the promotion of efficient use of road space. Four applications are developed in CVIS: i) Cooperative Network Management: Optimum area traffic management by using vehicle/driver destination and other characteristics, and individualized route guidance, ii) Cooperative Area Routing: Intersection controllers signal momentary disturbances in traffic flow in their area of control, and give individual, destination-based and appropriate rerouting advice to approaching vehicles, iii) Cooperative Local Traffic Control: Enhanced local intersection traffic control that cooperates with the approaching vehicles, gives control and traffic state related information to the driver and supports and creates green waves through speed recommendations (profiles) for the drivers and data exchange with neighbouring intersections, and iv) Cooperative Flexible Lane Allocation: To increase the capacity of the road infrastructure, a dedicated bus lane is made available to 'licensed' and CVIS-equipped vehicles, travelling in the same direction, allowing them to use the lane when and where it would not be a nuisance to public transport and the arguments of speed, punctuality and economy would not be compromised.

- *CINT – Cooperative Inter-urban Applications*: Aims to enable cooperation and communication between the vehicle and the infrastructure on inter-urban highways. It will develop and validate cooperative services to improve the efficiency, safety and environmental friendliness of traffic on the inter-urban road network and offer a safe and comfortable journey to drivers and passengers. Two applications are developed in CVIS: i) Enhanced Driver Awareness (EDA): This application focuses on safety and will Inform vehicle drivers within 5 seconds by communication from the roadside or even nearby motorists, about relevant aspects of the dynamic traffic situation, current speed and other regulations, road and weather conditions downstream, also offering the possibility of enhancing the effectiveness of in car systems for driver assistance, and ii) Cooperative Travelers Assistance (CTA): This application focuses on assistance of the drivers. It increases the transparency of the evolving traffic situation downstream on the road network, personalizes the information to travelers, enables them to make optimal use of the road network and assists the traveler in making the right choice when navigating through the road network, based upon full cooperation between roadside systems, in-vehicle sensors, traffic managers and service providers. This system will provide information to the driver within 15 seconds about a major congestion incident, and 15 seconds later they receive a recommendation about an alternative route.

- *CF&F – Cooperative Freight & Fleet*: Aims to increase the safety of dangerous goods transport and optimize transport companies' delivery logistics. The aim is to develop innovative cooperative systems for commercial vehicles where information about the current positions, the cargo types and the destinations of freight transport vehicles are given to the regional public authorities in order to increase: efficiency, safety, security and environmentally friendliness of cargo movements. The cooperation approach will be shown in three different application areas: i) monitoring and guidance of dangerous goods, ii) parking zone management, and iii) access control to sensitive infrastructures. The driver can have more precise and up-to-date information on: local traffic conditions and regulations/limitations affecting his journey, available parking zones for goods loading/unloading and resting, and suitable routes for the specific goods being transported.

To provide these cooperative applications with monitoring data anywhere and at any time, CVIS project defines a Cooperative Monitoring (COMO) block. It is placed as a central basic service inside the CVIS framework and will cooperate closely with CURB, CINT and CF&F applications to capture their particular requirements about monitoring of traffic and environmental information. The high data volume generated by fixed and mobile sensors requires new, innovative approaches to achieve fast response times and a reasonably (in-)expensive data communication between the vehicles, the roadside units and the centres. Since fully centralized systems might not be able to serve all of these goals, the aim is a cooperative system environment in which COMO is implementing applications for data collection, data fusion and (potentially) other applications. Hence, COMO aims to develop specifications and prototypes for the collection, integration and delivery of extended real-time information on individual and collective vehicle movements and on the state of the road network.

Particularly appealing examples that will use this cooperative monitoring block are IFTIS [39] and the parking zone management SmartPark [40]. IFTIS is a completely distributed and infrastructure-free mechanism for road density estimation. IFTIS is based on a distributed exchange and maintenance of traffic information between cooperating vehicles traversing the routes. It provides cooperative urban applications with real time information about the traffic within city roads. The idea of SmartPark is to solve the parking problem in a completely cooperative and decentralized fashion.

In the following, we give details of some of these applications and quantify performance gains due to relaying and cooperation.

5.3.4.1 Traffic Density Estimation

One of the most important components of the Intelligent Transportation System is the road traffic information handling (monitoring, transmission, processing and communication). The existing traditional ITS traffic information systems are based on a centralized structure in which sensors and cameras along the roadside *monitor* traffic density and *transmit* the result to a central unit for further *processing*. The results will then be *communicated* to road users via broadcast service or alternatively on demand via cellular phones. The centralized approaches are dependent on fixed infrastructures which demand public investments from government agencies or other relevant operators to build, maintain and manage such infrastructure: a large number of sensors are needed to be deployed in order to monitor the traffic situation. The traffic information service is then limited to streets where sensors are integrated. Besides, centralized designs have the disadvantage of being rigid, difficult to maintain and upgrade, require substantial computing/communications capabilities, and are susceptible to catastrophic events (sabotage or system failures). Moreover, such systems are characterized by long reaction times and thus are not useable by all the applications requiring reliable decision making based on accurate and prompt road traffic awareness.

We proposed a completely decentralized mechanism for the estimation of vehicular traffic density in city-roads IFTS (Infrastructure-Free Traffic Information System) [39]. The decentralized approach is based on the traffic information exchanged, updated and maintained among vehicles in the roads and revolves around the core idea of information relaying between groups of vehicles rather than individual vehicles. More precisely, the vehicles are arranged into location-based groups. For that, each road (section of street between two intersections) is dissected into small fixed area cells, each defining a group. Note that the cell size depends on the transmission range of vehicles and the coordinates of the cell centre gives the cell a unique identifier. Cells, and hence groups, overlap in such a way that any vehicle moving from one cell to the next belongs at least to one group. Among vehicles within the zone leader,[1] the closest vehicle to the cell centre is considered as the group leader for a given duration. Note that the overlapping zone is so small that it is not possible that a vehicle is considered to be group leader of both adjacent cells.

As shown in Figure 5.4, local density information is then computed by each group leader and relayed between groups using Cell Density Packet (CDP). The CDP gathers the density[2] of a given road (that is, all its cells). When initiating the CDP, a vehicle records the road ID, the transmission time[3] and a list of anchors through which the packet has to pass while travelling to the other intersection, and then, sends the packet in the backward direction. The

---

[1] A small area around a cell centre where a vehicle is elected as a group leader.
[2] By density, we mean the number of vehicles within the cell.
[3] Note that all the vehicles are synchronized by GPS.

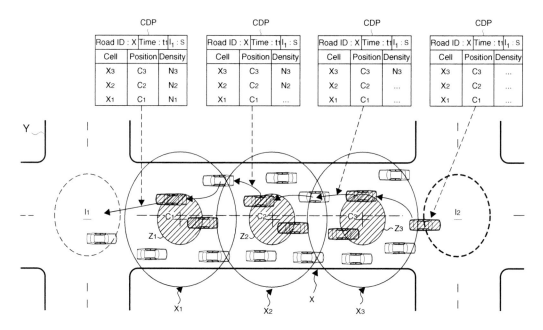

Figure 5.4 Relaying local density information between groups.

CDP header includes a limited list of anchors corresponding to the position of the cells' centres. Then, the CDP is forwarded towards the first anchor on the basis of greedy forwarding. Once the message is received by a group leader (the closest vehicle to the cell centre), this later updates it by including the density of the corresponding cell (the number of its neighbours) and then forwards it towards the next anchor. This is repeated until the CDP is completed while arriving to the destination intersection. After the last anchor (the destination intersection) is reached, the CDP is propagated to vehicles which are around the intersection so that all vehicles traversing through the intersection will receive it. These vehicles analyze the packet content and calculate the density for the respective road from which the CDP was received. This analysis is done by computing i) the average number of vehicles per cell N_{avg} and ii) the standard deviation of cells densities σ. Note that the standard deviation indicates how much variation there is away from the N_{avg}: a large standard deviation indicates that the cells densities are far from the mean and a small standard deviation indicates that they are clustered closely around the mean.

The performance analysis of the proposed mechanism depicted the accuracy of IFTIS and the promptness of information delivery based on delay analysis at the road traffic intersections. This is done in a distributed manner and based only on the cooperation between vehicles.

5.3.4.2 Smart Parking

The main goal of SmartPark [40] is to collect information about parking space availability and to coordinate drivers in order to guide them to free parking spots. To this extent, at every parking spot a wireless mote is deployed which tracks the occupancy and cooperates with other nearby motes and vehicles. Each vehicle is equipped with a wireless communication device that provides a driver with information about parking space availability and guides them eventually using turn-by-turn instructions (see Figure 5.5).

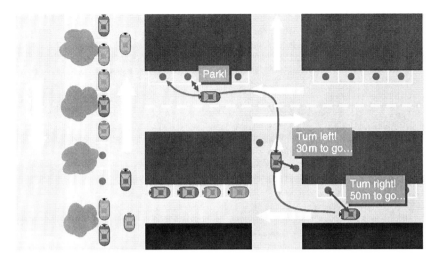

Figure 5.5 SmartPark system consists of sensor nodes embedded in parking spaces and on-board units inside vehicles.

5.4 Conclusion

Thanks to the advances in wireless technologies, vehicular networks have emerged as a new type of autonomous network allowing for vehicle-to-infrastructure and vehicle-to-vehicle communication. Applications in vehicular networks range from road safety applications oriented to the vehicle or to the driver, to entertainment and commercial applications for passengers, making use of a plethora of cooperating technologies. Passing traffic information such as travel times or warning messages about accidents or sloppy roads are only a few examples of the potentials created by equipping vehicles and roads with appropriate communication capabilities. The greater number of vehicles on the road increases significantly the unpredictable events outside vehicles. In fact, accidents arrive rarely from vehicles themselves and mainly originate from on-road dynamics. This means that cooperation using vehicular networks must be introduced into transportation networks to improve overall safety and network efficiency, and to reduce the environmental impact of road transport. Moreover, cooperation is crucial in entertainment applications to allow reliable services' access through the multihop communication during vehicles' mobility.

Cooperative techniques will likely survive in scenarios which are independent of users (no selfishness) but rather depending on machines or operator-programmed decision engines. Examples are machine-to-machine applications, such as vehicular networks. In this chapter we explored cooperation issues in autonomous vehicular networks at different levels. We notice that high-level services should be built following a cooperative model that depends exclusively on the participation of contributing vehicles and the existing infrastructure. We also notice that vehicular networks scenarios relying on an infrastructure (that could be eventually limited infrastructure) could satisfy cooperation through resolving several technological issues. Such scenarios are promising for real deployment of vehicular networks in a public context of generalized mobility.

References

[1] J. Blum, A. Eskandarian and L. Hoffmman, 'Challenges of intervehicle ad hoc networks' *IEEE Transactions on Intelligent Transportation Systems*, Vol. 5, No 4, pp. 347–351, 2004.

[2] M. Dohler, D.-E. Meddour, S.-M. Senouci, H. Moustafa, 'Cooperative Communication System Architectures for Cellular Networks' in M. Uysal (Ed.) 'Cooperative Communications for Improved Wireless Network Transmission: Frameworks for Virtual Antenna Array Applications,' *IGI-Global*, ISBN: 978-1-60566-665-5, July 2009.

[3] T. Taleb, A. Benslimane and K. Ben Letaief, 'Towards an Effective Risk-conscious and Collaborative Vehicular Collision Avoidance System,' *IEEE Trans. On Vehicular Technology*, Vol. 59, No. 3, Mar. 2010. pp. 1474–1486.

[4] G. M. T. Abdalla, M. Ali Abu -Rgheff, S. M. Senouci, 'Space -Orthogonal Frequency -Time Medium Access Control (SOFT MAC) for VANET', *IEEE GIIS'2009*, Hammamet, Tunisia, June 23–25, 2009.

[5] M. Jerbi, S. M. Senouci, T.M. Rasheed, Y. Ghamri-Doudane, 'Towards Efficient Geographic Routing in Urban Vehicular Networks,' *IEEE Transactions on Vehicular Technology*, Vol. 58, Issue 9, pp. 5048–5059, November 2009.

[6] M. Jerbi, S. M. Senouci, R. Meraihi and Y. Ghamri-Doudane, 'An Improved Vehicular Ad Hoc Routing Protocol for City Environments', *IEEE ICC'2007*, Glasgow, Scotland, UK, 24–28 June 2007.

[7] M. Cherif, S. M. Senouci, and B. Ducourthial, 'Vehicular Network Self-Organizing Architectures', *IEEE GCC'2009*, Kuwait, March 17–19, 2009.

[8] I. Salhi, S. M. Senouci, and M. Cherif, 'A New Framework for Data Collection in Vehicular Networks', *IEEE ICC'2009*, Dresden, Germany, June 14–18, 2009.

[9] C. Bettstetter, H. Adam, S. M. Senouci, 'A Multi-Hop-Aware Cooperative Relaying', *IEEE VTC'2009 Spring*, Barcelona, Spain, 26–29 April.

[10] H. Adam, C. Bettstetter, S. M. Senouci, 'Adaptive Relay Selection in Cooperative Wireless Networks', *IEEE PIMRC2008*, Cannes, France, 15–18 September 2008.

[11] G. Korkmaz, E. Ekici and F. Özgüner, 'Urban multihop broadcast protocol for inter-vehicle communication systems', *IEEE Transactions on Vehicular Technology*, Vol. 55, No. 3, pp. 865–875, 2006.

[12] H. Wu, R. Fujimoto, R. Guensler and M. Hunter, 'MDDV: a mobility centric data dissemination algorithm for vehicular networks', *In Proceedings of the 1st ACM Workshop on Vehicular Ad Hoc Networks (VANET'04)*, Philadelphia, PA, USA, Sep. 2004.

[13] B. Karp and H. T. Kung, 'GPSR: Greedy perimeter stateless routing for wireless networks' *In Proceedings of the 6th ACM/IEEE International Annual Conference on Mobile Computing and Networking (MOBICOM'00)*, Boston, MA, USA, August 2000.

[14] D. Niculescu and B. Nath, 'Trajectory based forwarding and its applications' *In Proceedings of the 9th ACM International Annual Conference on Mobile Computing and Networking (MOBICOM'03)*, San Diego, USA, 2003.

[15] C. Lochert, H. Hartenstein, J. Tian, D. Herrmann, H. Füßler, M. Mauve, 'A Routing Strategy for Vehicular Ad Hoc Networks in City Environments' In *Proceedings of IEEE Intelligent Vehicles Symposium (IV'03)*, Columbus, OH, USA, Jun. 2003.

[16] B.-C. Seet, G. Liu, B.-S. Lee, C. H. Foh, K. J. Wong, K.-K. Lee, 'A-STAR: A Mobile Ad Hoc Routing Strategy for Metropolis Vehicular Communications' In *Proceedings of the 3rd IFIP International Conferences on Networking (NETWORKING'04)*, Athens, Greece, May, 2004.

[17] J. Zhao and G. Cao, 'VADD: Vehicle-assisted data delivery in vehicular ad hoc networks' *IEEE Transactions on Vehicular Technology*, Vol. 57, No. 3, May, 2008.

[18] J. Davis, A. Fagg and B. Levine, 'Wearable Computers as Packet Transport Mechanisms in Highly-Partitioned Ad-Hoc Networks' *in Proceedings of the 5th International Symposium on Wearable Computing (ISWC'01)*, Zurich, Switzerland, Oct. 2001.

[19] S. Barghi, A. Benslimane and C. Assi, 'Connecting Vehicular Networks to the Internet: A Life Time-based Routing Protocol', *10th IEEE International Symposium World of Wireless, Mobile and Multimedia Networks & Workshops (WoWMoM 2009)*, Kos Greece, pp. 1–9, 15–19 June 2009.

[20] A. Benslimane, 'Optimized Dissemination of Alarm Messages in vehicular ad-hoc networks (VANET)', HSNMC, 2004, LNCS 3079, pp. 655–666.

[21] J. Nzoonta and C. Borcea, STEID: A protocol for emergency information dissemination in vehicular networks, Draft, 2006.

[22] G. Korkmaz, E. Ekici, F. Özgüner, and Ü. Özgüner, 'Urban multihop broadcast protocol for inter-vehicle communication systems,' in VANET '04: Proceedings of the 1st ACM International Workshop on Vehicular Ad Hoc Networks. Philadelphia, PA, USA: ACM Press, Sept. 2004, pp. 76–85.

[23] M. Jerbi, S. M. Senouci, A. L. Beylot, Y. Ghamri, 'Geo-localized Virtual Infrastructure for VANETs: Design and Analysis', *IEEE Globecom 2008*, New Orleans, LA, USA, 30 November – 4 December 2008.

[24] R.H. Frenkiel, B.R. Badrinath, J. Borras and R. Yates, 'The Infostations Challenge: Balancing Cost and Uniquity in Delivering Wireless Data,' in IEEE Personal Communications, April 2000.

[25] Tamer ElBatt, et al., 'Cooperative Collision Warning Using Dedicated Short Range Wireless Communications', *VANET'06*, September 29, 2006, Los Angeles, California, USA.

[26] R. Tatchikou, et al., 'Cooperative vehicle collision avoidance using inter-vehicle packet forwarding', Globecom 2005, pp. 2762–2766.

[27] Yao H. Ho, et al., 'Cooperation Enforcement in Vehicular Networks', Conference on Communication Theory, Reliability, and Quality of Service (CTRQ), 2008, pp. 7–12.

[28] X. Yang, et al., 'A Vehicle-to-Vehicle Communication Protocol for Cooperative Collision Warning', Mobiquitous 2004, pp. 114–123.

[29] B. Liang and Z.J. Haas, 'Virtual backbone generation and maintenance in ad hoc network mobility management', *IEEE INFOCOM 2000*, pp. 1293–1302, Tel-Aviv, Israel, March 2000.

[30] B. Chen, K. Jamieson, H. Balakrishnan and R. Morris, 'Span: An energy-efficient coordination algorithm for topology maintenance in ad hoc wireless networks', *ACM Wireless Networks Journal*, vol. 8, n°5, pp. 481–494, September 2002.

[31] H. Moustafa, G. Bourdon and Y. Gourhant, 'Authentication, authorization, and accounting (AAA) in hybrid ad hoc hotspots' environments,' ACM WMASH, 2006.

[32] H. Moustafa, J. Forestier and M. Chaari, 'Distributed Authentication for Services Commercialization in Ad hoc Networks,' ACM Mobility Conference 2009.

[33] D. Forsberg, O. Ohba, B. Patil, H. Tschofenig and A. Yegin, 'Protocol for carrying authentication and network access (PANA),'. RFC 5193, May 2008.

[34] S. Marti, T. J. Giuli, K. Lai and M. Baker, 'Mitigating routing misbehavior in mobile ad hoc Networks,' ACM Mobicom, 2000.

[35] S. Buchegger, and J. Y. le Boudec, 'Performance analysis of the CONFIDANT protocol,' ACM MobiHoc, 2002.

[36] R. Mahajan, M. Rodrig, D. Wetherall and J. Zahorjan, 'Sustaining cooperation in multihop wireless networks,' ACM NSDI, 2005.

[37] M. Raya, J. P. Hubaux and I. Aad, 'Domino: A system to detect greedy behavior in IEEE 802.11 Hotspots,' ACM MobiSys 2004.

[38] CVIS project, http://www.cvisproject.org/

[39] M. Jerbi, S. M. Senouci, T. Rasheed, Y. Ghamri-Doudane, 'An Infrastructure-Free Traffic Information System for Vehicular Networks', *IEEE WiVeC 2007*, Baltimore, USA, 30 September–1 October 2007.

[40] SmartPark, http://smartpark.epfl.ch/

6

Cooperative Overlay Networking for Streaming Media Content

F. Wang[1], J. Liu[1] and K. Wu[2]

[1]*School of Computing Science, Simon Fraser University, Burnaby, Canada*
[2]*Department of Computer Science, University of Victoria, Victoria, Canada*

6.1 Introduction

For the past two decades, we have witnessed a huge increase in streaming multimedia content over the Internet, which has been identified as one of the most popular network applications. Its large-scale deployment, however, is still in the early stages due to its unique characteristics. Compared with other applications such as webpage browsing and file downloading, streaming media content may consume enormous bandwidth from the media streaming source. Also, the media content itself and user behaviours may bring many dynamics. In addition, streaming media content has stringent time and bandwidth requirements as data are delivered simultaneously while the playback continues and data arriving after the playback deadline would become useless.

To tackle these challenges and enable streaming media content as a common Internet utility in a manner that any publisher can broadcast media content to any set of receivers, much research effort has had to be devoted. For much of the 1990s, the research and industrial community investigated the supports for such applications using the IP Multicast architecture [18]. However, serious concerns regarding its scalability, the support for higher-level functionality, and the difficulty in its deployment have dogged IP Multicast. The sparse deployment of IP Multicast, and the high cost of bandwidth required for server-based solutions or Content Delivery Networks (CDNs) are the two main factors that have limited the broadcast service to only a subset of Internet content publishers.

To reduce server/network loads, an effective approach is to cache frequently used data at proxies that are close to clients [28], [40]. Streaming media content could also benefit and have a significant performance improvement from proxy caching due to the temporal locality of client requests for streaming media content. In general, proxies grouped together can cooperate with each other to increase the aggregate cache space, balance loads, improve system scalability and thus the overall performance of streaming media content. While these aforementioned technologies enable many network service providers to provide IPTV services, which deliver quality media content to their own subscribers using packet switching, there remains a need for

Cooperative Networking, First Edition. Edited by Mohammad S. Obaidat and Sudip Misra.

cost-effective, ubiquitous support for Internet-wide media streaming, and the solutions will certainly be beneficial to IPTV as well.

Recently there have been significant interests in the use of peer-to-peer technologies for Internet media streaming. Two key driving forces make the approach attractive. First, this technology does not require the support from Internet routers or network infrastructure, and consequently it is extremely cost-effective and easy to deploy. Second, with such a technology, a participant who tunes into a broadcast not only downloads a media stream, but also uploads it to other participants watching the media content. Consequently, such an approach has the potential to scale well with group size, as greater demand also generates more resources.

In this chapter, we investigate various solutions proposed to advocate peer-to-peer media streaming. These solutions can be roughly divided into two categories, tree-based and mesh-based approaches. We found that both approaches may suffer from some performance degradation due to the vulnerability caused by dynamic end-hosts or the efficiency-latency tradeoff. Based on this observation, we present a novel approach toward a cooperative mesh-tree overlay design, called *mTreebone*. The key idea is to identify a set of stable hosts to construct a tree-based backbone, called *treebone*, with most of the data being pushed over this backbone. These stable hosts, together with others, are further organized through an auxiliary mesh overlay, which facilitates the treebone to accommodate dynamic hosts and also to fully explore the available bandwidth between host pairs.

In this cooperative mesh-tree design, however, a series of unique and important issues need to be addressed. First, we have to identify the stable hosts in an overlay, and gradually build up the treebone; second, we need to reconcile the push and pull data delivery used in the treebone and mesh overlays, respectively. They should work complementarily to maximize the joint efficiency in the presence of autonomous hosts. To this end, we derive an optimal age threshold for identifying stable hosts, which maximizes their expected service times in the treebone. We then propose a set of overlay construction and evolution algorithms to enable seamless treebone/mesh collaboration with minimized control overhead and transmission delay. Finally, we present a buffer partitioning and scheduling algorithm, which coordinates push/pull operations and avoids data redundancy.

We extensively evaluate the performance of mTreebone and compare it with existing mesh- and tree-based solutions. The simulation results demonstrate the superior efficiency and robustness of this hybrid solution. Such results are reaffirmed by our experimental results of an mTreebone prototype over the PlanetLab network [6].

The remainder of this chapter is organized as follows. Section 6.2 briefly discusses the architectural choices for streaming media content over the Internet. In Section 6.3, we highlight the key differences between media streaming and traditional peer-to-peer applications. In addition, we taxonomize the existing approaches for peer-to-peer media streaming. Section 6.4 gives an overview of the proposed cooperative mTreebone design. Details about the treebone's evolution and its interactions with the mesh are discussed in Sections 6.5 and 6.6, respectively. We evaluate the performance of mTreebone in Section 6.7. Finally, Section 6.8 concludes the chapter and offers potential future research directions.

6.2   Architectural Choices for Streaming Media Content over the Internet

We first review the architectural choices that support broadcasting/multicasting media streaming over the Internet (see Figure 6.1). There are subtle differences between broadcast and

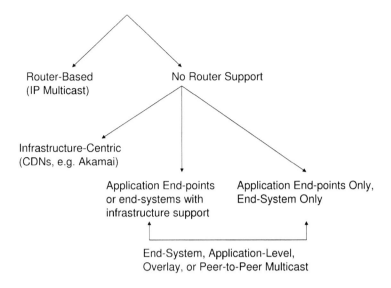

Figure 6.1 Taxonomy of architectures for Internet broadcast.

multicast: the former is to all the destinations and the latter is to a group of destinations only. While broadcast is possible in air, cable networks, or local area networks, it simply cannot be carried over the global Internet. Nevertheless, given the popular use of this term in radio and TV industries, in this chapter, we do not distinguish it from multicast if the context is clear.

6.2.1 Router-Based Architectures: IP Multicast

In the Internet environment, the primary issue for broadcast/multicast is to determine at which layer it should be implemented. Two conflicting considerations need to be reconciled. According to the end-to-end argument, a functionality should be 1) pushed to higher layers if possible, unless 2) implementing it at the lower layers can achieve significant performance benefits that outweigh the cost of additional complexity. In his seminal work in 1989 [18], Deering argued that the second consideration should prevail and multicast should be implemented at the IP layer. This view has since been widely accepted, leading to the IP multicast model.

IP multicast is a loosely coupled model that reflects the basic design principles of the Internet. It retains the IP interface, and introduces the concept of open and dynamic groups, which greatly inspires later proposals. Given that the network topology is best-known at the network layer, multicast routing in this layer is also the most efficient. Unfortunately, despite the tremendous efforts in the past 15 years, today's IP multicast deployment remains limited in reach and scope. The reason is complex and involves not only technical obstacles, but also, more importantly, economic and political concerns. First, IP multicast requires routers to maintain per-group states, which not only violates the 'stateless' architectural principle, but also introduces high complexity and serious scaling constraints at the IP layer. Second, IP multicast is a best-effort service, and attempts to conform to the traditional separation of routing and transport that has worked well in the unicast context. However, providing higher-level features such as error control, flow control and congestion control has been shown to be more difficult than in the unicast case. Finally,

IP multicast calls for changes at the infrastructure level, and this slows down the pace of its deployment. In particular, there is a lack of incentive to install multicast-capable routers and to carry multicast traffic.

6.2.2   Architectures with Proxy Caching

The placement of the multicast functionality was revisited in the new millennium, and several researchers have advocated moving multicast functionality away from routers towards end systems [17], [21], [14], [43], which lead to several choices in instantiating such an architecture as shown in Figure 6.1. On one end of the spectrum is an *infrastructure-centric* architecture, where an organization that provides value-added services deploys proxies at strategic locations on the Internet. End systems attach themselves to nearby proxies, and receive data using plain unicast. Such an approach is also commonly referred to as Content Delivery Networks (CDNs), and has been employed by companies such as Akamai [1]. On the other end of the spectrum is a purely *application end-point* architecture, where functionality is pushed to the users actually participating in the multicast group. Administration, maintenance, responsibility for the operation of such a peer-to-peer system are distributed among the users, instead of being handled by a single entity.

At either end of the spectrum, multicast related features, such as group membership, multicast routing and packet duplication, are implemented at end systems, assuming only unicast IP service. End systems participate in multicast communication via an overlay structure, in the sense that each of its edges corresponds to a unicast path between two nodes in the underlying Internet.

Moving multicast functionality to end systems has the potential to address many of the problems associated with IP multicast. Since all packets are transmitted as unicast packets, deployment is accelerated. It maintains the stateless nature of the network by requiring end systems, which subscribe only to a small number of groups, to perform additional complex processing for any given group. In addition, solutions for supporting higher-layer features can be significantly simplified by leveraging well-understood unicast solutions, and by exploiting application-specific intelligence.

It must be noted that moving multicast functionality away from routers involves performance penalties. For example, it is impossible to completely prevent multiple overlay edges from traversing the same physical link and thus some redundant traffic on physical links is unavoidable. Further, communication between end systems involves traversing other end systems, potentially increasing latency. Hence, many research efforts have focused on addressing these performance concerns with overlays.

6.2.2.1   Combining Proxy Caching with Media Streaming

Like caching, multicasting also explores the temporal locality of client requests. Specifically, it allows a media server to accommodate concurrent client requests with shared channels. However, multicast delivery suffers from important deficiencies. For example, while IP multicast is enabled in virtually all local-area networks, its deployment over the global Internet remains limited in scope and reach. Hence, it is unlikely that a multicast streaming protocol can be used for geographically dispersed servers and clients.

Interestingly, such deficiencies can be alleviated through the use of proxies. Specifically, a request can be instantaneously served by a cached prefix while waiting for the data from a multicast channel [33], [38], and proxies can bridge unicast networks with multicast networks, that is, employing unicast for server-to-proxy delivery. Wang et al. [38] have shown that a careful coupling of caching and multicasting can produce significant cost savings over using the unicast service, even if IP multicast is supported only at local networks.

6.2.2.2   Cooperative Proxy Caching

In general, proxies grouped together can achieve better performance than independent standalone proxies. Specifically, the group of proxies can cooperate with each other to increase the aggregate cache space, balance loads and improve system scalability [8], [13]. A typical cooperative media caching architecture is MiddleMan [8], which operates a collection of proxies as a scalable cache cluster. Media objects are segmented into equal-sized segments and stored across multiple proxies, where they can be replaced at a granularity of a segment. There are also several local proxies responsible for answering client requests by locating and relaying the segments. Note that, in cooperative web caching, a critical issue is how to efficiently locate web pages with minimum communication costs among the proxies. This is, however, not a major concern for cooperative media caching, as the bandwidth consumption for streaming objects is of orders of magnitude higher than that for object indexing and discovering. Consequently, in MiddleMan, a centralized coordinator works well in keeping track of cache states. On the other hand, while segment-based caching across different proxies facilitates the distribution and balance of proxy loads, it incurs a significant amount of overhead for switching among proxies to reconstruct a media object. To reduce such effects as well as to achieve better load balance and fault tolerance, Chae et al. [13] suggested a Silo data layout, which partitions a media object into segments of increasing sizes, stores more copies for popular segments, but still guarantees at least one copy stored for each segment.

6.2.3   Peer-to-Peer Architectures

Recall that in Figure 6.1, besides the one end of the spectrum with an infrastructure-centric architecture, the other end is a purely application end-point architecture, where functionality is pushed to the users actually participating in the multicast group. Instead of being handled by a single entity, administration, maintenance and responsibility for the operation of such a peer-to-peer system are distributed among the users, and the research focuses on simultaneous media content broadcast using the application end-point architecture, referred to as *peer-to-peer broadcast/multicast*. Such similar terms as *end-system multicast, overlay multicast, application-layer multicast*, have also been used in the literature. In the purest form, such architectures rely exclusively on bandwidth resources at application end-points. However, one could also conceive of hybrid architectures that seek to use the bandwidth resources of application end-points to the extent possible, but may leverage infrastructure resources where available.

The motivation of using the peer-to-peer paradigm derives to a large extent from its ability to leverage the bandwidth resources of end systems that actually participate in the communication. Although the addition of new participants requires more bandwidth support, they in the meantime contribute additional bandwidth. In contrast, while an infrastructure-centric service can potentially deal with a smaller number of well-defined groups, it is unclear whether it can support the bandwidth requirements associated with deploying tens of thousands of high-bandwidth broadcast applications. Further, the application end-point architecture is instantaneous to deploy, and can enable support of applications with minimal setup overhead and cost.

While the application end-point architectures have the promise to enable ubiquitous deployment, the infrastructure-centric architecture can potentially provide more robust data delivery with dedicated, better provisioned and more reliable proxies placed at strategic locations. In contrast, the application end-point architectures potentially involve a wide range of autonomous users that may not provide good performance but easily fail or leave at will. Individual user joining and leaving have more significant impact on the system performance. Thus, the key challenge for application end-point architectures is to function, scale and self-organize with a highly transient population of users, without the need of a central server and the associated management overhead.

6.3   Peer-to-Peer Media Streaming

In this section, we discuss the special characteristics of media content streaming applications. We then discuss why these characteristics correspond to a very different domain requiring very different solutions unlike many other peer-to-peer applications.

6.3.1   Comparisons with Other Peer-to-Peer Applications

A media streaming system typically has a single dedicated source, which may be assumed not to fail, and is present throughout a broadcast session. The address of the source is known in advance, serving as a rendezvous for new users to join the session. There are several special characteristics of such a system:

- Large scale, potentially having tens of thousands of users simultaneously participating in the broadcast.
- Performance-demanding, involving bandwidth requirements of hundreds of kilobits per-second.
- Real-time constraints, requiring timely and continuously streaming delivery. While interactivity may not be critical and minor delays can be tolerated through buffering, it is nevertheless critical to get video uninterrupted.
- Gracefully degradable quality, enabling adaptive and flexible delivery that accommodates bandwidth heterogeneity and dynamics.

The above characteristics altogether yield a unique application scenario that differs from other typical peer-to-peer applications, including *on-demand streaming, audio/video conferencing*, and *file download* (see Table 6.1).

Among these applications, on-demand streaming and audio/video conferencing also have stringent delay and bandwidth requirements. However, in on-demand streaming, the users are asynchronous, and it thus belongs to a different problem domain. Audio/video conferencing applications differ from broadcast applications in that they are interactive with latency being even more critical, and they are multi-point, because any participant may become a source. However, such applications are typically of smaller scales, involving only a few hundred participants. Example systems of this kind include Skype [7] (limited to audio conversation), and research proposals such as Narada [16] and Gossamer [14].

Peer-to-peer file download applications such as BitTorrent [2], and EMule [3] involve information distribution to tens of thousands of participants. However, video broadcast is

Table 6.1   A Taxonomy of Typical Peer-to-Peer Applications

Category	Bandwidth-Sensitive	Delay-Sensitive	Scale
File download	No	No	Large
On-demand streaming	Yes	Yes	Large
Audio/video conferencing	Yes/No	Yes	Small
Simultaneous broadcast	Yes	Yes	Large

more challenging due to the stringent real-time and bandwidth requirements. For example, BitTorrent enables peers to exchange any segment of the content being distributed, but the order in which they arrive is not important. In contrast, such techniques are not feasible in streaming applications. Furthermore, given the requirements on timely delivery, streaming video applications typically must include techniques for graceful degradation of video quality rather than involving excessive delays.

Another key problem in peer-to-peer file download is to design techniques for efficient indexing and search, that is, to locate a massive number of files distributed among a large number of peers. Solutions in this space include Napster, Gnutella and Distributed Hashing Table (DHT) techniques [5], [4], [35]. While the design of overlays for efficient indexing and searching a large video repository poses several challenges, we are more interested in the efficiency of data communication in peer-to-peer video broadcast.

6.3.2 Design Issues

The key problem in a peer-to-peer media streaming system is to organize the peers into an overlay for disseminating the video stream. The following are the important criteria for overlay construction and maintenance.

- *Overlay efficiency*: The overlay constructed must be efficient from the perspectives of both the network and the applications. For video broadcast, high bandwidth and low latencies are simultaneously required. However, given that applications are real-time but not interactive, a startup delay of a few seconds can be tolerated.
- *Scalability and load balancing*: Since broadcast systems can scale to tens of thousands of receivers, the overlay must scale to support such large sizes, and the overhead associated must be reasonable even at large scales.
- *Self-organizing*: The construction of overlay must take place in a distributed fashion and must be robust to dynamic changes in group membership. Furthermore, the overlay must adapt to long-term variations in Internet path characteristics (such as bandwidth and latency), while being resilient to inaccuracies. The system must be self-improving in that the overlay should incrementally evolve into a better structure as more information becomes available.
- *Honor per-node bandwidth constraints*: Since the system relies on users to contribute bandwidth, it is important to ensure that the total bandwidth a user is required to contribute does not exceed its inherent access bandwidth capacity. On the other hand, users also have heterogeneous inbound bandwidth capacities, and it is desirable to have mechanisms to ensure they can receive different qualities of video, proportional to their capacity.
- *System Considerations*: In addition to the above algorithmic considerations, several important system issues must be addressed in the design of a complete broadcasting system. Examples include the choice of transport protocol and the interaction with video players. Furthermore, a key challenge of peer-to-peer systems is to take care of a large fraction of users behind NATs and firewalls – the connectivity restrictions posed by such peers may severely limit the overlay capacity.

6.3.3 Approaches for Overlay Construction

A large number of proposals have emerged in recent years for peer-to-peer video broadcast [21], [14], [17], [19], [43], [31], [10], [30], [12], [24], [26], [15], [36], [32], [42], [37]. While these proposals differ on a wide-range of dimensions, in this chapter, we focus on the approach of building the proper overlay structure used for data dissemination. In particular, the proposals can be broadly classified into two categories, namely, **tree-based** and **mesh-based** overlay construction, which we discuss below.

Tree-Based Approaches: The vast majority of the proposals to date can be categorized as a tree-based approach. In such an approach, peers are organized into tree structures for delivering data, with each data packet being disseminated using the same structure. Nodes on the structure have well-defined relationships, for example, 'parent-child' relationships in trees. Such approaches are typically push-based, that is, when a node receives a data packet, it also forwards copies of the packet to each of its children. Since all data packets follow this structure, it becomes critical to ensure that the structure is optimized to offer good performance to all receivers. Furthermore, the structure must be maintained, as nodes join and leave the group at will. In particular, if a node crashes or stops performing adequately, all of its offspring in the tree will stop receiving packets, and the tree must be repaired. Finally, when constructing tree-based structures, we must address the problem of loop avoidance.

Tree-based solutions are perhaps the most natural approach, and work well with widely available video codecs. However, one concern with tree-based approaches is that the failure of nodes, particularly those close to the root of the tree may disrupt delivery of data to a large number of users, and potentially result in poor transient performance. Furthermore, the out-going bandwidth of the leaf nodes is not utilized. In response to these concerns, researchers have been investigating more resilient structures for data delivery. In particular, one approach that has gained popularity is multi-tree based approaches, where multiple disjoint trees are employed with each of them delivering a sub-stream of the video [31], [12], [36]. While the multi-tree based approaches can remarkably improve resilience, it is more complex to construct and maintain multiple trees. Optimizing the multiple trees as a whole without violating the disjoint property can be quite difficult in the presence of dynamic nodes [11], [29].

Mesh-Based Approaches:[1] Recently, researchers have proposed mesh-based approaches for peer-to-peer broadcast [32], [42]. Mesh-based overlay designs sharply contrast with tree-based designs in that they do not construct and maintain an explicit structure for delivering data. The underlying argument is that, rather than constantly repair a structure in a highly dynamic peer-to-peer environment, we can use the availability of data to guide the data flow.

A naive approach to distributing data without explicitly maintaining a structure is to use gossip algorithms [20]. In a typical gossip algorithm, a node sends a newly generated message to a set of randomly selected nodes; these nodes act similarly in the next round, and so do other nodes until the message is spread to all. The random choice of gossip targets achieves resilience to random failures and enables decentralized operations. However, gossip cannot be used directly for video broadcast because its random push may cause significant redundancy with the high-bandwidth video. Further, without an explicit structure support, minimizing startup and transmission delays becomes a difficult problem.

To handle this, approaches such as Chainsaw [32] and CoolStreaming [42] adopt pull-based techniques. More explicitly, nodes maintain a set of partners, and periodically exchange data availability information with the partners. A node may then retrieve unavailable data from one or more partners, or supply available data to partners. Redundancy is avoided, as the node pulls data only if it does not already possess it. Further, since any segment may be available at multiple partners, the overlay is robust to failures – departure of a node simply means its partners will use other partners to receive data segments. Finally, the randomized partnerships imply that the potential bandwidth available between the peers can be fully utilized.

---

[1] Also referred to as *data-driven randomized* approach [27].

The mesh-based approach at first sight may appear similar to techniques used in solutions for file download like BitTorrent [2]. However, the crucial difference here is that the real-time constraints imply that segments must be obtained in a timely fashion. Thus, an important component of a mesh-based broadcast system is a scheduling algorithm, which schedules the segments downloaded from various partners to meet the playback deadlines.

A drastic difference between the tree-based and the mesh-based overlays lies in their data delivery strategies. In a tree or multi-tree, a video stream is basically pushed along well-defined routes, that is, from parents to their children. In a mesh, given that multiple and dynamic neighbours may have video data available to send, a node has to pull data to avoid significant redundancies. A mesh-based system is therefore more robust, but experiences longer delays and higher control overhead. More explicitly, there is an efficiency-latency tradeoff [37], [41]: if the mesh nodes choose to send notifications for every data block arrival, the overhead will be excessive; periodical notifications containing buffer maps within a sliding window, as suggested in [32], [42], reduce the overhead, but increase the latencies.

Both tree/multi-tree and mesh solutions have shown their success in practical deployment [27], and yet neither completely overcomes the challenges from the dynamic peer-to-peer environment. The advantage of using the data-driven mesh overlays is their robustness, but the lack of a well-ordered parent/children relation implies that data have to be pulled from neighbours, which suffers the efficiency-latency tradeoff as discussed before. The push delivery in a tree is efficient, but has to face data outage in descendants when an internal node fails. The pre-defined flow direction also prevents the overlay from fully utilizing the bandwidth between node pairs, for example, that between two leaf nodes. Given the pros and cons of the two approaches, a natural question is whether we can combine them to realize a hybrid overlay that is both efficient and resilient.

In this chapter, we present a novel approach toward a hybrid overlay design, called mTreebone. The key idea is to identify a set of stable nodes to construct a tree-based backbone, called treebone, with most of the data being pushed over this backbone. These stable nodes, together with others, are further organized through an auxiliary mesh overlay, which facilitates the treebone to accommodate dynamic nodes and also to fully explore the available bandwidth between node pairs.

6.4   Overview of mTreebone

We consider a live media content streaming system using application-layer multicast. There is a single dedicated source node, which is persistent during the streaming session. The video stream, originated from the source, is divided into equal-length blocks. The clients that are interested in the video form an overlay network rooted at the source, and each overlay node, except for the source, acts both as a receiver and, if needed, an application-layer relay that forwards data blocks. In addition, these nodes can join and leave the overlay at will, or crash without notification.

The primary issue for building such a system lies in the construction of a resilient overlay structure with low overhead and short delay. In our proposed *mTreebone*, we advocate a hybrid tree/mesh overlay design that combines the best features of both approaches to meet such demands.

6.4.1   Treebone: A Stable Tree-Based Backbone

The core of our design is a tree-based backbone, referred to as *treebone*. Unlike existing tree-based approaches, this backbone consists of only a subset of the nodes, in particular, the stable nodes. Other non-stable nodes are attached to the backbone as outskirts. Most of the streaming data are pushed through the treebone and eventually reach the outskirts, as shown in Figure 6.2.

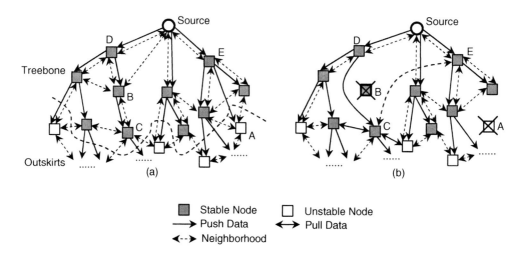

Figure 6.2   mTreebone framework. (a) A hybrid overlay. (b) Handling node dynamics.

It is worth noting that even a small set of stable nodes is sufficient to support the entire overlay. As an illustration, consider a simple K-ary tree of height H. The fraction of its internal nodes, that is, those belonging to the backbone, is no more than $1/K$ if the tree is complete and balanced ($\sum_{i=0}^{H-2} K^i / \sum_{i=0}^{H-1} K^i < \frac{1}{K}$). As such, the construction and maintenance overheads for the treebone are relatively low, particularly considering its nodes are stable, while the data delivery is efficient.

The critical question here is thus how to identify stable nodes. Recent studies have found that, in overlay multicast systems, nodes with a higher age tend to stay longer [11]. This fact offers a hint to identify stable nodes by periodically examining their ages. With this hint, we devise an optimal threshold-based method, which is discussed in Section 6.5.

6.4.2   Mesh: An Adaptive Auxiliary Overlay

The treebone, however, cannot completely eliminate repairing operations because the nodes are not absolutely persistent. In addition, it ignores the potential bandwidth between the unstable nodes.

To improve the resilience and efficiency of the treebone, we further organize all the nodes into a mesh overlay. Similar to CoolStreaming [42], in this auxiliary mesh overlay, each node keeps a partial list of the active overlay nodes and their status. This local list facilitates the node to locate a set of mesh neighbours as well as its dedicated treebone parent. To keep the list updated, the nodes use a lightweight, scalable, random gossip algorithm, SCAMP [22], to periodically exchange their status information. The mesh neighbours also periodically exchange their buffer maps. Unlike existing data-driven systems, a node will not actively schedule to fetch data blocks from its neighbours using data availability information. A fetch is invoked only if data outage occurs in the treebone.

Figure 6.1(a) illustrates this hybrid mTreebone design. When an unstable node, such as node A fails or leaves, it will not affect the data pushed along the treebone. On the other hand, the treebone nodes are stable and seldom leave; even if a leave happens, the impact can be remarkably mitigated with the help from the mesh overlay. For example, assume that node

B leaves, as shown in Figure 6.2(b). While node C is affected, it can easily pull the missing data from its mesh neighbours before it re-attaches to the treebone.

6.5 Treebone Construction and Optimization

To realize such a hybrid overlay for media content streaming, a series of unique and important issues have to be addressed. First, we have to identify the stable nodes in the overlay; second, we have to position the stable nodes to form the treebone, which should also evolve to optimize its data delivery; third, we have to reconcile the treebone and the mesh overlays, so as to fully explore their potentials. In this section, we present our solutions for the construction and optimization of the treebone. In the next section, we will give the details of its interactions with the mesh. For ease of exposition, we summarize the major notations in Table 6.2.

6.5.1 Optimal Stable Node Identification

Intuitively, the stability of a node is proportional to its duration in the overlay, which unfortunately cannot be known before the node actually leaves. We thus resort to a practical prediction using a node's age in the session, that is, the time elapsed since its arrival. As mentioned earlier, existing studies have shown that the nodes already with higher ages tend to stay longer [11]; hence, a node's age partially reflects its stability, and if its age is above a certain threshold, we consider it as a stable node and move it into the treebone. Once a stable node is in the treebone, it remains there until it leaves or the session ends.

The effectiveness of the treebone clearly depends on the age threshold. If the threshold is too low, many unstable nodes would be included in the treebone; on the other hand, if it is too high, few nodes could be considered stable. Our objective is thus to optimize the Expected Service Time (EST) of a treebone node by selecting an appropriate age threshold.

Let $f(x)$ be the probability distribution function (PDF) of node duration, and L be the length of the session. Since a node starts serving in the treebone when its age exceeds the corresponding threshold, for a treebone node arriving at time t, its expected service time $EST(t)$ can be calculated as the expected duration minus the corresponding age threshold, $T(t)$, that is,

$$EST(t) = \frac{\int_{T(t)}^{L-t} x f(x)\, dx + \int_{L-t}^{\infty} (L-t) f(x)\, dx}{\int_{T(t)}^{\infty} f(x)\, dx} - T(t)$$

Previous studies on video client behaviour have suggested that node durations generally follow a heavy-tailed distribution [9], [34], [39], in particular, the Pareto distribution with parameters

Table 6.2 List of Notations

Notation	Description
L	Session length
T	Node arrival time
$L - t$	Residual session length at time t
s	Node age
$EST(t)$	Expected service time in treebone for a node arriving at t
$T(t)$	Age threshold for a node arriving at t
$f(x)$	Probability distribution function of node duration
k, x_m	Shape and location parameters of Pareto distribution

k and x_m (k is a shape parameter that determines how skew the distribution is, and x_m is a location parameter that determines where the distribution starts). Given this model, we have the following expression,

$$EST(t) = \frac{\int_{T(t)}^{L-t} x \frac{kx_m^k}{x^{k+1}} dx + \int_{L-t}^{\infty} (L-t) \frac{kx_m^k}{x^{k+1}} dx}{\int_{T(t)}^{\infty} \frac{kx_m^k}{x^{k+1}} dx} - T(t)$$

$$= \frac{\frac{kx_m^k}{k-1}\left(\frac{1}{T^{(k-1)}(t)} - \frac{1}{(L-t)^{k-1}}\right) + \frac{x_m^k}{(L-t)^{k-1}}}{\frac{x_m^k}{T^k(t)}} - T(t)$$

$$= \frac{T(t)}{k-1}\left[1 - \left(\frac{T(t)}{L-t}\right)^{k-1}\right]$$

To maximize $EST(t)$ with respect to $T(t)$, we have

$$EST(t)'_{T(t)} = \frac{1}{k-1} - \frac{k}{k-1}\left(\frac{T(t)}{L-t}\right)^{k-1} = 0,$$

which follows that $T(t) = (L-t)\left(\frac{1}{k}\right)^{\frac{1}{k-1}}$.

For the typical k value close to 1, $EST(t)$ is maximized when $T(t)$ is roughly about $0.3(L-t)$. In other words, the age threshold for a node arriving at time t is 30% of the residual session length. This is the default setting used in our experiments. In practice, we can also online estimate k and adjust the threshold accordingly.

6.5.2 Treebone Bootstrapping and Evolution

Given the optimal age threshold, we now discuss how the nodes evolve into a stable treebone. We assume that initially only the source node is in the treebone. Each newly joined node obtains L and t from the source, as well as a partial list of existing overlay nodes, at least one of which is in the treebone. The new node then attaches itself to one of the treebone nodes and locates mesh neighbours using the list.

If a node is not in the treebone, it will periodically check its own age in the overlay. Once its age exceeds the threshold $T(t)$, it will promote itself as a treebone node. Figure 6.3 shows an example, where the numeric label of each node is its arrival time.

In this basic promotion method, before time $T(0)$, no node but the source is included in the treebone, which reduces the efficiency of data delivery in this period. To alleviate this problem, we introduce a randomized promotion for the initial period of the session. For a node arriving at time t, the algorithm achieves a probability $s/T(t)$ for the node to be in the treebone when its age is up to s ($0 \le s \le T(t)$). Specifically, each non-treebone node independently checks its status per unit time; for the s-th check (that is, at time $t+s$), it will be promoted to the treebone with probability $1/(T(t) - s + 1)$ (and 0 for $s = 0$). Such an early promotion will speed up the establishment of the treebone. And it is fully distributed with no extra message exchange among the overlay nodes. In addition, as suggested by observations from [23], the built-in randomness of the promotion will also reduce the churn of the treebone.

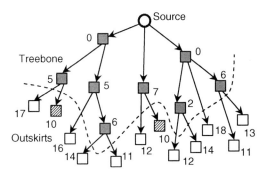

Figure 6.3 An example of treebone evolution. Numeric labels indicate the node arrival times. The shaded nodes will promote themselves into the treebone as their ages become longer than the threshold.

6.5.3 Treebone Optimization

The treebone constructed by the above basic algorithm does not necessarily minimize the latency for data delivery. In particular, two non-optimal substructures could exist, as shown in Figure 6.4 and 6.5, respectively. In the first case, a node has more children than its parent, and a swap of them can reduce the average depth of the treebone nodes. In the second case, a treebone node closer to the source may still be able to accept new children; a node can use this chance to reduce its depth in the treebone. We now introduce two localized algorithms that implement such optimizations.

High-Degree-Preemption. Each treebone node x periodically checks whether it has more children than a node that is closer to the source in the treebone. Such a node, referred to as y, could either be the parent of x in the treebone, or a node known from x's local node list. If so, node x will then preempt y's position in the treebone, and y will re-join the treebone. In practice, y can simply attach itself to x, as illustrated in Figure 6.4.

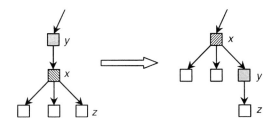

Figure 6.4 An illustration of *high-degree-preemption*.

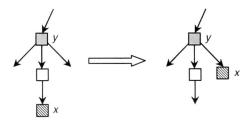

Figure 6.5 An illustration of *low-delay-jump*.

Low-Delay-Jump. Each treebone node x periodically checks whether there are nodes closer to the source than its parent. If so and one such node, say y, has enough bandwidth to support a new child, node x will leave its original parent and attach itself to y as a child, as illustrated in Figure 6.5.

The above two algorithms will be executed by the overlay nodes iteratively until no node can further locate candidates for swapping. The average depth of the treebone nodes is monotonically decreasing in the iterations, leading to a minimal average depth. More explicitly, we have the following theorem to show the minimal average depth is actually minimum:

Theorem 6.1: The average depth of the treebone is minimized when *high-degree-preemption* and *low-delay-jump* terminate at all treebone nodes.

Proof: We prove this in two steps: First, we show that the two properties mentioned above hold for any tree with the minimum average depth. Second, based on this observation, we prove that both our tree and a tree with the minimum depth have the same average depth.

We prove the first by contradiction. Assume there exists a tree A that does not have the two properties but is of minimum average depth. Apparently, A must be a balanced tree; otherwise we can use *low-delay-jump* to reduce its average depth. Now consider that A violates the second property, that is, there must be at least one node x whose out degree is smaller than another node y but is closer to the root. We first consider the case that y is a descendant of x, as shown in Figure 6.6. In this case, we can swap nodes x and y, with y still serving its other children (node z in Figure 6.6). This operation reduces the average depth, which contradicts to the assumption that A has minimized average depth. For the case that y is not x's descendant, we can first swap y with one of x's descendants at the same level. This operation does not change the average depth, but y now becomes x's descendant; so we can then apply the swapping as in the previous case to reduce A's average depth and find the contradiction.

Next, we define a unification operation that swaps nodes only at the same level if the left one has lower out-degree than the right one. Obviously, this unification does not change the average depth. Applying this operation iteratively to a tree until no swap can be found, the unified tree will have the following new property: at each level, the out-degrees of the nodes are non-increasing from left to right. It is easy to show that, given our treebone and any tree with the minimum average depth, the two trees after the unification are isomorphic, and thus of the same average depth. In other words, our treebone achieves the minimum average depth after *high-degree-preemption* and *low-delay-jump* terminate.

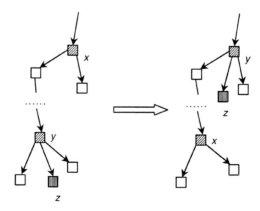

Figure 6.6   A more general example for *high-degree-preemption*.

6.6 Collaborative Mesh-Tree Data Delivery

We next discuss the collaboration between the treebone and the mesh within the mTreebone framework. Such collaboration reflects in two aspects, namely, delivering video data and handling node dynamics.

6.6.1 Seamless Push/Pull Switching

In mTreebone, the data blocks are delivered by two means. In general, they are pushed over the treebone. And if a gap appears in the stream received by a node, due to either temporal capacity fluctuation in the treebone or node dynamics as discussed later, the node may pull the missed blocks through the mesh overlay. We introduce a seamless push/pull buffer that coordinates the treebone and the mesh to make data delivery efficient yet resilient against failure.

Figure 6.7 illustrates the push/pull switching, where a *tree-push* pointer is used to indicate the latest data block delivered by the push method, and a *mesh-pull* window facilitates the pull delivery. When a node is temporarily disconnected from the treebone, its tree-push pointer will be disabled and only the mesh-pull window works to fetch data from its mesh neighbours. When it connects to the treebone again, the tree-push pointer will be re-activated. The mesh-pull window is always kept behind the tree-push pointer so as not to request data currently being delivered by the treebone. Therefore, no duplicated data blocks are received from both treebone and mesh.

6.6.2 Handling Host Dynamics

A node may gracefully leave the overlay, or abruptly fail without any notification. In the former, the node will proactively inform its mesh neighbours and its treebone children if it resides in the treebone. In the latter, the abrupt leave can be detected by the mesh neighbours after a silent period with no control message exchange, or by the children in the treebone after observing persistent losses. In either case, its mesh neighbours need to re-establish neighbourships with other known nodes in their local node lists, and, if the node is in the treebone, its children have to relocate their parents.

If the affected child is an unstable node in the outskirts of the treebone, it will check its local node list and directly attach to one node that is nearest to the source with enough available bandwidth. On the other hand, if it is a stable node, it has to re-join the treebone. To this end, the node will first locate a treebone node with enough available bandwidth and then attach itself to the node. If no such treebone node is known, the node needs to preempt the position of an unstable node that is currently a child of a treebone node.

In the above process, mesh overlay will temporarily take over data delivery before the treebone is repaired, as discussed previously. Our simulation and experimental results demonstrate that this two-tier overlay effectively reduces data losses in the tree repairing process. Meanwhile,

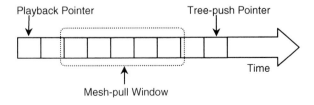

Figure 6.7 Design of the push/pull switch buffer.

its control overhead is kept at a reasonable level, which is lower than that of the purely mesh-based approaches.

6.7 Performance Evaluation

To evaluate mTreebone, we have conducted extensive simulations and PlanetLab-based experiments. We have also implemented two state-of-the-art application layer multicast systems for comparison, namely, CoolStreaming [42] and ChunkySpread [37]. The former is a typical data-driven mesh system, and the latter is a multi-tree system, which also adopts an auxiliary neighbouring graph to facilitate tree construction.

In our evaluation and comparison, we use the following three typical metrics, which altogether reflect the quality of service experienced by overlay nodes.

Startup latency, which is the time taken by a node from requesting to join the session to receiving enough data blocks for playback.

Transmission delay, which is the time to deliver a data block from the source to the node. Due to the buffer-and-relay nature in overlay networks, the end-to-end transmission delay is orders of magnitude higher than that in IP multicast, thus we use *second* as the unit, as for startup latency.

Data loss rate, which is defined as the fraction of the data blocks missing their playback deadlines, that is, either lost during transmission or experienced excessive delays.

We adopt a dynamic scenario for the evaluation, where the overlay nodes arrive at different times and may also leave or fail before the session ends. As discussed in Section 6.5, we use the Pareto distribution to model the durations of the nodes in a session, which is also suggested by many empirical trace studies [9], [34]. The default parameters of the distribution are adopted from the recently modeling work for peer-to-peer streaming in [11] as well as trace studies of PPLive, a popular commercial peer-to-peer streaming system [25], [39]. We also investigate the impact of other parameter settings, in particular, the skew factor k, which reflects the churn rate of the clients.

6.7.1 Large-Scale Simulations

Unless otherwise specified, the following default parameters are used in our simulation, most of which follow the typical values reported in [9], [25], [39]. The session length L is set to 6000 seconds and each data block is of 1-second video; there are 5000 overlay nodes; the maximum end-to-end delay is 1000 ms between two overlay nodes, and the maximum upload bandwidth is uniformly distributed from 4 to 12 times of the bandwidth required for a full streaming. For comparison, the number of partners in CoolStreaming is set to 5 and the number of substreams in ChunkySpread is set to 16. The details about these two parameters and their setting guidelines can be found in [42], [37], respectively.

We set the default age threshold $T(t)$ to 30% of the residual session length, which corresponds to the optimal setting (see Section 6.5.1) when k is close to 1. We will also examine other settings of $T(t)$ as well as the impact of k.

Figure 6.8 shows the CDF of the startup latency. We can see that mTreebone has the lowest latency, followed by CoolStreaming and then by ChunkySpread. This is because, when two events (join and leave) occur closely, the tree repair process in ChunkySpread may delay the newly joined node from receiving data. In addition, the new node needs to receive sub-streams across multiple trees to assemble the original stream, which further increases the delay. A Cool-Streaming node, however, has to locate multiple mesh neighbours and establish relations, and

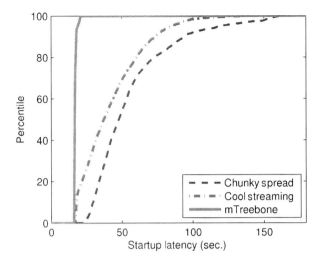

Figure 6.8   CDF of startup latency (simulation).

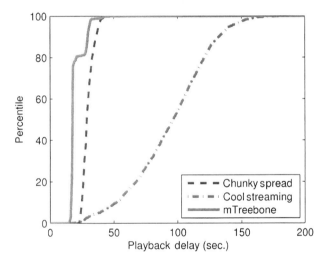

Figure 6.9   CDF of transmission delay (simulation).

the data pull operation also slows down the process. In contrast, a new node in mTreebone can receive data in the tree structure even before establishing the mesh neighbourship, and the *high-degree-preemption* and *low-delay-jump* operations minimize the delay of the treebone. Similar reasons also explain the results of transmission delay in Figure 6.9. The data loss rates for the three systems are given by Figure 6.10. mTreebone also outperforms CoolStreaming and ChunkySpread, validating the advantage of the hybrid design.

Figure 6.11 shows the data loss rate in mTreebone with different age thresholds. Although the data loss rates are generally below 1% for a wide range of thresholds, the minimal appears at 30% of the residual session length, which corresponds to the optimal threshold setting. Also, we have conducted simulations with different k values. In general, a larger k value means the overlay has a higher churn rate, that is, more dynamic. We show the data loss rate of mTreebone

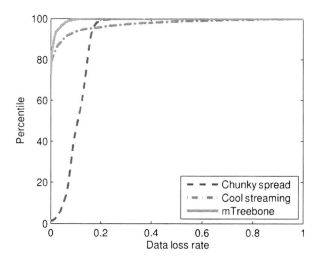

Figure 6.10   CDF of data loss (simulation).

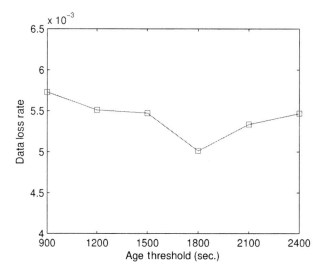

Figure 6.11   Avg. data loss with different $T(t)$ (simulation).

as a function of k in Figure 6.12. When k is no greater than 1.5, our mTreebone is fairly stable, and the default value of 1 is thus representative for performance evaluation. It is worth noting that, although the data loss rate noticeably increases for k over 1.5, it remains less than 2%, implying that our hybrid design resists to node dynamics well.

6.7.2   PlanetLab-Based Experiments

To further investigate the performance of mTreebone, we have implemented a prototype and conducted experiments on the PlanetLab. Besides the three typical metrics for overlay nodes, we also examine the cost of the whole system in such a real network, that is, the message

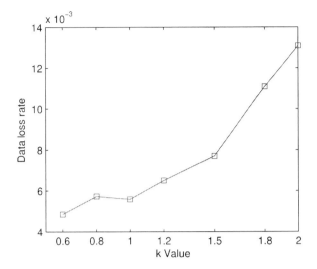

Figure 6.12   Avg. data loss with different k (simulation).

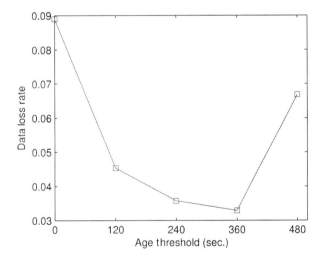

Figure 6.13   Avg. data loss with different $T(t)$ (PlanetLab).

overhead to construct, maintain and repair the overlay. When a mesh is applied, the overhead also includes the messages to exchange data availability and to request blocks from neighbours. Such messages are generally of small sizes, but their excessive total number could still impose heavy load to routers. Hence, in our evaluation, we use the total number of the control messages as the metric to measure their impact.

In each experiment, we let 200 PlanetLab nodes stay in the session following the Pareto distribution. To accumulate enough join and leave events for evaluation, we also allow a node to re-join the overlay after it leaves the overlay for a while. We first vary the age threshold to see its impact on data loss rate. The result is shown in Figure 6.13, where the threshold at 30% again leads to the minimum loss rate. Hence, we still use it as the default setting in mTreebone.

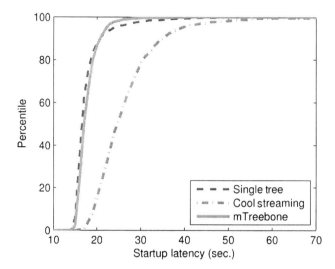

Figure 6.14   CDF of startup latency (PlanetLab).

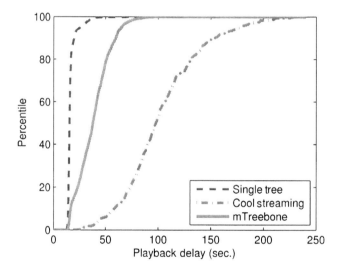

Figure 6.15   CDF of transmission delay (PlanetLab).

Figures 6.14–6.16 give the results of startup latency, transmission delay and data loss rate of mTreebone, respectively. For comparison, we also implement CoolStreaming and a single tree based system, and present their respective results in the figures. From these results, we can see that the startup latency of mTreebone is very close to that of the tree-based approach, and is much lower than that of CoolStreaming. Similar observation applies to transmission delay. In contrast, mTreebone and CoolStreaming both have much lower data loss rates, as compared to the single tree. Comparing with the simulation results, the data loss rates on the PlanetLab are generally higher. This is due to the influences of background traffic.

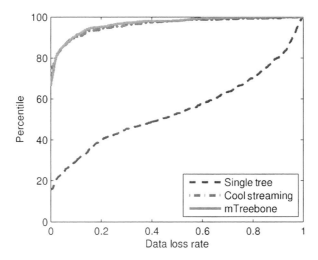

Figure 6.16 CDF of data loss (PlanetLab).

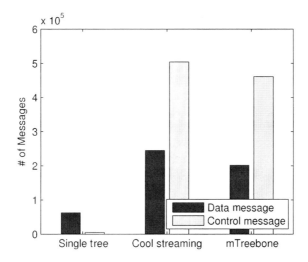

Figure 6.17 Message overhead (PlanetLab).

Figure 6.17 compares the control overhead of the three systems. It is not surprising that mTreebone has a higher overhead than the single tree; but, even though mTreebone has to maintain two overlays, its overhead is still lower than CoolStreaming. This result suggests that the stable treebone is the major delivery path in mTreebone. Also the total traffic volume of the (small) control messages are indeed quite low, which is generally less than 1% of the total data traffic of mTreebone in our experiments. We thus believe that the control overhead of mTreebone is acceptable.

To further demonstrate the resilience of mTreebone against node dynamics, Figure 6.18 shows the data loss rates for different values of k. The results reaffirm that mTreebone is quite stable for a wide range of churn levels.

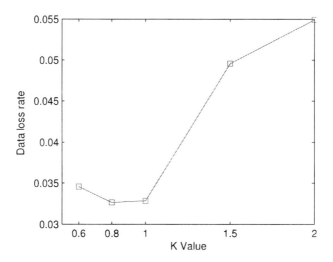

Figure 6.18   Avg. data loss with different k (PlanetLab).

6.8   Conclusion and Future Work

In this chapter, we reviewed various solutions proposed to advocate peer-to-peer media streaming. These solutions can largely be divided into two categories, namely, tree-based and mesh-based approaches, both of which may suffer performance inefficiencies either due to the vulnerability caused by dynamic end-hosts or the efficiency-latency tradeoff. To this end, we explored the opportunity to leverage both tree and mesh approaches cooperatively within a hybrid framework, *mTreebone*. We presented effective and coherent solutions to reconcile the two overlays in the mTreebone framework. Specifically, we derived an optimal age threshold to identify stable nodes, which maximizes their expected service time in the treebone. We designed a set of overlay construction and evolution algorithms, which minimize the startup and transmission delays. Finally, we gave a buffer partitioning and scheduling algorithm, which enables seamless treebone/mesh collaboration in data delivery.

We extensively evaluated the performance of mTreebone and compared it with existing mesh and tree based solutions. The simulation results demonstrated the superior efficiency and robustness of this hybrid solution, which were further validated by our experiments of an mTreebone prototype over the PlanetLab network.

It is worth emphasizing that diverse optimization techniques could be carried over the treebone, for its relatively smaller scale and inherent stability. In our future work, we plan to examine advanced node organization/re-organization methods to further improve its efficiency and its interactions with the mesh. We will also explore the use of multi-tree-based backbone, which may lead to better balanced load and finer-grained bandwidth control. Finally, we are interested in experiments of larger scales with our prototype as well as real deployment over the global Internet.

References

[1] Akamai. [Online]. Available: http://www.akamai.com
[2] BitTorrent. [Online]. Available: http://www.bittorrent.com
[3] EMule. [Online]. Available: http://www.emule-project.net
[4] Gnutella. [Online]. Available: http://www.gnutella.com

[5] Napster. [Online]. Available: http://www.napster.com

[6] PlanetLab. [Online]. Available: http://www.planet-lab.org/

[7] Skype. [Online]. Available: http://www.skype.com

[8] S. Acharya and B. C. Smith, 'Middleman: A Video Caching Proxy Server,' in *Proc. NOSSDAV*, June 2000.

[9] K. C. Almeroth and M. H. Ammar, 'Collecting and Modeling the Join/Leave Behavior of Multicast Group Members in the MBone,' in *Proc. IEEE HPDC*, August 1996.

[10] S. Banerjee, B. Bhattacharjee and C. Kommareddy, 'Scalable Application Layer Multicast,' in *Proc. ACM SIGCOMM*, August 2002.

[11] M. Bishop, S. Rao and K. Sripanidkulchai, 'Considering Priority in Overlay Multicast Protocols under Heterogeneous Environments,' in *Proc. IEEE INFOCOM*, April 2006.

[12] M. Castro, P. Druschel, A.-M. Kermarrec, A. Nandi, A. Rowstron and A. Singh, 'SplitStream: High-Bandwidth Multicast in Cooperative Environments,' in *Proc. ACM SOSP*, October 2003.

[13] Y. Chae, K. Guo, M. M. Buddhikot, S. Suri and E. W. Zegura, 'Silo, Rainbow, and Caching Token: Schemes for Scalable, Fault Tolerant Stream Caching,' *IEEE JSAC*, vol. 20, no. 7, pp. 1328–1344, September 2002.

[14] Y. Chawathe, S. McCanne and E. A. Brewer, 'An Architecture for Internet Content Distribution as an Infrastructure Service,' http://yatin.chawathe.com/-yafin/papers/scattercast.ps.

[15] Y.-H. Chu, A. Ganjam, T. E. Ng, S. G. Rao, K. Sripanidkulchai, J. Zhan and H. Zhang, 'Early Deployment Experience with an Overlay Based Internet Broadcasting System,' in *Proc. USENIX Annual Technical Conference*, June 2004.

[16] Y.-H. Chu, S. G. Rao, S. Seshan and H. Zhang, 'Enabling Conferencing Applications on the Internet Using an Overlay Multicast Architecture,' in *Proc. ACM SIGCOMM*, August 2001.

[17] Y.-H. Chu, S. G. Rao and H. Zhang, 'A Case for End System Multicast,' in *Proc. ACM SIGMETRICS*, June 2000.

[18] S. Deering and D. Cheriton, 'Multicast Routing in Datagram Internetworks and Extended LANs,' *ACM Trans. Computer Systems*, vol. 8, no. 2, pp. 85–110, May 1990.

[19] H. Deshpande, M. Bawa and H. Garcia-Molina, 'Streaming Live Media over Peer-to-Peer Network,' Stanford University, Tech. Rep., 2001.

[20] P. Eugster, R. Guerraoui, A.-M. Kermarrec and L. Massoulie, 'From Epidemics to Distributed Computing,' *IEEE Computer*, 2004.

[21] P. Francis, 'Yoid: Extending the Interent Multicast Architecture,' http://www.icir.org/yoid/.

[22] A. J. Ganesh, A. M. Kermarrec and L. Massoulie, 'Peer-to-Peer Membership Management for Gossip-based Protocols,' *IEEE Trans. Computers*, vol. 52, no. 2, pp. 139–149, February 2003.

[23] P. B. Godfrey, S. Shenker and I. Stoica, 'Minimizing Churn in Distributed Systems,' in *Proc. ACM SIGCOMM*, September 2006.

[24] M. Heffeeda, A. Habib, B. Botev, D. Xu and B. Bhargava, 'PROMISE: Peer-to-Peer Media Streaming Using CollectCast,' in *Proc. ACM Multimedia*, November 2003.

[25] X. Hei, C. Liang, J. Liang, Y. Liu and K. Ross, 'Insights into PPLive: A Measurement Study of a Large-scale P2P IPTV System,' in *Proc. Workshop on IPTV services over World Wide Web*, May 2006.

[26] D. Kostic, A. Rodriguez, J. Albrecht and A. Vahdat, 'Bullet: High Bandwidth Data Dissemination Using an Overlay Mesh,' in *Proc. ACM SOSP*, October 2003.

[27] J. Liu, S. G. Rao, B. Li and H. Zhang, 'Opportunities and Challenges of Peer-to-Peer Internet Video Broadcast,' *Proceedings of the IEEE*, vol. 96, no. 1, pp. 11–24, January 2008.

[28] J. Liu and J. Xu, 'Proxy Caching for Media Streaming over the Internet,' *IEEE Communications*, vol. 42, no. 8, pp. 88–94, August 2004.

[29] N. Magharei, R. Rejaie and Y. Guo, 'Mesh or Multiple-Tree: A Comparative Study of Live P2P Streaming Approaches,' in *Proc. IEEE INFOCOM*, May 2007.

[30] V. N. Padmanabhan, H. Wang and P. Chou, 'Resilient Peer-to-Peer Streaming,' in *Proc. IEEE ICNP*, November 2003.

[31] V. N. Padmanabhan, H. J. Wang, P. A. Chou and K. Sripanidkulchai, 'Distributing Streaming Media Content Using Cooperative Networking,' in *Proc. NOSSDAV*, May 2002.

[32] V. Pai, K. Tamilmani, V. Sambamurthy, K. Kumar and A. Mohr, 'Chainsaw: Eliminating Trees from Overlay Multicast,' in *Proc. The 4th International Workshop on Peer-to-Peer Systems (IPTPS)*, February 2005.

[33] S. Ramesh, I. Rhee and K. Guo, 'Multicast with Cache (Mcache): An Adaptive Zero-Delay Video-on-Demand Service,' in *Proc. IEEE INFOCOM*, April 2001.

[34] K. Sripanidkulchai, B. Maggs and H. Zhang, 'An Analysis of Live Streaming Workloads on the Internet,' in *Proc. ACM IMC*, October 2004.

[35] I. Stoica, R. Morris, D. Karger, M. F. Kaashoek and H. Balakrishnan, 'Chord: A Scalable Peer-to-Peer Lookup Service for Internet Applications,' in *Proc. of ACM SIGCOMM*, August 2001.

[36] R. Tian, Q. Zhang, Z. Xiang, Y. Xiong, X. Li and W. Zhu, 'Robust and Efficient Path Diversity in Application-Layer Multicast for Video Streaming,' *IEEE Trans. Circuits and Systems for Video Technology*, vol. 15, no. 8, pp. 961–972, August 2005.

[37] V. Venkataraman, P. Francis and J. Calandrino, 'ChunkySpread: Multi-Tree Unstructured Peer-to-Peer Multicast,' in *Proc. The 5th International Workshop on Peer-to-Peer Systems (IPTPS)*, February 2006.

[38] B. Wang, S. Sen, M. Adler and D. Towsley, 'Optimal Proxy Cache Allocation for Efficient Streaming Media Distribution,' in *Proc. IEEE INFOCOM*, June 2002.

[39] F. Wang and J. Liu, 'A Trace-Based Analysis of Packet Flows in Data-Driven Overlay Networks,' Simon Fraser University, Tech. Rep., 2006.

[40] J. Wang, 'A Survey of Web Caching Schemes for the Internet,' *ACM Computer Communication Review (CCR)*, vol. 29, no. 5, pp. 36–46, October 1999.

[41] M. Zhang, J.-G. Luo, L. Zhao and S.-Q. Yang, 'A Peer-to-Peer Network for Live Media Streaming Using a Push-Pull Approach,' in *Proc. of ACM Multimedia*, November 2005.

[42] X. Zhang, J. Liu, B. Li and T.-S. P. Yum, 'CoolStreaming/DONet: A Data-driven Overlay Network for Peer-to-Peer Live Media Streaming,' in *Proc. IEEE INFOCOM*, March 2005.

[43] S. Q. Zhuang, B. Y. Zhao and A. D. Joseph, 'Bayeux: An Architecture for Scalable and Fault-Tolerant Wide-Area Data Dissemination,' in *Proc. NOSSDAV*, June 2001.

7

Cooperation in DTN-Based Network Architectures

Vasco N. G. J. Soares[1,2] and Joel J. P. C. Rodrigues[1]

[1] *Instituto de Telecomunicações, University of Beira Interior, Portugal*
[2] *Superior School of Technology, Polytechnic Institute of Castelo Branco, Portugal*

7.1 Introduction

Available Internet protocols perform transmission control protocol/Internet protocol (TCP/IP), making implicit assumptions of continuous, bi-directional end-to-end paths, short round-trip times (RTT), high transmission reliability and symmetric data rates. However, a wide range of emerging networks (outside the Internet) usually referred to as opportunistic networks, intermittently connected networks, or episodic networks violate these assumptions. These networks fall into the general category of delay-tolerant networks (DTNs). Sparse connectivity, frequent network partitioning, intermittent connectivity, large or variable delays, asymmetric data rates, and high packet loss belong to the main characteristics of DTNs. More importantly, end-to-end connection cannot be assumed to be available on these networks.

Delay-tolerant networking [1, 2] is a network research topic focused on the design, implementation, evaluation and application of architectures and protocols that intend to enable data communication among heterogeneous networks in extreme environments. In order to answer these challenges, DTN introduces a bundle layer, creating a store-and-forward overlay network above the transport layers of underlying networks.

Vehicular delay-tolerant networking (VDTN) [3] appears as a novel network architecture proposal based on the delay-tolerant network architecture, that aims to provide innovative solutions for challenged vehicular communications. VDTN assumes out-of-band signaling, with separation between control and data planes, and considers IP over VDTN approach, placing the DTN-based layer over the data link layer.

Effective operation of DTNs and VDTNs relies on the cooperation of network nodes to store-carry-and-forward data over partitioned and challenged network environments. Until recently, most of the research on these network architectures has assumed that nodes are fully cooperative. However, this assumption may not be realistic. For instance, nodes may decide not to cooperate in order to conserve their resources (for example, energy), or may maliciously consume the resources contributed by other nodes without contributing any resources in return. This selfish behaviour can degrade the overall network performance significantly.

This chapter introduces and studies the concept of cooperation and its importance on DTN and VDTN network architectures. The remainder of this chapter is organized as follows. Section 7.2 starts with an overview of DTNs focusing on its architecture and application scenarios. Then, the cooperation concept is introduced, and related work on cooperation in DTNs is described. Section 7.3 briefly describes VDTN network architecture and its innovative approach. Then, the VDTN cooperation functions at the control plane and data plane level are discussed. At the end of this section, the impact of node cooperation on the performance of data plane with two DTN routing protocols is analyzed. Finally, Section 7.4 concludes the chapter providing a final summary of the study.

7.2 Delay-Tolerant Networks

The DTN architecture [1] incorporates a store-carry-and-forward paradigm by overlaying a protocol layer, called bundle layer, that provides internetworking on heterogeneous networks (regions) operating on different transmission media. DTN gateways forward asynchronous messages called bundles (that represent entire blocks of application-program user data) between two or more DTN regions. Therefore, gateways are responsible for mapping data from the lower-layer protocols used in each region they span. This approach allows data to traverse multiple regions, via region gateways, to reach a destination region and, finally, a host within that region.

DTN store-carry-and-forward paradigm avoids the need for constant connectivity. It is used to move bundles across a region, exploiting node mobility. DTN store-carry-and-forward operating principle can be described as follows. A source node originates a bundle and stores it (using some form of persistent storage), until an appropriate communication opportunity becomes available. The bundle will be forwarded when the source node is in contact with an intermediate node that is expected to be more close to the destination node. Afterwards, the intermediate node stores the bundle and carries it while no new contact is available. This process is repeated and the bundle will be relayed hop by hop until (eventually) reaching its destination. This process is illustrated in Figure 7.1.

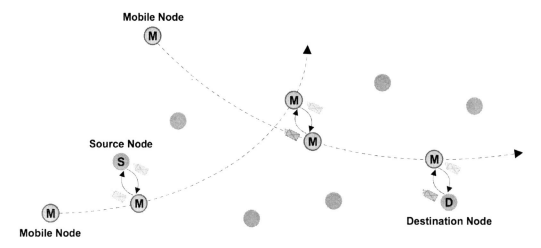

Figure 7.1 DTN store-carry-and-forward paradigm.

The contact schedule between network nodes depends on the network environment [4], and contacts can be classified as scheduled, predicted, or opportunistic. If nodes move along predictable paths, they can predict or receive time schedules of their future positions. Thus, communication sessions will be scheduled. Predicted contacts require analyzing previous observed contacts to predict the opportunities to transmit data. In opportunistic contacts, communication opportunities happen unexpectedly.

Numerous routing strategies have been presented in the literature for different types of DTNs [4–6]. They are used according to the contact characteristics and the knowledge available about the network. In general, routing strategies can be classified as either forwarding-based or flooding-based. Forwarding strategies use network topology information. They maintain a single copy of a bundle in the network that is forwarded using the best path. On the contrary, flooding strategies do not need to have information about the network. They replicate bundles at contact opportunities hoping to find a path to a destination. Bundle replication contributes to the delivery probability improvement and minimizes the delivery delay. The downside is that it increases the contention for network resources (for example, bandwidth and storage), potentially leading to poor overall network performance [7].

7.2.1 DTN Application Domains

The DTN concept was initially developed for interplanetary networking [8]. However, over the last years, a number of real-world environments where DTN techniques are required have been emerging in the literature. Underwater networks enable applications for oceanographic data collection, pollution monitoring, offshore exploration, disaster prevention, assisted navigation and tactical surveillance applications [9, 10]. Wildlife tracking networks are designed to support wildlife tracking for biology research [11, 12]. Data MULEs are used for data retrieval in the context of sensor network applications [13]. Transient networks provide connectivity to remote and underdeveloped communities [14–18]. Disaster recovery networks can support communications in catastrophe hit areas lacking a functioning communication infrastructure [19]. People networks can explore transfer opportunities between mobile wireless devices carried by humans [20, 21].

Integrating DTN concepts to military tactical networks can ease communications in hostile environments (battlefields) subject to frequent disruption of end-to-end connectivity [22, 23]. Vehicular networks can also benefit from DTN approach and have several application scenarios, including networks to disseminate information advertisements (for example, marketing data) or safety related information (for example, emergency notification, traffic condition and collision avoidance) [24], and monitoring networks to collect data (for example, pollution data, road pavement defects) [25].

7.2.2 Cooperation in Delay-Tolerant Networks

It is important to notice that limited storage capacity, limited network bandwidth and limited energy influence the performance and the capacity of delay-tolerant networks [7, 26]. Furthermore, these problems are reinforced by the long or variable propagation delay, low node density, low transmission reliability, node mobility and disruption, commonly observed in these networks.

Cooperation is a key issue to the success of data communication in DTNs, where nodes use their storage, bandwidth and energy resources to mutually enhance the overall network performance. In a cooperative environment (shown in Figure 7.1), network nodes collaborate with each other, storing and distributing bundles not only in their own interest, but also in the interest

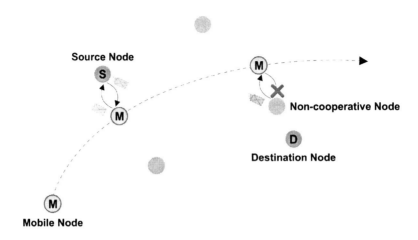

Figure 7.2   Non-cooperative network scenario.

of other nodes. Such a behaviour increases the number of possible transmission paths improving the robustness to failure of individual nodes. In a non-cooperative environment, network nodes exhibit a selfish behaviour. For instance, as illustrated in Figure 7.2, a network node may deny storing and distributing the other nodes' bundles and, at the same time, exploit the other nodes resources to disseminate its data. This behaviour can be caused by several reasons, such as, resource limitations (for example, storage, energy) or rogue operation (malicious behaviour). More importantly, it may lead to overall degradation of the network performance.

Although cooperation is crucial for improving the limited capability of network nodes and, consequently, for increasing the overall network performance, to the best of our knowledge, little research has been done in this field. Panagakis et al. [27] state that the literature related to the performance of DTN routing protocols usually assumes fully cooperative environments, which can be an unrealistic assumption, since network nodes may be unable to cooperate due to resource constraints, to a selfish behaviour or to a common strategy. The authors do not propose any mechanisms to detect non-cooperative behaviour or to enforce cooperation among network nodes. They define cooperation as the probability of a node to forward message copies or drop them on their arrival, and evaluate its effect in terms of delivery delay and transmission overhead on Epidemic [28], Two-Hop [29] and Binary Spray and Wait [30] DTN routing algorithms. The results presented in this study reveal the importance of considering cooperation effects on the network performance. Algorithms that perform better under a fully cooperative environment can be outperformed in non-cooperative environments. In general, Two-Hop routing algorithm is more resilient to less cooperative node behaviour.

Buttyán et al. [31] studied the problem of selfish node behaviour in DTNs used for personal wireless communications. In this work, a mechanism to discourage selfish node behaviour during message exchange based on the principles of barter is proposed. This barter-based approach is analyzed with a game-theoretic model. Simulation results show that the approach indeed stimulates cooperation of the network nodes. Shevade et al. [32] also demonstrate the degradation of a DTN network performance due to selfish node behaviour. In their opinion, this motivates the need to introduce an incentive mechanism to stimulate cooperation. In this sense, an incentive-aware DTN routing scheme is proposed based on the use of pair-wise tit-for-tat (TFT) incentive mechanism. TFT is based on the principle that every node forwards as much traffic for a neighbour as the neighbour forwards for it. This routing strategy allows selfish users to optimize their performance without significantly deteriorating the system-wide performance. A simulation-based study shows that the proposed routing protocol effectively stimulates cooperation among selfish network nodes, and thus improves the overall delivery ratio.

Morillo-Pozo et al. [33, 34] propose a variation of the cooperative ARQ (C-ARQ) scheme to be used in delay-tolerant vehicular networks. The principal objective of this scheme is to reduce packet losses in transmissions between fixed access points placed on the roads and passing-by vehicles. Vehicles buffer all data sent by access points. Cooperation between vehicles is established in areas where there is no connectivity to the access points. In these areas, vehicles request packets incorrectly received or lost (from the access point) to near-by vehicles. This scheme intends to allow network nodes to work cooperatively in order to increase their delivery rate. The performance of the proposed scheme is evaluated through an experimental prototype running in a real urban environment. The obtained results have shown that using this cooperation scheme packet loss can be halved for transmissions between vehicles and access points. This study is extended in [35] by the same authors. They propose a new cooperative ARQ protocol scheme named DC-ARQ (Delayed Cooperative ARQ). Contrary to C-ARQ cooperation that occurs in a packet-by-packet basis, in DC-ARQ the cooperation is delayed until vehicles are out of the range of the access point. DC-ARQ scheme performance is evaluated through simulations.

Following the previous works [33–35], Trullols-Cruces et al. [36] present a vehicular framework that benefits from two cooperative mechanisms. A DC-ARQ scheme is used to reduce packet losses in the transmissions between network nodes (vehicles and access points). In addition, vehicle route prediction together with a carry-and-forward paradigm is used to improve throughput, delay and the number of access points. Through simulation, the authors show that sparse vehicle scenarios benefit more from the carry-and-forward paradigm, while dense vehicle scenarios benefit from the DC-ARQ scheme.

Resta et al. [37] present a theoretical framework for evaluating the effects of different degrees of node cooperation on the performance of DTN routing protocols. The first part of the work assumes a fully cooperative node behaviour, and presents an analytical characterization of the performance of Epidemic and Two-Hop routing in terms of packet delivery ratio. These results are then used in the second part of the work to analytically characterize the performance of Epidemic under different degrees of node cooperation. The third part of the work presents a simulation study, which evaluates the performance of Epidemic, Two-Hop and Binary Spray and Wait protocols under different degrees of node cooperation. The observed results show that Binary Spray and Wait has the better resilience to lower node cooperation, while presenting the best compromise between packet delivery ratio and message overhead.

Altman [38] analyzes competitive and cooperative operation in DTNs considering a Two-Hop routing strategy. The effect of competition between network nodes is studied in a game theoretical setting. Insights into the structure of equilibrium policies are provided as well as a comparison to a cooperative scenario. Solis et al. [39] present a work related to this area of research. The work introduces the concept of 'resource hog' as a DTN network node that, on average, attempts to send more of its own data and possibly forward less peer data than a typical well-behaved node. Moreover, a 'malicious resource hog' may accept all incoming messages sent by the peer nodes, but immediately drop them. A performance evaluation through simulation reveals that the delivery ratio of well-behaved nodes decreases significantly in the presence of a reduced number of nodes acting as resource hogs. The work also proposes and evaluates resource management solutions to deal with the 'resource hogs' problem.

7.3   Vehicular Delay-Tolerant Networks

Vehicular delay-tolerant networking (VDTN) has emerged as a new DTN-based network architecture, which assumes the separation between the control and data plane, and locates the bundle layer below the network layer, introducing an IP over VDTN approach [3]. The control plane is responsible for exchanging signaling information and resources allocation. This information is carried out-of-band and it is used to set up a data plane connection to transmit data bundles between network nodes. Bundles are defined as the protocol data unit at the VDTN bundle layer, and represent aggregates of Internet protocol (IP) packets with common characteristics, such as the same destination node. Invariably, VDTN uses the store-carry-and-forward DTN model for bundle delivery. This long-term storage paradigm may be combined with bundle replication schemes to increase the probability of successfully delivery and to decrease delivery delay.

In VDTNs, the mobility of vehicles (mobile nodes) is exploited by setting them to operate as relays according to the store-carry-and-forward paradigm. Additionally, stationary relay nodes are introduced in the network to increase contact opportunities in scenarios with low vehicle density. These special nodes are fixed devices located at road intersections with store-and-forward capabilities. They allow vehicles passing by to pickup and deposit data on them, contributing to increase the bundles delivery ratio while decreasing their delivery delay [40]. Figure 7.3 illustrates the interactions between these network nodes – relay and mobile nodes, and between mobile nodes.

VDTNs are characterized by high mobility, which results in frequent changes on the network topology and its corresponding partition. The vehicles' high speed causes short contact durations between network nodes, which restricts the volume of data transfer that is already restricted by the finite bandwidth and the transmission range limitations. In addition, environment physical obstacles, and interferences, also contribute to intermittent connectivity and high error rates.

7.3.1   Cooperation in Vehicular-Delay Tolerant Networks

Following delay-tolerant networks, vehicular delay-tolerant networks also take advantage of the benefits introduced by the cooperative behaviour of network nodes, in order to obtain significant enhancement of the network performance. In particular, all VDTN network functions are based on the principle of cooperation between network nodes. For example, this encompasses the strategies for signaling and resources reservation (for example, storage and bandwidth).

As above-mentioned VDTN architecture assumes control and data planes separation. The control plane provides signaling functionalities for node discovery and resources reservation.

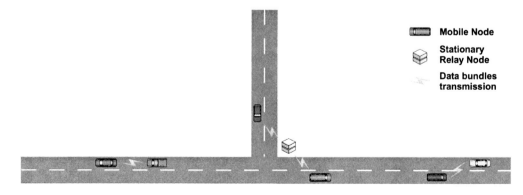

Figure 7.3   Illustration of a vehicular-delay tolerant network.

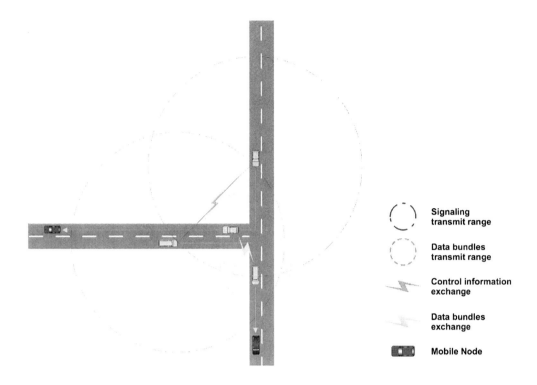

Figure 7.4 Control information and data bundles exchange.

It uses a low-powered, long-range, low bandwidth link, and is responsible for carrying information to configure and control the data plane. Data bundles are transmitted in the data plane using
a high-powered, short-range and high bandwidth link connection.

When network nodes encounter each other (Figure 7.4), they exchange signaling information.
This includes, among others, information about the node type, geographical location, current
path and velocity, energy and buffer status, and data plane link rate and transmission range.

The cooperative exchange of node localization information (geographical location, current
path, and velocity) among network nodes, allows predicting the period of time during which the
nodes' data plane links will be in range with each other [41]. Then, these link connections can be
configured to be active only during that time. This results in saving energy, which is important for
network nodes with limited energy resources like stationary relay nodes. Moreover, it will also
be possible to predict the maximum number of bytes that can be transmitted during the contact
opportunity. This allows improving and optimizing the data plane link bandwidth utilization,
since it will be possible to avoid incomplete data bundle transmissions and consequently the
waste of link capacity [42].

The cooperative exchange of signaling information can also be used to decide whether to
ignore or accept a contact opportunity. For instance, a contact opportunity may be ignored if a
node has energy constraints, if the predicted contact duration is too short for a successful bundle
transmission, or to prevent buffer overflow and thus avoid bundle drops and unnecessary energy
consumption. Other kinds of information can be cooperatively exchanged and used to improve
the overall network performance, like routing state information or bundle delivery notifications,
that can be used to free essential buffer space.

In vehicular delay-tolerant networks, cooperation is not restricted only to the control plane. At the data plane level, network nodes rely on mutual cooperation (and node mobility) to relay bundles between source and destination. However, network nodes are constrained with limited data plane resources, such as storage and link bandwidth. In this sense, a fully cooperative behaviour, such as unconditionally store and forward bundles for others, cannot be taken for granted. Nodes might not be willing to unconditionally store all bundles sent by other network nodes, in order to save buffer resources for their own bundles. The same applies to scheduling bundle forwarding. For instance, network nodes might give preference to scheduling first all their own bundles for transmission. This may limit severely the transmission of the other bundles due to short-lived links and finite bandwidth.

In a previous work [41], we have demonstrated the performance gains that can be achieved by node cooperation at the control plane. We focused on using node localization information to improve the overall performance of a VDTN. In this work, we are interested in studying the impact of node cooperation at the data plane level. More concretely, at data plane, we assume that nodes' resources (for example, storage space and bandwidth) will be divided into two parts. One part of the resources is reserved to store, carry and forward the bundles hosted on nodes, whereas the other part is used for cooperation purposes. We evaluate the impact of different amounts of data plane resources reserved for node cooperation on the overall network performance.

7.3.2 Performance Assessment of Node Cooperation

This section investigates the impact of node cooperation (at the data plane level) on the performance of VDTN networks. The study was conducted by simulation using a modified version of the Opportunistic Network Environment (ONE) simulator [43]. ONE was modified to support the VDTN layered architecture model proposed in [3]. Additional modules were developed to implement the buffer management scheme, the scheduling and drop policies used in the context of this work. The next subsections describe the simulation scenario and the corresponding performance analysis.

7.3.2.1   Network Setup

For the simulation scenario, we use a map-based model of a small part of the city of Helsinki presented in Figure 7.5. Stationary relay nodes were placed at five road intersections presented in the figure, each one with a 500 Mbytes message buffer. During a 12 hours period of time (for example, from 8:00 to 20:00), 100 mobile nodes (vehicles) move on the map roads at an average speed of 30 km/h, between random locations, with random pause times between 5 and 15 minutes. Different storage constraints are introduced by changing the mobile nodes buffer size between 25, 50, 75 and 100 Megabytes, across the simulations. Network nodes use a data plane link connection with a constant transmission data rate of 6 Mbps and an omni-directional transmission range of 30 meters.

Data bundles are generated using an inter-bundle creation interval that is uniformly distributed in the range of [15, 30] (seconds), and have random source and destination vehicles. Data bundles size is uniformly distributed in the range of [250 K, 2 M] (Bytes), and they have a time-to-live (TTL) of 180 minutes. Data bundles are discarded in cases of buffer congestion or when TTL expires.

Epidemic [28] and Spray and Wait [30] are used as the underlying routing schemes. Epidemic is a flooding-based routing protocol that replicates bundles at contact opportunities. It does not impose any limit on the amount of replications, per bundle, that a node can make. On the

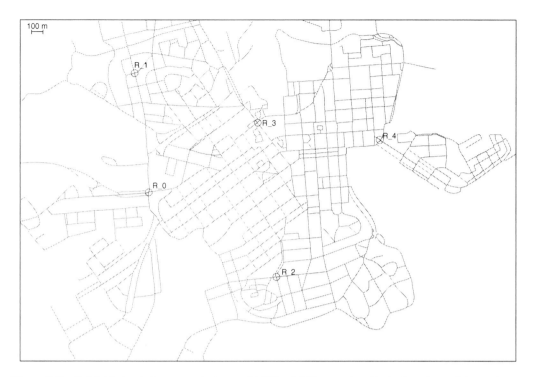

Figure 7.5 Helsinki simulation scenario (area of 4500 × 3400 meters), with the locations of the stationary relay nodes (R).

contrary, Spray and Wait routing limits the number of bundle replicas (copies), and it assumes two main phases. In the 'spray phase', for each original bundle, L bundle copies are spread to L distinct relays. At the 'wait phase', using direct transmission, it waits until any of the L relays finds the destination node. This work considers a binary spraying method, where the source node starts with a number of copies N (assuming 8, in this study) to be transmitted ('sprayed') per bundle. Any node A that has more than 1 bundle copies and encounters any other node B that does not have a copy, forwards to B $N/2$ bundle copies and keeps the rest of the copies. When a node carries only 1 copy left, it only forwards it to the final destination.

Performance metrics considered in this study are both the overall bundle delivery probability (measured as the relation of the number of unique delivered bundles to the number of bundles sent), as well as the overall bundles delivery delay (measured as the time between bundles creation and delivery). We analyze the different performance results when mobile nodes employ 10, 20, 30, 40, or 50% of their buffer capacity and bandwidth resources, to cooperate in bundle relay.

7.3.2.2 Performance Analysis of Epidemic Routing Protocol

We start the results analysis with the delivery ratio observed for Epidemic routing protocol. Figure 7.6 shows the effect of node cooperation percentage, from 10% to 50%, on the delivery probability. It can be observed that this routing protocol registers low delivery probability values across all simulations. These results are due to Epidemic's pure flooding approach that wastes network resources and severely degrades the overall network performance when resources are scarce. Increasing the cooperation percentage at network nodes from 10% to 20% results in

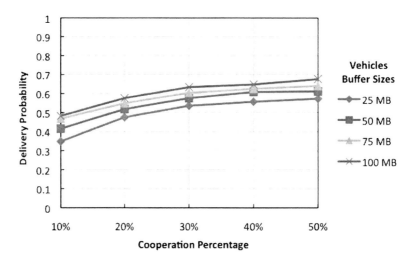

Figure 7.6 Effect of node cooperation percentage on the delivery probability for Epidemic routing protocol.

allowing 20% of the full buffer capacity to be used for storing bundles relayed by other nodes while the remaining 80% will be always available to store bundles generated at the nodes. In addition, at a contact opportunity, 20% of the transmission link bandwidth will be used to relay the other nodes bundles, instead of the previous 10%. Hence network nodes will be able to store, carry and forward more bundles generated by other nodes. As expected, this behaviour results in improving the overall network performance in terms of delivery ratio. For example, when network nodes use a 25 MB buffer, increasing the cooperation percentage from 10% to 20%, results in improving the overall delivery ratio in 13%.

Increasing the buffer size of network nodes attenuates the high buffer occupancy utilization problem of this routing protocol. Moreover, the analysis of Figure 7.6 also allows concluding that a better delivery ratio can be obtained using a 25 MB buffer size and a 30% cooperation percentage, instead of a 100 MB buffer with a 10% cooperation percentage.

As discussed above, by increasing the mobile nodes buffer size, these nodes will be able to store, carry and exchange more bundles during longer periods of time, before being dropped due to buffer overflow or TTL expiration. This contributes to the increase of the delivery ratio (Figure 7.6), but also increases the average delivery delay, as can be observed in Figure 7.7. This effect is reinforced by the increase of the nodes' cooperation percentage that also increases the average time that bundles spend in buffers before being delivered.

An analysis of Figure 7.7 reveals that these conclusions are valid for mobile nodes buffer sizes equal to or great than 50 MB. Using only a 25 MB buffer, two distinct behaviours can be observed. When cooperation percentage is lower than 20%, the storage space available for cooperating in the bundle relay process is very close to the average bundle size, which results in a very low number of cooperative bundles stored, and leads to frequent drops of such bundles. This behaviour is very pronounced at 10% cooperation percentage. As a consequence, most of the delivered bundles are transmitted from the source mobile nodes themselves, which results in an average delay greater than the one registered for the other buffer sizes.

7.3.2.3 Performance Analysis for Spray and Wait Routing Protocol

Spray and Wait routing strategy reduces transmissions by bounding the total number of copies/transmissions per bundle. Hence, it succeeds in limiting some of the overhead of the

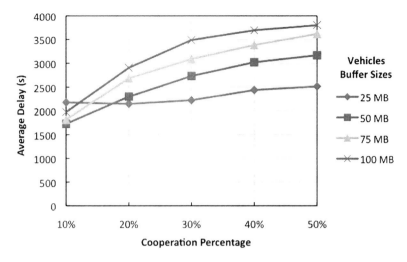

Figure 7.7   Effect of node cooperation percentage on the delivery delay for Epidemic routing protocol.

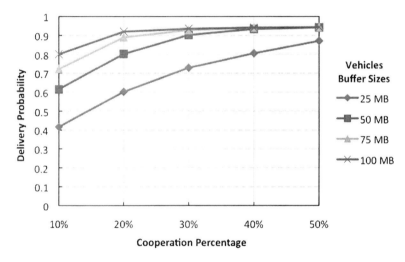

Figure 7.8   Effect of node cooperation percentage on the delivery probability for Spray and Wait routing protocol.

pure epidemic diffusion. Therefore, as expected, Spray and Wait performs better than Epidemic, presenting greater delivery ratios across all simulations, in this resource constrained network scenario (Figure 7.6 and Figure 7.8).

The effect of cooperation, as an effective strategy to increase the overall network bundle delivery, is even more pronounced in this routing protocol. Figure 7.8 shows that when mobile nodes have a 25 MB buffer size, changing the cooperation percentage from 10% to 50%, increases the bundles delivery ratio in approximately 19%, 12%, 8%, and 7%, respectively. In addition, as can be seen in this figure, employing a 100 MB buffer with a 10% cooperation percentage results in a similar delivery probability to a 25 MB buffer with a 40% cooperation percentage. The analysis of this figure also allows us to conclude that increasing cooperation above 20%

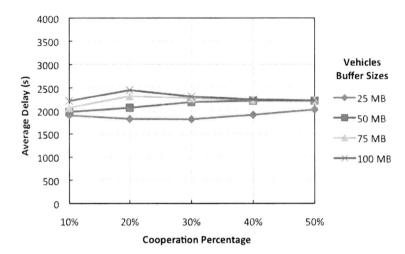

Figure 7.9 Effect of node cooperation percentage on the delivery delay for Spray and Wait routing protocol.

when 75 MB or 100 MB buffer sizes are used, does not improve the delivery ratio significantly, for this simulation scenario.

Spray and Wait routing scheme not only registers better delivery ratios (Figure 7.8), but also achieves better delivery delays (Figure 7.9) than previously shown for Epidemic flooding-based routing scheme (Figure 7.7). Moreover, as it can be observed from the analyses of Figure 7.8 and Figure 7.9, increasing the buffer size of mobile nodes and/or the cooperation percentage, it increases the delivery ratio. However, it has no significant effect on the average delay registered for Spray and Wait. Figure 7.9 also shows that similar average delays are registered for buffer sizes greater than 25 MB and cooperation percentages greater than 20%. This behaviour was expected since similar delivery probabilities are observed in these cases.

7.4 Conclusions

Delay tolerant networking (DTN) is a timely topic that addresses communication in challenged network environments including interplanetary networks, underwater networks, wildlife tracking networks, remote area networks, social networks, military networks and vehicular networks. These networks face unconventional difficulties on data communications due to sparse connectivity, network partitioning, intermittent connectivity, long delays, asymmetric data rates, high error rates, and even no end-to-end connectivity. DTNs rely on cooperative behaviour to help deliver bundles across sporadically connected nodes. Non-cooperative behaviours adversely affect the network operation and performance.

This chapter presented an overview of the delay-tolerant networking paradigm that serves as the basis for an innovative network architecture for vehicular communications, called vehicular delay-tolerant network (VDTN). This study addressed the research problem of node cooperation in DTN-based network architectures. The related literature about this topic was surveyed. In addition, a study to evaluate the impact of cooperation on the performance of VDTNs using two DTN routing protocols (Epidemic and Spray and Wait) was driven. It was shown that Spray and Wait protocol performs better than Epidemic and nodes' cooperation improves the overall bundle delivery, corresponding to a better performance of VDTN networks. We expect that this work will provide a deep understanding of the implications of node cooperation over the performance of DTN-based network architectures, leading to insights for further research studies.

7.5 Acknowledgements

Part of this work has been supported by the *Instituto de Telecomunicações*, Portugal, in the framework of the Project VDTN@Lab, and by the Euro-NF Network of Excellence from the Seventh Framework Programme of EU, in the framework of Project VDTN.

References

[1] V. Cerf, S. Burleigh, A. Hooke, L. Torgerson, R. Durst, K. Scott, K. Fall, and H. Weiss, 'Delay-Tolerant Networking Architecture,' RFC 4838, April 2007, [Online]. Available: http://www.rfc-editor.org/rfc/rfc4838.txt.

[2] K. Scott and S. Burleigh, 'Bundle Protocol Specification,' RFC 5050, November 2007, [Online]. Available: http://www.rfc-editor.org/rfc/rfc5050.txt.

[3] V. N. G. J. Soares, F. Farahmand and J. J. P. C. Rodrigues, 'A Layered Architecture for Vehicular Delay-Tolerant Networks,' in *Fourteenth IEEE Symposium on Computers and Communications (ISCC'09)*, Sousse, Tunisia, July 5–8, 2009, pp. 122–127.

[4] E. P. C. Jones and P. A. S. Ward, 'Routing Strategies for Delay-Tolerant Networks,' *Submitted to ACM Computer Communication Review (CCR)*, 2006.

[5] J. Shen, S. Moh and I. Chung, 'Routing Protocols in Delay Tolerant Networks: A Comparative Survey,' in *The 23rd International Technical Conference on Circuits/Systems, Computers and Communications (ITC-CSCC 2008)*, Shimonoseki City, Yamaguchi-Pref., Japan, July 6–9, 2008, pp. 1577–1580.

[6] Z. Zhang, 'Routing in Intermittently Connected Mobile Ad Hoc Networks and Delay Tolerant Networks: Overview and Challenges,' *IEEE Communications Surveys & Tutorials*, vol. 8, no. 1, pp. 24–37, 2006.

[7] A. Balasubramanian, B. N. Levine and A. Venkataramani, 'DTN Routing as a Resource Allocation Problem,' in *ACM SIGCOMM 2007*, Kyoto, Japan, August 27–31, 2007, pp. 373–384.

[8] S. Burleigh, A. Hooke, L. Torgerson, K. Fall, V. Cerf, B. Durst, K. Scott and H. Weiss, 'Delay-Tolerant Networking: An Approach to Interplanetary Internet,' in *IEEE Communications Magazine*, vol. 41, 2003, pp. 128–136.

[9] J. Partan, J. Kurose and B. N. Levine, 'A Survey of Practical Issues in Underwater Networks,' in *1st ACM International Workshop on Underwater Networks, in conjunction with ACM MobiCom 2006*, Los Angeles, California, USA, Sep. 25, 2006, pp. 17–24.

[10] I. Katz, 'A Delay-Tolerant Networking Framework for Mobile Underwater Acoustic Networks,' in *Fifteenth International Symposium on Unmanned Untethered Submersible Technology (UUST'07)*, Durham, NH, USA, August, 2007.

[11] P. Juang, H. Oki, Y. Wang, M. Martonosi, L. S. Peh and D. Rubenstein, 'Energy-Efficient Computing for Wildlife Tracking: Design Tradeoffs and Early Experiences with ZebraNet,' *ACM SIGOPS Operating Systems Review*, vol. 36, no. 5, pp. 96–107, 2002.

[12] UMass Diverse Outdoor Mobile Environment (DOME) project, 'UMass TurtleNet,' [Online]. Available: http://prisms.cs.umass.edu/dome/turtlenet [Accessed: November, 2009].

[13] S. Jain, R. Shah, W. Brunette, G. Borriello and S. Roy, 'Exploiting Mobility for Energy Efficient Data Collection in Wireless Sensor Networks,' *ACM/Kluwer Mobile Networks and Applications (MONET)*, vol. 11, no. 3, pp. 327–339, June 2006.

[14] N4C and eINCLUSION, 'Networking for Communications Challenged Communities: Architecture, Test Beds and Innovative Alliances,' [Online]. Available: http://www.n4c.eu/ [Accessed: January, 2009].

[15] Wizzy Digital Courier, 'Wizzy Digital Courier – leveraging locality,' [Online]. Available: http://www.wizzy.org.za/ [Accessed: January, 2008].

[16] A. Pentland, R. Fletcher and A. Hasson, 'DakNet: Rethinking Connectivity in Developing Nations,' in *IEEE Computer*, vol. 37, 2004, pp. 78–83.

[17] A. Doria, M. Uden and D. P. Pandey, 'Providing Connectivity to the Saami Nomadic Community,' in *2nd International Conference on Open Collaborative Design for Sustainable Innovation*, Bangalore, India, December, 2002.

[18] A. Seth, D. Kroeker, M. Zaharia, S. Guo and S. Keshav, 'Low-cost Communication for Rural Internet Kiosks Using Mechanical Backhaul,' in *12th ACM International Conference on Mobile Computing and Networking (MobiCom 2006)*, Los Angeles, CA, USA, September 24–29, 2006, pp. 334–345.

[19] M. Asplund, S. Nadjm-Tehrani and J. Sigholm, 'Emerging Information Infrastructures: Cooperation in Disasters,' in *Lecture Notes in Computer Science, Critical Information Infrastructure Security*, vol. 5508/2009: Springer Berlin/ Heidelberg, 2009, pp. 258–270.

[20] N. Glance, D. Snowdon and J.-L. Meunier, 'Pollen: Using People as a Communication Medium,' *Computer Networks: The International Journal of Computer and Telecommunications Networking*, vol. 35, no. 4, pp. 429–442, March 2001.

[21] P. Hui, A. Chaintreau, J. Scott, R. Gass, J. Crowcroft and C. Diot, 'Pocket Switched Networks and Human Mobility in Conference Environments,' in *ACM SIGCOMM 2005 – Workshop on Delay Tolerant Networking and Related Networks (WDTN-05)*, Philadelphia, Pennsylvania, USA August 22–26, 2005, pp. 244–251.

[22] T. Olajide and A. N. Washington, 'Epidemic Modeling of Military Networks using Group and Entity Mobility Models,' in *5th International Conference on Information Technology: New Generations (ITNG 2008)*, Las Vegas, Nevada, USA, April 7–9, 2008, pp. 1303–1304.

[23] J. Jormakka, H. Jormakka and J. Väre, 'A Lightweight Management System for a Military Ad Hoc Network,' in *Lecture Notes in Computer Science, Information Networking. Towards Ubiquitous Networking and Services*, vol. 5200/2008: Springer Berlin / Heidelberg, 2008, pp. 533–543.

[24] R. Tatchikou, S. Biswas and F. Dion, 'Cooperative Vehicle Collision Avoidance using Inter-vehicle Packet Forwarding,' in *IEEE Global Telecommunications Conference (IEEE GLOBECOM 2005)*, St. Louis, MO, USA, 28 Nov. – 2 Dec., 2005.

[25] L. Franck and F. Gil-Castineira, 'Using Delay Tolerant Networks for Car2Car Communications,' in *IEEE International Symposium on Industrial Electronics 2007 (ISIE 2007)*, Vigo, Spain, 4–7 June, 2007, pp. 2573–2578.

[26] Z. J. Haas and T. Small, 'Evaluating the Capacity of Resource-Constrained DTNs,' in *International Wireless Communications & Mobile Computing Conference (IWCMC 2006) – International Workshop on Delay Tolerant Mobile Networks (DTMN)*, Vancouver, Canada, July 3–6, 2006, pp. 545–550.

[27] A. Panagakis, A. Vaios and I. Stavrakakis, 'On the Effects of Cooperation in DTNs,' in *2nd International Conference on Communication Systems Software and Middleware (COMSWARE 2007)*, Bangalore, India, January 7–12, 2007, pp. 1–6.

[28] A. Vahdat and D. Becker, 'Epidemic Routing for Partially-Connected Ad Hoc Networks,' Duke University, Technical Report, CS-2000-06, April, 2000.

[29] M. Grossglauser and D. N. C. Tse, 'Mobility Increases the Capacity of Ad Hoc Wireless Networks,' *IEEE/ACM Transactions on Networking*, vol. 10, no. 4, pp. 477–486, August 2002.

[30] T. Spyropoulos, K. Psounis and C. S. Raghavendra, 'Spray and Wait: An Efficient Routing Scheme for Intermittently Connected Mobile Networks,' in *ACM SIGCOMM 2005 – Workshop on Delay Tolerant Networking and Related Networks (WDTN-05)*, Philadelphia, PA, USA, August 22–26, 2005, pp. 252–259.

[31] L. Buttyán, L. Dóra, M. Félegyházi and I. Vajda, 'Barter-Based Cooperation in Delay-Tolerant Personal Wireless Networks,' in *IEEE International Symposium on a World of Wireless, Mobile and Multimedia Networks (WOWMOM 2007)*, Helsinki, Finland, 18–21 June, 2007, pp. 1–6.

[32] U. Shevade, H. H. Song, L. Qiu and Y. Zhang, 'Incentive-Aware Routing in DTNs,' in *The 16th IEEE International Conference on Network Protocols (ICNP 2008)*, Orlando, Florida, USA, October 19–22, 2008, pp. 238–247.

[33] J. Morillo-Pozo, J. M. Barcelo-Ordinas, O. Trullos-Cruces and J. Garcia-Vidal, 'Applying Cooperation for Delay Tolerant Vehicular Networks,' in *Fourth EuroFGI Workshop on Wireless and Mobility*, Barcelona, Spain, January 16–18, 2008.

[34] J. M. Pozo, O. Trullols, J. M. Barceló and J. G. Vidal, 'A Cooperative ARQ for Delay-Tolerant Vehicular Networks,' in *The 28th International Conference on Distributed Computing Systems (ICDCS 2008)*, Beijing, China, June 17–20, 2008, pp. 192–197.

[35] J. Morillo-Pozo, Ó. Trullols-Cruces, J. M. Barceló-Ordinas and J. García-Vidal, 'Evaluation of a Cooperative ARQ Protocol for Delay-Tolerant Vehicular Networks,' in *Lecture Notes in Computer Science, Wireless Systems and Mobility in Next Generation Internet*, vol. 5122/2008: Springer Berlin / Heidelberg, 2008, pp. 157–166.

[36] O. Trullols-Cruces, J. Morillo-Pozo, J. M. Barcelo and J. Garcia-Vidal, 'A Cooperative Vehicular Network Framework,' in *IEEE International Conference on Communications (ICC'09)*, Dresden, Germany, June 14–18, 2009.

[37] G. Resta and P. Santi, 'The Effects of Node Cooperation Level on Routing Performance in Delay Tolerant Networks,' in *Sixth Annual IEEE Communications Society Conference on Sensor, Mesh and Ad Hoc Communications and Networks (SECON 2009)*, Rome, Italy, June 22–26, 2009.

[38] E. Altman, 'Competition and Cooperation Between Nodes in Delay Tolerant Networks with Two Hop Routing,' in *Lecture Notes in Computer Science, Network Control and Optimization*, vol. 5894/2009: Springer Berlin / Heidelberg, 2009, pp. 264–278.

[39] J. Solis, N. Asokan, K. Kostiainen, P. Ginzboorg and J. Ott, 'Controlling Resource Hogs in Mobile Delay-Tolerant Networks,' *Computer Communications*, Elsevier, 2009.

[40] V. N. G. J. Soares, F. Farahmand and J. J. P. C. Rodrigues, 'Improving Vehicular Delay-Tolerant Network Performance with Relay Nodes,' in *5th Euro-NGI Conference on Next Generation Internet Networks (NGI 2009)*, Aveiro, Portugal, July 1–3, 2009, pp. 1–5.

[41] V. N. G. J. Soares, J. J. P. C. Rodrigues, F. Farahmand and M. Denko, 'Exploiting Node Localization for Performance Improvement of Vehicular Delay-Tolerant Networks,' in *2010 IEEE International Conference on Communications (IEEE ICC 2010) – General Symposium on Selected Areas in Communications (ICC'10 SAS)*, Cape Town, South Africa, May 23–27, 2010.

[42] M. Pitkänen, A. Keränen and J. Ott, 'Message Fragmentation in Opportunistic DTNs,' in *Second International IEEE WoWMoM Workshop on Autonomic and Opportunistic Communications (AOC 2008)*, Newport Beach, California, USA, 23 June, 2008, pp. 1–7.

[43] A. Keränen, J. Ott and T. Kärkkäinen, 'The ONE Simulator for DTN Protocol Evaluation,' in *Second International Conference on Simulation Tools and Techniques (SIMUTools 2009)*, Rome, March 2–6, 2009.

8

Access Selection and Cooperation in Ambient Networks

Ramón Agüero
University of Cantabria, Spain

8.1 Leveraging the Cooperation in Heterogeneous Wireless Networks

The recent proliferation of various radio access technologies (RAT), likely motivated by the advances which have been seen in the electronic design and the miniaturization processes, has brought about the presence, which is becoming more noticeable everyday, of communication devices equipped with various wireless interfaces. It is reasonable to be looking at a not too distant future where not very advanced terminals will be able to use different RATs. On the other hand, the spring of different technologies and the appearance of various elements to access the network are also intrinsic elements of the so-called new wireless communications generation. Under these circumstances, current solutions to select one connection alternative are rather rudimentary, requiring, in many cases, the direct intervention from the end-user or the establishment of non-flexible policies.

Besides, this type of heterogeneous network deployment is not only beneficial for the end-user, but also to the operators, which will be able to benefit from the multi-access systems, considering, above all, the network deployment costs. Needless to say, each technology presents particular characteristics regarding its capacity and coverage, as well as the required deployment investment. For a network with multiple access technologies, the characteristics of each of them can be used so as to establish the most appropriate planning strategy. The cooperation between them entails the possibility of using load balancing techniques, thus increasing the available capacity margin, which yields a better quality of service for the end users.

It has to be considered that the aforementioned heterogeneity does not only encompass the subjacent technologies, but it also extends to upper level entities, for example, mobility solutions, new transport layer protocols, or the operators themselves. Hence, the growing tendency for communications diversity leads to the need to tackle a novel design of terminals, algorithms and protocols, so as to be used over heterogeneous network environments, strengthening the need for novel multi-access, mobility architectures, without jeopardizing the correct operation of the currently existing solutions.

Cooperative Networking, First Edition. Edited by Mohammad S. Obaidat and Sudip Misra.
© 2011 John Wiley & Sons, Ltd. Published 2011 by John Wiley & Sons, Ltd.

As has been previously mentioned, the relevance of multi-access systems has recently gathered relevant attention from the scientific community, and there are numerous examples in the available literature proposing solutions for heterogeneous network deployments [1–7]. There are a few which focus on the integration of particular technologies, for example, 3G and WLAN, studying mobility solutions between them [8–14]. There are other works which analyze the implications of using other technologies, for example, WiMax and WLAN [15], and so on.

In fact, even standardization bodies have understood the relevance that these types of network deployment will have in the near future and have already analyzed aspects related to the multi-access paradigm. In this sense 3GPP has analyzed the possibility of incorporating different technologies (most notably, WiMax) to their reference scenarios [16]. On the other hand, the IEEE 802.21 group [17] is working towards the definition of technology agnostic mechanisms so as to improve the handover procedures between heterogeneous networks. This approach presents two main problems: first, it focuses on event generation from the lower layers, without considering potential situations which might come from the upper layers; on the other hand, it just establishes the way to disseminate generic information, without specifying the way it could be used. Considering this reasoning, none of these initiatives is able to address the level of flexibility and dynamicity which might be required to face the future network deployment challenges, although they must be taken into account so as to favour an adequate migration procedure.

8.2   The Ambient Networks Philosophy

The Ambient Networks architecture [18, 19] was devised with the main goal of being both scalable and flexible. One of its main characteristics is to enable the cooperation between networks making use of the composition process, by means of which, it is possible to establish negotiations and spontaneous service level agreements between different administrative domains [20]. Figure 8.1 shows a high level vision of such architecture, which is discussed with a greater level of detail in [21]. The following main elements can be highlighted.

- The Ambient Control Space (ACS) [21, 22] encompasses a set of control functionalities, such as mobility and security management. It has been designed so that just a few functions are

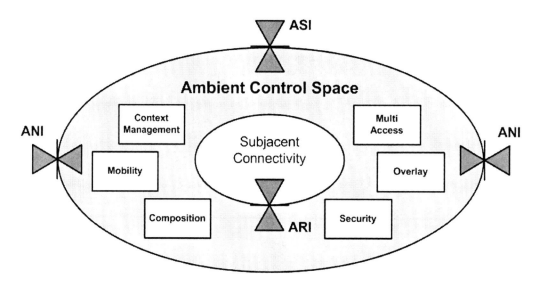

Figure 8.1   Ambient Control Space high level architecture.

compulsory and, furthermore, adding or deleting elements are relatively easy tasks. These control functions have been structured as a set of Functional Entities (FE).

- The abstraction layer of the subjacent connectivity allows the ACS to have a uniform and generic vision of the resources that a terminal may have at a particular moment.
- Three main interfaces are defined [23]:
  - The connection between elements belonging to different ACSs is done by means of the Ambient Networks Interface (ANI), over which the composition procedure takes place.
  - The interface between the ACS and the subjacent connectivity resources is the Ambient Resource Interface (ARI), which provides a homogeneous mechanism to manage them, especially those concerning the access technologies.
  - In order to let services and applications benefit from the set of functionalities offered by the ACS, the Ambient Service Interface (ASI) is defined.

As has been previously said, the focal point of the Ambient Networks philosophy comes from the capacity of executing dynamic composition between networks [24–26], which brings about the possibility of establishing a common control plane for all of them, so that the functionalities provided by any could be used by the others. In this sense, the composition process goes beyond the current interaction between control planes, allowing, amongst many other features, the establishment of certain degrees of cooperation between networks operated by different entities [27], the improvement of terminal mobility management and the optimized usage of multiple interface terminals.

The Multi-Radio Access (MRA) architecture [28–35] which was defined within the framework of the Ambient Networks project entails two main entities: the Multi-Radio Resource Management (MRRM) [36] and the Generic Link Layer (GLL) [37, 38]; obviously, they are part of the overall ACS, which provides them with a greater capacity, since it allows the interaction with other elements with some influence in the process of access selection, most notably with entities managing mobility of the terminal [39–42]. Figure 8.2 shows how both entities are included within the ACS, as well as the interfaces which have been established between them and other entities.

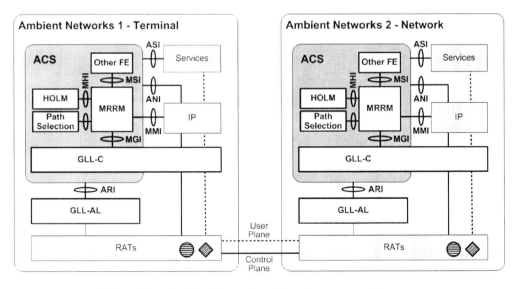

Figure 8.2 Ambient Networks multi-radio access architecture.

8.2.1 Generic Link Layer

The GLL allows, on a transparent and dynamic level, the set of wireless interfaces which a terminal might be using at any time to be managed. In this sense, it offers information about the available resources, so that other entities may use them, especially the MRRM when executing its access selection algorithms. The GLL abstracts the information it obtains from the corresponding RATs, so that it is independent of the technology which is being used, so as to facilitate the fair comparison between them [43, 44]. It incorporates a function which maps the 'link quality' over a generic metric, independently from the particular characteristics of the subjacent technology. On the other hand, the GLL is also able to abstract information about the current load and available capacity of the different technologies, which could be later compared by the MRRM mechanisms.

The GLL functionality is divided into two parts: the GLL_C is in charge of interacting with the rest of ACS entities, especially the MRRM, activating the services provided by the GLL Abstraction Layer (GLL_AL) which, as can be seen in Figure 8.2, is placed outside the ACS, while being a fundamental element to guarantee the required abstraction level in the subjacent connectivity layer. In this sense, the GLL_AL implements a relevant set of functionalities offered by the ARI, as shown in Figure 8.3.

8.2.2 Management of Heterogeneous Wireless Resources

The MRRM is the main control entity of the multi-access Ambient Networks architecture. Its main goal is the joint management of the radio resources within heterogeneous network environments. In this way, the MRRM is in charge of selecting the optimum access each time, as well as balancing the load between the different network connection elements. Thanks to the services provided by the GLL, the MRRM monitors the available networks, collects information about the status of the existing links (in terms of their quality and the availability of resources), and correlates it with the restrictions imposed from the upper layers. Based on this information, the MRRM executes the access selection algorithms, which might eventually derive in handovers; access selection could be triggered by different types of events, as the request for a new session, a change to an existing one, the detection of a new radio access, and so on. On the other hand,

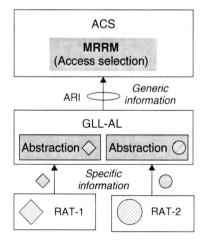

Figure 8.3 Abstraction mechanisms provided by the GLL.

the algorithms which are executed allow the selection of the most appropriate choice, depending on various parameters, as the QoS requirements from the services, the characteristics of the available accesses, and so on.

8.2.3 Additional Functional Entities

Although the multi-access Ambient Networks architecture does only embrace the two previously described components, there are a set of additional elements with which both the GLL and the MRRM heavily interact. We refer to the FEs which have been designed in order to manage end user device mobility; amongst them, we can highlight: the Triggering FE, the Handover and Locator Management (HOLM) FE and the Path Selection FE. Below each of these entities is briefly described.

8.2.3.1 Triggering FE

The Triggering (TRG) FE [45, 46] is in charge of collecting events related to, mostly, node mobility, generating information elements, referred to as triggers, and which are delivered to all the elements of the ACS which were interested in receiving such information.

In order to better understand the operation of the TRG FE, the following entities must be introduced.

- The sources from which the different events come from are called producers.
- Those FEs which are interested in receiving notifications or certain types of information from the TRG FE and subscribe to the corresponding service are known as TRG-Subscribers.

The events are generated according to the information provided by the TRG-Subscriber, establishing a set of rules and policies which allow a filtering and grouping of the information which arrives at the TRG FE to be carried out.

Taking into account the intrinsic functionalities which are provided by both the GLL and the MRRM, it is sensible assuming that the combination with the TRG FE might become very beneficial for a large number of use cases. In this sense, in [47] the Ambient Networks Heterogeneous Access Selection Architecture (ANHASA) is described, as the natural and straightforward integration and cooperation between the three FEs. Finally the TRG, which has some relationship with the work which is being carried out within the IEEE 802.21 framework, appears as a key element in the architecture which was defined to manage context in the Ambient Networks framework [48, 49].

8.2.3.2 Handover and Locator Management FE

The HOLM FE is the focal piece from the point of view of mobility within the ACS. It is in charge of managing the connectivity at the IP level, as well as the changes in the corresponding locators (addresses) during the handover processes [41, 50]. The HOLM provides a set of tools which can be executed on a dynamic way, based on the particular needs of each of the cases. In this sense, any time the MRRM decides that a handover is needed, it will make use of the services provided by the HOLM so as to select the protocols which should be employed for each of the situations, activate the adequate functionalities and successfully finish the handover, ensuring that all the actions required so that the terminal continues connected in its new location (change of address, routing information updates, and so on) are executed.

8.2.3.3 Path Selection FE

As has been previously discussed, the main role of the MRRM is to execute the access selection algorithms, applying a number of restrictions. Obviously, the information which is managed by the GLL, or the general policies which an end-user might use (preferred operator, costs, and so on) are types of data which are locally managed within a terminal and thus, by the MRRM itself. There are, though, a set of additional information elements which are not directly accessible by the MRRM, especially when it comes to the establishment of an end-to-end communication. Amongst them, we could mention the characteristics of the whole route towards the final destination (note that the MRRM is only aware of the wireless access part), the mobility protocols which could be activated in each particular case, and so on.

The ACS architecture advocates the use of an additional FE, namely Path Selection [51], whose main goal is to establish the conditions and characteristics which affect the elements which are being evaluated, on those aspects in which the impact of the MRRM is not enough (when the radio access part is not affected).

8.2.4 Multi-Access Functions and Procedures

The main goal of the MRRM, assisted by the information and the services provided by the GLL, is to execute access selection procedures. For that a set of information elements are defined, since they allow managing the available resources, as well as their current status. These objects become fundamental for all the procedures which have been specified in the MRRM operation. As can be seen in Figure 8.4, four different elements are defined, which are briefly introduced below [52].

- Detected Set (DS). It contains the access elements which are detected by a terminal. It is basically built based on the information provided by the GLL (by means of search procedures),

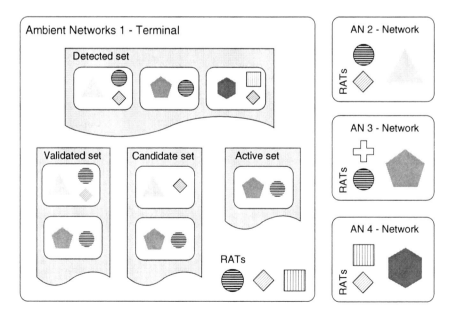

Figure 8.4 Information elements managed by the MRRM during the communication establishment procedures.

although it is possible using dedicated signalling messages by MRRM entities of the network, by means of which terminals are informed about the presence of additional networks, saving the extra overhead and energy consumption which might be required by the search processes of the GLL. Obviously there exists a different DS per terminal. In the example depicted in the figure, the terminal detects all the available networks, but it is not able to use one of the RATs of AN-3.

- Validated Set (VS). Once the MRRM is aware of the available resources, these are validated against a number of general policies from the end-user: preferred operator, cost and security preferences, and so on, without considering the specific requirements from the services. As it was for the DS, there is a different VS per terminal. In the example which is being used to illustrate the whole process, we can see how the terminal deletes AN-4 in the VS, most likely because it is not within the list of possible operators of the end-user.

- Candidate Set (CS). It is a subset of the VS, so that its elements can cope with the particular requirements, in terms of capacity, QoS, and so on, of a concrete service. In this case, there might be a CS for each of the active applications a node might have at a particular time. Furthermore, if the needs of the service change, the CS elements could be reconfigured accordingly. In the figure, we can see that one of the RATs of AN-2 is deleted in the DS, possibly due to the low quality of the corresponding radio link.

- Active Set (AS). These are the resources which are assigned to each information flow. From the elements which are within the CS, the MRRM selects those considered to be more appropriate, taking into account the access selection algorithm which it might be using, so as to satisfy the particular needs of the application. Although in most of the cases the AS comprises a single element, the possibility of assigning more than one resource to a single flow is also envisaged. In this latter case, the GLL is the entity in charge of making use of the most likely available capacity, by multiplexing and demultiplexing the frames accordingly. In the illustrative example of Figure 8.4, we can see that the MRRM only selects a RAT of AN-3 to handle the communication.

8.2.4.1 Announcement and Discovery

The first step to be taken when a terminal is switched on is to build the DS. For that, the MRRM instructs the GLL so that it performs periodic search procedures of the networks which are within its coverage area; thanks to these, the MRRM would become aware of new access alternatives, as well as the disappearance of those which were previously part of the DS, and relevant changes on the conditions and characteristics of each of their elements. Besides, it has to be taken into account that the GLL provides all this information abstracting the characteristics of the subjacent technology, so that the MRRM can compare the alternatives in a fair way.

Search and detection networks are usually based on the generation, at the network elements, of Beacon packets which announce their presence and indicate certain additional characteristics of the performance they offer. In the case of the Ambient Networks approach these packets incorporate complementary information elements, which can be used when taking the decision of the network to connect to [53, 54].

Some of the main drawbacks which are usually attributed to broadcast and dissemination processes are the additional overhead and delay they impose, as well as their energy consumption. In order to limit the effect of these adverse aspects, in certain cases, the active connection of a terminal (through one of its RATs) could be used so that the network MRRM entity could inform it about the presence of other accesses in its surroundings, to initiate, only if it was appropriate, a new search process (previously knowing the availability of potentially interesting access elements) or to locate different users [55]. The MRRM, to tackle the scalability problems which are attributed to traditional network discovery procedures, incorporates an additional functionality by means of which any terminal can interrogate the network about the presence of other access alternatives around its current position.

Last, it is worth highlighting the role that the TRG plays in these procedures. Apart from the MRRM, there might be other entities within the ACS interested in receiving information about the presence of new available networks or the disappearance of those which were previously present. In order to avoid the MRRM and/or the GLL had the burden of keeping track of those entities, they benefit from the functionality of the TRG, which will be in charge of disseminating the corresponding information, according to the particular necessities of the different subscribers.

Albeit not having any ongoing communication or service, there might be cases when the terminal establishes a connection with the network, so as to carry out the exchange of signalling information and to ensure that it can be paged from the outside.

8.2.4.2 Connection Establishment

The establishment of the connection is the process by means of which an application requests the ACS a flow over which to transport the end-user information. Once a service uses the ASI to set up a communication with any other node, the ACS establishes the access characteristics required to satisfy the QoS demands, information which is sent to the MRRM at the terminal. This, in cooperation with the MRRM at the network uses the VS elements to build the CS (set of available resources which can cope with the service requirements). In this case, it also uses an additional functionality provided by the Path Selection FE, so as to apply a number of additional restrictions to all the resources which belong to the VS, identifying those which are able to satisfy the particular service requirements.

Afterwards, taking into account the characteristics of the access selection algorithm, the AS is built. It usually comprises a single element, which will be used to establish the communication. It has to be considered that it is feasible to provide more influence to either the network or the terminal in the decision process.

Once an access element has been selected, the MRRM sends this information to the HOLM, which is in charge of carrying out the required actions to identify and assign the locator associated to the new connection, updating the state of the node from the point of view of the mobility solution. Finally, the MRRM informs the GLL the frequency on which it would like to receive information from the AS element, so as to be able to trigger a handover process in case it would be required (for example, when the quality of the corresponding radio link drops below a predefined threshold).

8.2.4.3 Handover

There might be different actions leading the MRRM to initiate a handover process for any active service at any particular moment [56]. In this sense, and using the information provided by the GLL, the MRRM realizes that the current AS stops being appropriate for the service and, thus, needs to identify a better choice. Besides, there might be other additional events which may derive into that circumstance, for example, a change on the service requirements, user preferences, and so on.

After determining that a handover needs to be executed, the procedure is rather similar to the one which has been previously described for the communication establishment. The MRRM should build the DS, the VS and the CS, applying for this latter one, the restrictions gathered by the Path Selection FE. The access selection algorithm would be executed again, establishing the AS. From that moment onwards the HOLM will be used in order to carry out the required functions from the point of view of mobility.

8.3   Related Work

Although the approach presented before appears as one of the most relevant ones, there are other similar proposals or works which are worth mentioning.

For instance, [2] describes the CRRM, which shares some key functions with the MRRM, but it does not rely on an abstraction mechanism, as the one which is provided by the GLL and it was conceived to address a set of particular technologies; in addition, it does not tackle multi-operator issues, since the initial assumption is that the CRRM is managed by a single administrative domain. [3] depicts the JRRM, which again does not propose a framework as generic as the one which was advocated by the MRRM; the work is interesting, since it studies how the information which might be managed at such levels could be used to foster an optimum access selection. Finally [57] presents the CARM, an evolution of the CRRM, which adds some functionality to facilitate the coordination between the wireless access part and the IP network. It follows the conventions and approach of the 3GPP efforts and does not consider the cooperation between different operators.

Furthermore, there are some works focused on particular technologies [9], [10], [58] which analyze the integration between WLAN and 3G, studying different mobility schemes; [59] discusses the benefits of employing various metrics to determine the optimum access, although they do not consider the possibility of changing one access once it has been selected. The work presented in [60] follows the guidelines proposed by the 3GPP [17] group and focuses on the integration of UMTS and WLAN technologies.

In addition, and as has been briefly mentioned before, another relevant aspect would be the study of different cooperation models between networks. This might become really significant, due to the increasing number of factors which are behind different networks. The analysis of different business models and cooperation strategies among operators has also gathered the attention of some researchers, see for example [61].

8.4   Outlook

So far we have discussed the barebones of an architecture which will foster the cooperation between the network elements so as to ensure the most appropriate access selection to the end user, based on different aspects, including the current characteristics of the available resources, the general policies of the end-user as well as the particular requirements of the ongoing services.

There are other aspects which have recently gathered interest from the scientific community and which should be included in the previously described framework, so as to broaden its possibilities and functionalities. Two of the most relevant ones are the cognition capability and the use of mesh topologies. Their common denominator is that they rely on a close cooperation between the different entities, as is briefly discussed below.

8.4.1   Cognition

One of the most stringent requirements is the need to bring about an efficient usage of the available resources. This calls for a departure from the current centralized approach, to a more flexible and distributed architecture, where the intelligence is shared between the different devices and involved entities. Under these assumptions, the use of cognition and cooperation springs as a key issue to be addressed.

If well understood, cognition must span across the different layers of the protocol stack. As previously said, all entities need to cooperate between them, in order to acquire the required knowledge about the Ambient Context, so as to be able to take smart decisions about the usage of the networks resources, adapting it to the specific needs of the current services and vice versa, modifying the current services' characteristics depending on the situation of the subjacent resources.

Although cognitive radio has gathered notable interest from the scientific community since it was originally proposed by Mitola in [62], there is still a large gap to be filled, the particular relevance of which is the lack of a holistic approach towards the cognition problem. It is clear that an efficient usage of the spectrum is mandatory, and this has become even more relevant now that there will be, in the short term, some resources which will be eventually freed, for example, television white spaces. However, this can not and must not be done without appropriate participation from the various involved entities. This entails a greater relevance when multi-operator environments are considered, since cooperative and competitive use of the available spectrum would even pose more challenges, due to the property profiles of the resources. In this sense, the frameworks which have been previously described must be able to deal with the capacity of managing those resources on a highly dynamic way.

Haykin, in [63] provides a very interesting and exhaustive summary of some of the aspects which should be considered in the cognitive radio (CR). The paper discusses three different tasks, namely the analysis of the radio scene, the estimation and prediction of the channel estate and the dynamic management of the spectrum and transmission power. The paper provides a very good insight into some of the problems which are usually attributed to CR, but most (not to say all) of them deal with PHY layer issues. Some of these techniques, based on the availability to sense the available spectrum, allow the same device to use regulated and not regulated frequency bands [64], taking advantage of its own capacity to be aware of its environment: devices that surround it (collaboration/competition), or accessing the network using different technologies [65]. For this reason, new algorithms that maximize the opportunities derived from the knowledge about the current use of the spectrum and minimize the delay in locating an available channel are being studied and analyzed [66].

The IEEE 802.22 working group is currently researching in this field, and they have defined the network WRAN (Wireless Regional Area Network), in which heterogeneous devices coexist with different networks and access technologies, and terminals must be aware of their environment so as to promote a more efficient use of the resources [67]. Furthermore, with respect to the standardization efforts, in April, 2007 the SCC41 (Standards Coordinating Committee 41) was created, supported by the IEEE Communications Society and the EMC Society, (originally, it was known as IEEE 1900 Committee). It develops the standards related to the DySPAN networks (Dynamic Spectrum Networks Access), which aims at improving the spectrum usage by means of new techniques and mechanisms to manage the interference, and to coordinate wireless technologies, including network management and information sharing.

Other works have also analyzed additional cognitive aspects mostly concerning the upper layers. However, in order to get full advantage from the cognitive capacity, a holistic approach must be followed, in which the cognition mechanisms are embedded on a wider framework, as the one which has been presented before. In this sense, the role of cognitive managers and aspects such as whether a centralized or a distributed approach is more appropriate need to be considered; obviously, this has a strong influence over the multi-radio access architecture which has been presented before, so it is sensible to think that the *cognitive* manager will become a new functional entity of such frameworks. Distributed architectures may have several advantages, but they will also impose a number of challenges, especially considering the situations where multiple networks (both technology-wise and operator) are being used. Another aspect which has a clear impact is the location of the knowledge; a distributed architecture for storing such information is deemed necessary, and the framework which was designed in the Ambient Networks project to manage context information might also appear as a sensible approach.

8.4.2 Mesh Topologies

The importance of wireless multi-hop networks has experienced a tremendous growth in the last few years. Although the usage of this type of topology has been traditionally attributed to rather particular circumstances (for example, natural disasters or emergency situations), leading to the so-called MANET realm, it is now believed that the use of multi hop communications may also bring about additional benefits [68]. In this sense, traditional operators may use this sort of deployment, so as to broaden their coverage area, in a cost efficient way, if, for example, some users relay traffic to/from others that would be otherwise unable to connect to the network or by deploying fixed relays (mesh networks), thus providing low cost and flexible network extensions. As a consequence, a strong interest in the use of such type of topologies has recently appeared in the relevant standardization bodies, for example, IEEE 802.11s [69] or IEEE 802.16j [70]. Furthermore, the use of multi-hop network deployments is also being analyzed in the context of future cellular technologies, such as LTE [71]. In many of the aforementioned use cases, the focus is on the improvements that may be brought about by using cooperative techniques, for example, in terms of spectral efficiency.

The use of mesh topologies has a twofold benefit: they provide a remarkable increase of the coverage compared to more traditional network infrastructures, without deploying additional base stations and, thus, greatly reducing the required investment [72, 73]; and, using cooperative techniques, the capacity could be strongly improved [74].

The use of mesh topologies in the access part of the network may impose new challenges to the multi-radio access. For instance, some works apply Network coding techniques in order to optimize the performance [75]. The idea is to take advantage of the space diversity derived from forwarding packets over different routes (in this sense, the cooperative relaying techniques can be somehow related to the multiple antennas systems – MIMO [76]). Traditionally, these mechanisms are classified in three main groups: amplify & forward, decode & forward, decode & codify [77]. The intermediate nodes of a multi-hop network smartly combine the contents of the packets before relaying them. These techniques take advantage of the intrinsic broadcast nature of the wireless channel, and therefore, it also allows optimizing multicast transmissions. In short, the Network Coding techniques which were originally proposed to be used over traditional wired networks [78] are postulated to be applied in mobile environments.

8.5 Conclusions

This chapter has presented the approach which was followed in the Ambient Networks framework to deal with heterogeneous access networks. It was based on the cooperation between a set of functional entities, in which the two more relevant are the GLL and the MRRM.

The former plays a key role, since it allows a fair comparison between different radio access technologies, on a uniform way, alleviating the decision engine from the burden of assessing the specific characteristics for all the involved technologies. It also helps the migration process, as far as incorporating a new technology into the framework is guaranteed if the corresponding abstraction mechanisms are implemented and the ARI specification is respected.

The intelligence of the proposal resides within the MRRM. It is an entity which gathers pieces of information from various sources and takes decisions based on such knowledge about the most appropriate access to handle a communication flow. Amongst the elements which it considers, it is worth highlighting the role played by the information elements gathered from the GLL. Although this approach is similar to other frameworks, the degree of flexibility of the MRRM must be remarked.

This flexibility entails the MRRM to include new functionalities, which are deemed necessary, due to the recent advances which have been recently seen. For instance, the capacity of managing available resources on a cognitive way will certainly appear as a focal aspect in the forthcoming wireless communication architectures. We advocate that the framework provided by the MRRM brings about the possibility of taking advantage from this cognitive capacity on a holistic way, since it provides a means to share all the information from the various layers, ranging from the lower ones up to the services, since their requirements will be considered when taking decisions.

Another aspect which should be also included in any framework aimed at the smart selection of access networks comes from the non-conventional extensions which might become popular in the near future. The role of multi-hop (mesh) topologies is believed to become more relevant and therefore their particular characteristics should be taken into consideration. In this case, the focus should be given to the difficulties which may arise when trying to ensure certain degree of quality of service over a route embracing more than two hops, since it might be impossible to control other sources of interference. However, the cooperation between the involved nodes may bring about several advantages to the end-user, who will perceive a better quality of service and to the operators, who will be able to extend their coverage in a relatively economic way.

References

[1] A. Tolli, P. Hakalin and H. Holma. 'Performance evaluation of common radio resource management (CCRM)'. Proceedings of the IEEE International Conference on Communications, ICC. April 2002.

[2] J. Perez-Romero, O. Sallent, R. Agusti, P. Karlsson, A. Barbaresi, L. Wang, F. Casadevall, M. Dohler, H. Gonzalez and F. Cabral-Pinto. 'Common radio resource management: Functional models and implementation requirements'. Proceedings of the 16th IEEE International Symposium on Personal Indoor and Mobile Radio Communication, PIMRC. September 2005.

[3] L. Giupponi, R. Agusti, J. Perez-Romero and O. Sallent. Joint radio resource management algorithm for multi-RAT networks. Proceedings of the IEEE Global Telecommunications Conference GLOBECOM. November 2005.

[4] A. Banchs. The DAIDALOS architecture for QoS over heterogeneous wireless networks. Proceedings of the 2nd edition workshop in trends in radio resource management. November 2005.

[5] J. Perez-Romero, O. Sallent and R. Agusti. A novel algorithm for radio access technology, selection in heterogeneous B3G networks. Proceedings of the 63rd IEEE Vehicular Technology Conference, VTC. April 2006.

[6] J. Perez-Romero, O. Sallent and R. Agusti. A generalized framework for multi-RAT scenarios characterization. Proceedings of the 65th IEEE Vehicular Technology Conference, VTC. April 2007.

[7] T. Melia, D. Corujo, A. Oliva, A. Vidal, R. Aguiar and I. Soto. Impact of heterogeneous network controlled handovers on multi-mode mobile device design. Proceedings on Wireless Communications and Networking Conference, WCNC. March 2007.

[8] Q. Zhang, C. Guo, Z. Guo and W. Zhu. Efficient mobility management for vertical handovers between WWAN and WLAN. *IEEE Communications Magazine*, 41(11): 16–28, November 2003.

[9] Salkintzis. Interworking techniques and architectures for WLAN/3G integration toward 4G mobile data networks. *IEEE Wireless Communications*, 11(3): 50–61, June 2004.

[10] L. Ma, F. Yu, V. Leung and T. Randhawa. A new method to support UMTS/WLAN vertical handover using SCTP. *IEEE Wireless Communications*, 11(4): 44–51, August 2004.

[11] W. Wei, N. Banerjee, K. Basu and S. Das. SIP-based vertical handover between WWANs and WLANs. *IEEE Wireless Communications*, 12(3): 66–72, June 2005.

[12] N. Shenoy and R. Montalvo. A framework for seamless roaming across cellular and wireless local area networks. *IEEE Wireless Communications*, 12(3): 50–57, June 2005.

[13] T. Melia, A. Oliva, I. Soto, C. Bernardos and A. Vidal. Analysis of the effect of mobile terminal speed on WLAN/3G vertical handovers. Proceedings of the IEEE Global Telecommunications Conference, GLOBECOM. November 2006.

[14] R. Good and N. Ventura. A multilayered hybrid architecture to support vertical handover between IEEE 802.11 and UMTS. Proceedings of the 2006 international conference on Wireless communications and mobile computing, IWCMC. Julio 2006.

[15] D. Niyato and E. Hossain. Integration of IEEE 802.11 WLANs with IEEE 802.16-based multihop infrastructure mesh/relay networks: a game-theoretic approach to radio resource management. *IEEE Network*, 21(3): 6–14, May 2007.

[16] 3rd Generation Partnership Project 3GPP TS 23.402. 3GPP System Architecture Evolution: Architecture Enhancements for non-3GPP accesses (Release 8), January 2007.

[17] IEEE Computer Society LAN MAN Standards Committee. IEEE 802.21: Media Independent Handover, January 2009.

[18] N. Niebert, A. Schieder, H. Abramowicz, G. Malmgren, J. Sachs, U. Horn, C. Prehofer and H. Karl. Ambient Networks: An architecture for communication networks beyond 3G. *IEEE Wireless Communications*, 11(2): 14–22, April 2004.

[19] N. Niebert, A. Schieder, J. Zander and R. Hancock. *Ambient Networks: Co-operative Mobile Networking for the Wireless World*. Wiley, 2007. ISBN 978-0-470-51092-6.

[20] J. Markendahl, P. Poyhonen and O. Strandberg. Impact of operator cooperation on traffic load distribution and user experience in ambient networks business scenarios. Proceedings of the LA Global Mobility Roundtable. June 2007.

[21] M. Johnsson, B. Ohlman, A. Surtees, R. Hancock, P. Schoo, K. Ahmed, F. Pittmann, R. Rembarz y M. Brunner. A future-proof network architecture. Proceedings of the 16th IST Mobile and Wireless Communications Summit. July 2007.

[22] N. Charkani, M. Cano, S.-W. Svaet y M. Johnsson. Migration approach of the Ambient Control Space. En Proceedings of the 16th IST Mobile and Wireless Communications Summit. July 2007.

[23] A. Schieder, P. Schoo, J. Gebert, D. Zhou, K. Balos and M. Cano. The reference points of an Ambient Network. Proceedings of the 16th IST Mobile and Wireless Communications Summit. July 2007.

[24] M. Johnsson, A. Mehes, G. Selander, N. Papadoglou, M. Priestley, P. Poyhonen, R. Aguero, C. Kappler and J. Markendahl. Network composition. Proceedings of the 15th Wireless World Research Forum meeting, WWRF. December 2005.

[25] N. Akhtar, C. Kappler, P. Schefczik, L. Tionardi and D. Zhou. Network composition: A framework for dynamic interworking between networks. Proceedings of the International Conference on Communications and Networking in China (CHINACOM). August 2007.

[26] C. Kappler, P. Poyhonen, M. Johnsson and S. Schmid. Dynamic network composition for beyond 3G networks: a 3GPP viewpoint. *IEEE Network*, 21(1): 47–52, January-February 2007

[27] P. Poyhonen, J. Tounonen, H. Tang and O. Strandberg. Study of handover strategies for multi-service and multi-operator Ambient Networks. Proceedings of the Second International Conference on Communications and Networking in China, CHINACOM. August 2007.

[28] J. Sachs, L. Munoz, R. Aguero, J. Choque, G. Koudouridis, R. Karimi, L. Jorguseski, J. Gebert, F. Meago and F. Berggren. Future wireless communication based on multi-radio access. Proceedings of the Wireless World Research Forum 11, WWRF. June 2004.

[29] J. Lundsjo, R. Aguero, E. Alexandri, F. Berggren, C. Cedervall, K. Dimou, J. Gebert, R. Jennen, L. Jorguseski, R. Karimi, F. Meago, H. Tang and R. Veronesi. A multi-radio access architecture for ambient networking. En Proceedings of the 14th IST Mobile and Wireless Communications Summit. June 2005.

[30] G. Koudouridis, R. Aguero, E. Alexandri, M. Berg, A. Bria, J. Gebert, L. Jorguseski, R. Karimi, I. Karla, P. Karlsson, J. LundsjÄo, P. Magnusson, F. Meago, M. Prytz and J. Sachs. Feasibility studies and architecture for multi-radio access in Ambient Networks. Proceedings of the 15th Wireless World Research Forum meeting, WWRF. December 2005.

[31] G. Koudouridis, P. Karlssson, J. Lundsjo, A. Bria, M. Berg, L. Jorguseski, F. Meago, R. Aguero, J. Sachs and R. Karimi. Multi-radio access in Ambient Networks. Proceedings of the 2nd edition workshop in trends in radio resource management. November 2005.

[32] M. Prytz, P. Karlsson, C. Cedervall, A. Bria and Karla. Infrastructure cost benefits of Ambient Networks multi-radio access. Proceedings of the 63rd IEEE Vehicular Technology Conference, VTC. April 2006.

[33] J. Sachs, R. Aguero, M. Berg, J. Gebert, L. Jorguseski, I. Karla, P. Karlsson, G. Koudouridis, J. Lundsjo, M. Prytz and O. Strandberg. Migration of existing access networks towards multi-radio access. Proceedings of the 64th IEEE Vehicular Technology Conference, VTC. Septembre 2006.

[34] M. Johnsson, J. Sachs, T. Rinta-aho and T. Jokikyyny. Ambient Networks: a framework for multi-access control in heterogeneous networks. Proceedings of the 64th IEEE Vehicular Technology Conference, VTC. September 2006.

[35] J. Gebert, R. Aguero, K. Daoud, G. Koudouridis, M. Prytz, T. Rinta-aho, J. Sachs and H. Tang. Access flow based multi-radio access connectivity. B3G and SRM Cluster Workshop on network detection and heterogeneous radio resource management. March 2007.

[36] P. Magnusson, F. Berggren, I. Karla, R. Litjens, F. Meago, H. Tang and R. Veronesi. Multi-radio resource management for communication networks beyond 3G. Proceedings of the 62nd IEEE Vehicular Technology Conference,
VTC. September 2005.

[37] G. Koudouridis, R. Aguero, E. Alexandri, J. Choque, K. Dimou, R. Karimi, H. Lederer, J. Sachs and R. Sigle.
Generic link layer functionality for multi-radio access networks. Proceedings of the 14th IST Mobile and Wireless
Communications Summit. June 2005.

[38] K. Dimou, R. AgÄuero, M. Bortnik, R. Karimi, G. Koudouridis, H. Lederer, J. Sachs and R. Sigle. Generic Link
Layer: a solution for multi-radio transmission diversity in communication networks beyond 3G. Proceedings of
the 62nd IEEE Vehicular Technology Conference, VTC. September 2005.

[39] J. Eisl, J. Holler, S. Uno and R. Aguero. Towards modular mobility management in Ambient Networks. Proceedings
of the 1st International ACM workshop on dynamic interconnection of networks, DIN. September 2005.

[40] R. Aguero, J. Eisl, V. Typpo and S. Uno. Analysis of mobility control functions in Ambient Networks. Proceedings
of the 6th IEE International Conference on 3G and Beyond, IEE 3G. November 2005.

[41] R. Aguero, A. Surtees, J. Eisl and M. Georgiades. Mobility management in Ambient Networks. Proceedings of
the 65th IEEE Vehicular Technology Conference, VTC. April 2007.

[42] J. Eisl, J. Mäakela, R. Agüero and S. Uno, ed. Ambient Networks: Co-operative Mobile Networking for the Wireless
World, chapter Ambient Networks Mobility Management. Wiley, 2007.

[43] J. Sachs, R. Aguero, K. Daoud, J. Gebert, G. Koudouridis, F. Meago, M. Prytz, T. Rinta-aho and H. Tang. Generic
abstraction of access performance and resources for multi-radio access management. Proceedings of the 16th IST
Mobile and Wireless Communications Summit. July 2007.

[44] F. Meago, J. Gebert, J. Sachs, J. Choque, R. Aguero and O. Blume. On capacity/load- based and availability-based
resource abstractions for multi-access networks. Proceedings of the 1st Ambient Networks Workshop on Mobility,
Multiaccess and Network Management, M2NM. October 2007.

[45] J. Makela and K. Pentikousis. Trigger management mechanisms. Proceedings of the 2nd International Symposium
on Wireless Pervasive Computing, ISWPC. February 2007.

[46] J. Makela. Towards seamless mobility support with cross-layer triggering. Proceedings of the 18th IEEE International Symposium on Personal Indoor and Mobile Radio Communication, PIMRC. September 2007.

[47] K. Pentikousis, R. Aguero, J. Gebert, J. A. Galache, O. Blume y P. Paakkonen. The Ambient Networks heterogeneous access selection architecture. Proceedings of the 1st Ambient Networks Workshop on Mobility, Multiaccess
and Network Management, M2NM. October 2007.

[48] R. Giaffreda, K. Pentikousis, E. Hepworth, R. Aguero and A. Galis. An information service infrastructure for
Ambient Networks. Proceedings of the IASTED International Conference on Parallel and Distributed Computing
and Networks, PDCN. February 2007.

[49] K. Pentikousis, R. Giaffreda, E. Hepworth, R. Aguero and A. Galis. Information management for dynamic networks.
Proceedings of the International Multi-Conference on Computing in the Global Information Technology, ICCGI.
March 2007.

[50] S. Uno, J. Eisl and R. Aguero. Study of mobility control functions in Ambient Networks. Proceedings of the 8th
International Conference in Advanced Communication Technology, ICACT. February 2006.

[51] A. Gunnar, B. Ahlgren, O. Blume, L. Burness, P. Eardley, E. Hepworth, J. Sachs and A. Surtees. Access and
path selection in Ambient Networks. Proceedings of the 16th IST Mobile and Wireless Communications Summit.
July 2007.

[52] G. Koudouridis, R. Aguero, K. Daoud, J. Gebert, M. Prytz, T. Rinta-aho, J. Sachs and H. Tang. Access flow based
multi-radio access connectivity. Proceedings of the 18th IEEE International Symposium on Personal Indoor and
Mobile Radio Communication, PIMRC. September 2007.

[53] T. Rinta-aho, R. Campos, A. Mehes, U. Meyer, J. Sachs and G. Selander. Ambient Network attachment. Proceedings
of the 16th IST Mobile and Wireless Communications Summit. July 2007.

[54] L. Ho, J. Markendahl and M. Berg. Business aspects of advertising and discovery concepts in Ambient Networks.
Proceedings of the 17th IEEE International Symposium on Personal Indoor and Mobile Radio Communication,
PIMRC. September 2006.

[55] H. Tang, P. Poyhonen, O. Strandberg, K. Pentikousis, J. Sachs, F. Meago, J. Tounonen and R. Aguero. Paging
issues and methods for multiaccess. Proceedings of the Second International Conference on Communications and
Networking in China, CHINACOM. August 2007.

[56] O. Blume, A. Surtees, R. Aguero, E. Perera and K. Pentikousis. A generic signalling framework for seamless
mobility in heterogenous wireless networks. Proceedings of the 1st Ambient Networks Workshop on Mobility,
Multiaccess and Network Management, M2NM. October 2007.

[57] J. Olmos, R. Ferrús, O. Sallent, J. Pérez-Romero, F. Casadevall. 'A Functional End-to-End QoS Architecture
Enabling Radio and IP Transport Coordination'. Proc. of IEEE WCNC 2007.

[58] W. Wu, N. Banerjee, K. Basu, and S. K. Das, 'SIP-Based Vertical Handoff between WWANs and WLANs,' IEEE
Wireless Communications, pp. 66–72, June 2005.

[59] F. Berggren, R. Litjens, 'Performance analysis of access selection and transmit diversity in multi-access networks,' Proc. 12th annual international conference on Mobile computing and networking, pp. 251–261, 2006.

[60] J. Olmos, R. Ferrús, O. Sallent, J. Pérez-Romero, F. Casadevall. 'A Functional End-to-End QoS Architecture Enabling Radio and IP Transport Coordination'. Proc. of the First Ambient Networks Workshop on Mobility, Multiaccess, and Network Management (M2NM 2007). Sidney (Australia) October 2007.

[61] D. Niyato, E. Hossain. 'A Game Theoretic analysis of service competition and pricing in heterogeneous wireless access networks'. *IEEE Transactions on Wireless Communications*. Vol. 7. No 12. December 2008.

[62] J. Mitola III, G. Maguire. 'Cognitive Radio: Making Software Radios More Personal'. IEEE Personal Communications, August 1999.

[63] S. Haykin, 'Cognitive Radio: Brain-Empowered Wireless Comm.,' *IEEE J. Selected Areas in Comm.*, vol. 23, no. 2, pp. 201–220, Feb. 2005.

[64] R. Rajbanshi, Q. Chen, A.M. Wyglinski, G.J. Minden, and J.B. Evans, 'Quantitative Comparison of Agile Modulation Techniques for Cognitive Radio Transceivers,' Proc. Fourth IEEE Consumer Comm. and Networking Conf. Workshop Cognitive Radio Networks, pp. 1144–1148, Jan. 2007.

[65] C. Cordeiro, K. Challapali, and M. Ghosh, 'Cognitive PHY and MAC Layers for Dynamic Spectrum Access and Sharing of TV Bands,' Proc. First Int'l Workshop Technology and Policy for Accessing Spectrum (TAPAS '06), Aug. 2006.

[66] H. Kim and K. G. Shin, 'Efficient Discovery of Spectrum Opportunities with MAC-Layer Sensing in Cognitive Radio Networks', *IEEE Trans. On Mobile Computing*, Vol. 7, No. 5, pp. 533–545, May 2008.

[67] IEEE Computer Society LAN MAN Standards Committee. Part 22.1: Standard to Enhance Harmful Interference Protection for Low Power Licensed Devices Operating in TV Broadcast Bands. Draft 8, July 2010.

[68] R. Pabst, B. Walke, D. Schultz, P. Herhold, H. Yanikomeroglu, S. Mukherjee, H. Viswanathan, M. Lott, W. Zirwas, M. Dohler, H. Aghvami, D. Falconer, and G. Fettweis, 'Relay-based deployment concepts for wireless and mobile broadband radio,' *IEEE Communications Magazine*, vol. 42, no. 9, pp. 80–89, September 2004.

[69] IEEE Computer Society LAN MAN Standards Committee. Part 11: Wireless LAN Medium Access Control (MAC) and Physical Layer (PHY) specifications-Amendment 10: Mesh Networking. Draft 8, December 2010.

[70] IEEE Computer Society LAN MAN Standards Committee. Part 16: Air Interface for Broadband Wireless Access Systems Amendment 1: Multiple Relay Specification. June 2009.

[71] R. Schoenen, R. Halfmann, B. H. Walke 'MAC Performance of a 3GPP-LTE Multihop Cellular Network'. Proc. of ICC 2008. Beijing (China). May 2008.

[72] R. Agüero, J. Choque, L. Muñoz. 'On the Relay-Based Coverage Extension for Non-Conventional Multi-Hop Wireless Networks'. Proc of ICC 07. Glasgow (Scotland). July 2007.

[73] R. Agüero, J. Choque, L. Muñoz. 'On the outage probability for multi-hop communications over array network deployments'. Proc of WCNC 08. Las Vegas (USA). March 2008.

[74] S. Song, K. Son, H.-W. Lee, and S. Chong, 'Opportunistic relaying in cellular network for capacity and fairness improvement,' in IEEE Global Telecommunications conference GLOBECOM, November 2007.

[75] S. Katti, H. Rahul, W. Hu, D. Katabi, M. Médard, J. Crowcroft. 'XORs in the Air: Practical Wireless Network Coding'. *IEEE/ACM Transactions on Networking*. Vol. 16. No 3. June 2008.

[76] M. Dohler, A. Gkelias, H. Aghvami, '2-Hop distributed MIMO communication system', IEEE Electron. Lett. (2003).

[77] P. Herhold, E. Zimmermann, G. Fettweis. 'Cooperative multi-hop transmission in wireless networks'. Computer Networks 49. 2005.

[78] R. Ahlswede, N. Cai, S. R. Li, and R. W. Yeung, 'Network information flow,' *IEEE Trans. Inf. Theory*, vol. 46, no. 4, pp. 1204–1216, Jul. 2000.

9

Cooperation in Intrusion Detection Networks

Carol Fung and Raouf Boutaba
Department of Computer Science, University of Waterloo, Waterloo, Ontario, Canada

9.1 Overview of Network Intrusions

In modern days almost every computer is connected to the Internet. Applications which rely on networks such as email, web-browsing, social networks, remote connections and online messaging are being used by billions of users every day. At the same time, network intrusions are becoming a severe threat to the privacy and safety of computer users. By definition, network intrusions are unwanted traffic or computer activities that may be malicious and destructive. The consequence of a network intrusion can be the degradation or termination of the system (denial of service), user identity information theft (ID theft), unsubscribed commercial emails (*Spam*), or fraud of legitimate websites to obtain sensitive information from users (*Phishing*). Network intrusions usually rely on the executing of malicious code (*Malware*) to achieve their attack goals. People who write or utilize Malware for attacking are called *hackers*. In recent years, network intrusions are becoming more sophisticated and organized. Hackers tend to control many attacking sources/hosts to launch cooperative attacks, such as Botnets (Section 9.1.2). In the following subsections, we briefly describe Malware and their potential damage to computers and their users.

9.1.1 Single-Host Intrusion and Malware

If the network intrusion traffic is from a single host, then we call it a single-host intrusion. Normally a network intrusion accomplishes its goals by executing malicious code on the victim machine or attacker machine. *Malware* is the common term for code designed to exploit or damage a computer, server, or network. Malware include Virus, Worms, Trojan and Spyware.

A *computer virus* is a computer program that can insert/copy itself into one or more files without the permission or knowledge of the user and perform some (possibly null) actions [10]. Although not all viruses are designed to be malicious, some viruses may cause a program to run abnormally or corrupt a computer's memory. A computer can be infected with virus through

Cooperative Networking, First Edition. Edited by Mohammad S. Obaidat and Sudip Misra.
© 2011 John Wiley & Sons, Ltd. Published 2011 by John Wiley & Sons, Ltd.

copying data from other computers or through using infected external data media such as a flash memory or a removable disk.

In general, computer viruses do not actively search for victims through network. Computer viruses which actively search for victims are called *worms*. A formal definition of a computer worm is a program which propagates itself through network automatically by exploiting security flaws in widely-used services [38]. Worms can cause the most extensive and widespread damage of all types of computer attacks due to their characteristic of automatic spreading. There have been a large number of different worms created in history. Some of the famous ones are Morris (1988), CodeRed (2001), SQL Slammer (2003), and the Conficker worm (2008).

The essential character of a computer virus is its self-replication. Some harmful software/code do not replicate themselves, such as Trojan horses (Trojan). A *Trojan* is a program with an overt (documented or known) effect and a covert (undocumented or unexpected) effect [10]. A Trojan appears to perform a desirable function, but in fact facilitates unauthorized access to the user's computer. A Trojan requires interactions with a hacker.

The last type of malware we mention in this chapter is Spyware. A *spyware* is a malware that is installed surreptitiously on a personal computer to collect information about the user, including their browsing habits without their informed consent. A Spyware can also be used by hackers to obtain user email address, identity information, credit card information and passwords.

9.1.2   Distributed Attacks and Botnets

In recent years, network intrusions are becoming more sophisticated and organized. Intruders tend to control a group of compromised computers to launch distributed attacks, for example, Distributed Denial of Service (*DDoS*) attack. Compromised nodes which may run on a Malware communicate with a master through a command and control server [37]. The group of compromised nodes with a master together forms a *Botnet*. The compromised nodes are called *Bot-nodes*. The master is called a *Bot-master*. Bot-nodes can be used to commit profit-driven cyber crimes such as DDOS attacks, Spam spreading, ID theft, or Phishing. CSI report [32] indicates that in 2008 financial loss caused by Bot computers in US enterprises stands the second highest, following the financial fraud. Another report [9] predicts that the percentage of compromised computers into Bot-nodes worldwide will keep rising up from 10% in 2008 to 15% in 2009. In a cooperative attack, compromised nodes can play different roles in the whole attack scenario. In the next subsection, we give an example of how Botnet is used to launch Phishing attacks.

9.1.3   Cooperative Attacks and Phishing

Phishing is the criminally fraudulent process of attempting to acquire sensitive information such as usernames, passwords and credit card details by masquerading as a trustworthy entity in an electronic communication. There are mainly two mechanisms used by Phishers. The first, also known as *social engineering*, makes use of spoofed emails purporting to be from legitimate businesses and agencies to lead consumers to counterfeit websites designed to trick recipients into divulging financial data such as usernames and passwords. The second mechanism, also known as *technical subterfuge*, plants Malware onto user PCs to steal credentials directly through intelligent key-loggers and/or corrupting browsers navigation to mislead customers to counterfeit websites. Gartner has estimated the cost of identity theft increased from 2.3 million to 3.2 billion in 2007 in the USA [21].

Large-scale phishing websites like any large-scale online service rely on online availability as a key parameter to good profit. Phishing websites, however, may be easily taken down if they use fixed IP address(s). This is not only specific to phishing websites. In fact, any illegal online organization which targets victims on a large scale requires high availability for the continuation of its operation.

Recently, *Fust-Flux Service Networks* [31] appeared to serve this requirement of high availability yet evasiveness of illegal websites. Fast-Flux Service Networks (FFSN) is a term coined in the anti-spam community to describe a decentralized Botnet used to host online criminal activities. FFSNs employ DNS techniques to establish a proxy network on the compromised machines. These compromised machines are used to host illegal online services like phishing websites, malware delivery sites, and so on, with a very high availability. An FFSN generally has hundreds or even thousands of IP addresses assigned to it. These IP addresses are swapped in and out of flux with extreme frequency, using a combination of round-robin IP addresses and a very short Time-To-Live (TTL) for any given particular DNS Resource Record (RR). Website hostnames may be associated with a new set of IP addresses as often as every 3 minutes [31], which makes it hard to take down the actual service launcher as the control node (*mothership*) is not known. The proxy agents do the actual work for the control node and these proxy agents change rapidly. ATLAS is a system from Arbor Networks which identifies and tracks new fast-flux networks [29].

When an FFSN is detected, the domain registrars can be contacted to shut the corresponding domain, hence removing the FFSN. Contrary to the fact that this mitigation technique sounds doable, it is often a tedious and time-consuming job given the fact that not all registrars respond to abuse complaints [28].

9.2   Intrusion Detection Systems

Intrusion Detection Systems (IDS) are software/hardware designed to monitor network traffic or computer activities and alert administrators for suspicious intrusions. Based on the model of detection, IDSs can be divided into signature-based and anomaly-based. Based on data sources, they can be host-based or network-based.

9.2.1   Signature-Based and Anomaly-Based IDSs

Signature-based (misuse) IDSs compare data packets with the signatures or attributes database of known intrusions to decide whether the observed traffic is malicious or not. A Signature-based IDS is efficient in detecting known intrusions with fixed signatures. However, it is not efficient in detecting unknown intrusions or intrusions with polymorphic signatures. *Anomaly-based* IDSs observe traffic or computer activities and detect intrusions by identifying activities distinct from a user's or a system's normal behaviour. Anomaly-based detection can detect unknown intrusions or new intrusions. However, it usually suffers from a high false positive rate problem. Most current IDSs employ both techniques to gain better detection ability.

9.2.2   Host-Based and Network-Based IDSs

A *Host-based* IDS (HIDS) runs on an individual host or device in the network (Figure 9.1). It monitors inbound/outbound traffic to/from a computer as well as the internal activities such as system calls and system logs. A HIDS can only monitor an individual device and may not be aware of the whole picture of the network activities. Some examples of HIDSs are OSSEC [4] and tripwire [7]. *Network-based* IDSs (NIDS) monitor network traffic packets, such as TCP-dump,

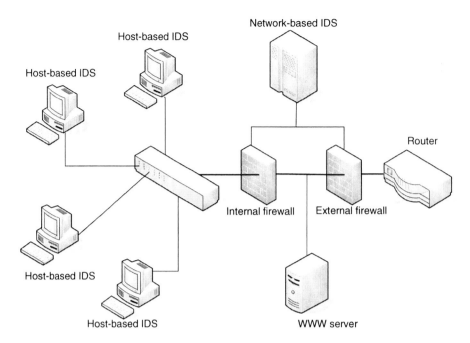

Figure 9.1   An example of host-based IDS and Network-based IDS.

to/from the network system. A NIDS contains sensors to sniff packets, and a data analyzer to process and correlate data. Alarms are raised whenever suspected intrusions are found. However, a NIDS does not have knowledge about internal activities of individual computers. Examples of NIDSs are Snort [6] and Bro [1].

9.3   Cooperation in Intrusion Detection Networks

Isolated intrusion detection systems do not have information about the whole environment and may be easily compromised by new intrusions. Cooperation in intrusion detection systems enables the system to use collective information from other IDSs to make more accurate intrusion detection locally or system wide. The network that connects IDSs to exchange information among them is a cooperative intrusion detection network (*CIDN*). Many different CIDNs have been proposed in the past few years. A typical CIDN collects data from distributed peer IDSs and to gain better intrusion detection accuracy. In this section, we propose a set of features (that is, cooperation topology, cooperation scope and specialization) to categorize CIDNs, and present the CIDs using a taxonomy.

9.3.1   Cooperation Topology

The cooperation topology of an IDN can be centralized or decentralized. In a centralized system, all the intrusion data from end nodes are forwarded to a central server for process and analysis. In general, a centralized system (for example, DShield [34] and CRIM [12]) can easily gather abundant data to make more accurate detection. However, the disadvantage is that it may cause traffic bottle-neck to the analyzing server and leads to a *single-point-of-failure*. On the contrary, a decentralized system can avoid the bottle-neck problem by sending intrusion data to many nodes for process and analysis. A decentralized system can be fully distributed or partially decentralized.

In a fully distributed system (for example, Indra [22], NetShield [11], and HBCIDS [16]), every node in the network is a contributor as well as an analyzer. The failure of a single node will have little impact on the functionality of the cooperation network. However, the lack of sufficient data in each analyzer may lead to less accurate intrusion detection. In a partially decentralized system, some (super) nodes may take the responsibility of analyzing data passed from their surrounding nodes, therefore, have heavier workload than peers which are only contributes. The network structure may be clustered (for example, ABDIAS [18]) or hierarchical (for example, DOMINO [39]). A partially decentralized system aims at finding a balanced solution between a centralized system and a fully distributed system.

9.3.2   Cooperation Scope

Another feature which can be used to categorize CIDNs is the cooperation scope. The cooperation scope of a CIDN can be local, global, or hybrid. In a local-scope CIDN (for example, Indra [22] and Gossip [14]), peers in the CIDN are usually assumed to be fully trusted. The privacy concern of exchanging packet payload is usually neglected since all nodes are in the same administrative boundary. Therefore, data packets can be in full disclosure and exchanged freely among IDSs. In a global CIDN (for example, DShield [34] and NetShield [11]), peers exchange intrusion information with other IDSs outside administration boundaries. Therefore only limited information can be shared due to the privacy concern. In this case, data payload (or IP addresses, and so on) is either digested or removed in the exchanged information. In a hybrid system (for example, DOMINO [39] and ABDIAS [18]), the network is divided into different trust zones. Each trust zone has its own customized data privacy policies.

9.3.3   Specialization

IDNs can be specialized to detect different intrusions such as worms (for example, NetShield [11], Gossip [14] and Worminator [26]), Spam (for example, ALPACA [40]), Botnet detection [41], or can be used to detect general intrusions (for example, Indra [22], CRIM [12], and HBCIDS [16]). We also adopt this feature to categorize the IDSs.

9.3.4   Cooperation Technologies and Algorithms

There are several technical components essential to CIDNs, namely, data correlation, trust management and load-balancing. In this section, we briefly describe each component and give some examples of solutions.

9.3.4.1   Data Correlation

IDSs are known to generate a large amount of alerts with high false positives. In practice, it is common to have tens of thousands alerts per day within a small enterprise network. With a sheer number of alerts to deal with, it becomes hard to decide in a timely way about which alert to deal with. The need for alert correlation becomes apparent in the presence of cooperative attacks as it helps in linking together different alerts that may be spaced in time and place. *Alert Correlation* is about the combination of fragmented information contained in the alert sequences and interpreting the whole flow of alerts. It is the process that analyzes alerts produced by one or more IDSs and provides a more succinct and high-level view of occurring or attempted intrusions [36]. In doing so, some alerts may get modified or cleared, new alerts may be generated and others delayed depending on the security policy in use.

Table 9.1 Classification of Cooperative Intrusion Detection Networks

IDN	Topology	Scope	Specialization	Technology and Algorithm
Indra	Distributed	Local	Worm	–
DOMINO	Decentralized	Hybrid	Worm	–
DShield	Centralized	Global	General	Data Correlation
NetShield	Distributed	Global	Worm	Load-balancing
Gossip	Distributed	Local	Worm	–
Worminator	–	Global	Worm	–
ABDIAS	Decentralized	Hybrid	General	Trust Management
CRIM	Centralized	Local	General	Data Correlation
HBCIDS	Distributed	Global	General	Trust Management
ALPACAS	Distributed	Global	Spam	Load-balancing
CDDHT	Decentralized	Local	General	–
SmartScreen	Centralized	Global	Phishing	–
FFCIDN	Centralized	Global	Botnet	Data correlation

Alert correlation techniques can be put into three broad categories. *Alert Clustering* is used to group alerts into clusters (or threads) based on some similarity measure, such as IP addresses or port numbers [35]. The second category relies on the pre-specification known attack sequences [15], [27]. The third category uses logical dependencies between alerts by matching prerequisites (of an attack) with its consequences [12], [30]. Attack graphs [33] are used in this regard to simplify the identification of attack patterns that are made up of multiple individual attacks.

9.3.4.2 Trust Management

Trust management is an important component for intrusion detection cooperation, especially when the participants cross administration boundaries. Without a trust model, dishonest nodes may degrade the efficiency of the IDN by providing false information. A trust management system can help to identify dishonest nodes by monitoring the past behaviour of participating nodes. Some CIDNs with trust management system are: Simple voting model [18], Simple linear model [16] and Bayesian statistic model [17].

9.3.4.3 Load-Balancing

When a CIDN needs to deal with a large amount of data and over the capacity of a single IDS processor, it is necessary to distribute workload into multiple servers to speed up the detection. Load-balancing algorithms help to distribute workload evenly on each IDS processor, therefore improve the overall efficiency of the system. Some examples of load-balancing algorithms are, signature distributing among collaborative IDSs group [25], data flow distributing among an IDS group [23], and data packets distributing among distributed IDSs [11].

9.3.5 Taxonomy

Based on the features we provide above, we categorize a list of selected CIDNs using the taxonomy as in Table 9.1.

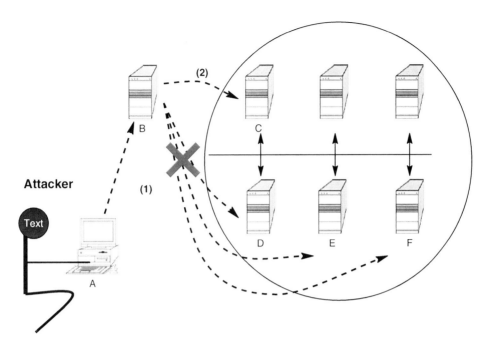

Figure 9.2 Indra architecture. (Adapted from [22].)

9.4   Selected Intrusion Detection Networks

9.4.1   Indra

Indra [22] is one of the earliest papers to propose a cooperative intrusion detection system (Figure 9.2). In the proposed system, host-based IDSs in a local area network take a pro-active approach and send warnings to other trusted nodes about the intruder through a peer-to-peer network. For example, as shown in Figure 9.2, if an attacker compromises a weak node B then launch attacks from B to hosts in the trusted network. Node C detects an attack from B and then multicasts a security warning to its trusted neighbours regarding B. Therefore, if B intends to attack other devices in the network, it will be repelled straight away by the forewarned nodes. Indra is a fully distributed system which targets local area networks.

9.4.2   DOMINO

DOMINO [39] is an IDS collaboration system which aims at monitoring Internet outbreaks for a large scale network. In the system architecture (Figure 9.3), heterogeneous IDSs located in diverse locations share their intrusion information with each other. There are typically three types of nodes, namely, axis nodes, satellite nodes and terrestrial contributors. *Satellite nodes* are organized hierarchically and are responsible for gathering intrusion data and sending them to parent nodes in the hierarchical tree. Parent nodes aggregate intrusion data and further forward data up to the tree till they reach axis nodes; *Axis nodes* analyze intrusion data, generate digested summary data and then multicast them to other axis nodes. Network-based IDSs and active sink nodes (such as Honeypot [13]) are integrated to axis nodes to monitor unused IP addresses for incoming worms; *Terrestrial contributors* do not follow DOMINO protocols but they can also contribute to the system through DOMINO access points. In DOMINO, heterogeneous nodes are

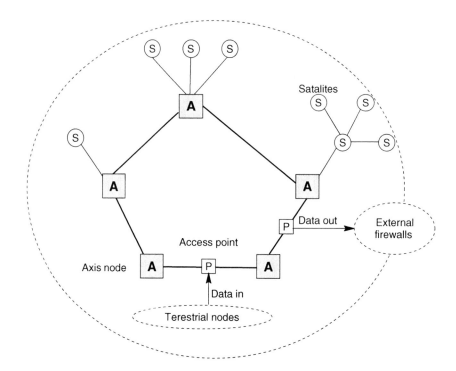

Figure 9.3   DOMINO architecture. (Adapted from [39].)

involved in the cooperation overlay. Information from axis nodes, satellite nodes and terrestrial contributors are distinguished by different trust levels. This feature enables DOMINO to be able to handle inter-administration-zone cooperation. DOMINO is a decentralized system with hierarchical structure. It is a global IDN with a good scalability.

9.4.3   DShield

DShield [34] is a community-based firewall log correlation system. The central server receives firewall logs from worldwide volunteers and then analyzes attack trends based on the information collected. Similar systems include myNetWatchMan [3] and CAIDA [2]. DShield is used as the data collection engine behind the SANS Internet Storm Center (ISC) [5]. Analysis provided by DShield has been used in the early detection of several worms, such as 'Code Red' and 'SQL Snake'. Due to the number of participants and volume of data collected, DShield is a very attractive resource and its data is used by researchers to analyze attack patterns. However, DShield is a centralized system and it does not provide real-time analysis or rule generation. Also due to the privacy issues, payload information and some headers can not be shared, which makes the classification of attacks often not possible.

9.4.4   NetShield

NetShield [11] is another IDN which uses the DHT Chord system to reduce communication overhead. In the system architecture (Figure 9.4), IDSs contribute and retrieve information from the system through a P2P overlay. Each IDS maintains a local prevalence table to record the number

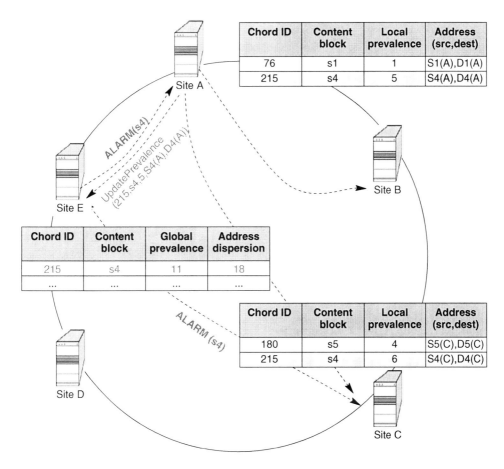

Figure 9.4   NetShield architecture. (Adapted from [11].)

of occurrences of each content block signature locally and its corresponding source address and destination address. An update will be triggered if the local prevalence of the content block exceeds a local threshold (for example, site A in Figure 9.4). If the global prevalence is higher than a threshold, and the address dispersion exceeds a certain threshold then an alarm will be raised regarding the corresponding content block. Netshield targets on epidemic worm outbreaks and DOS attacks. However, the limitation is that using content blocks as attack identification is not effective to polymorphic worms. Also the system assumes all participants are honest, which makes them vulnerable to the collusion attack of some malicious nodes.

9.4.5   Gossip

Denver et al. [14] proposed a collaborative worm detection system for enterprise level IDN for host-based INSs. A fully distributed model is adopted to avoid a single point of failure. In their system, a host-based IDSs (*local detector*) raises an alert only if the number of new-created connections per unit time exceeds a certain threshold. The alert will then be propagated to neighbours for aggregation. A Baysian network based alert aggregation model is used here for alert aggregation on *global detectors*. Their proposed system aims at detecting slow propagating

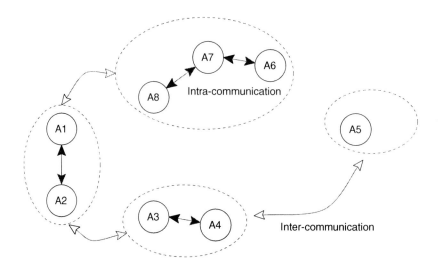

Figure 9.5   ABDIAS architecture. (Adapted from [18].)

worms in a local area network. However, their system only uses new connection rate as the sign of possible worm spreading. This is not effective to worms spread without connections, such as UDP worms.

9.4.6   Worminator

Worminator [26] was proposed to enable IDSs to share alert information with each other to detect worm propagation. Alert correlation is used to gain better detection accuracy. Different from most other systems, Worminator concerns about the privacy of exchanging alerts and proposed to use bloom filter to encode IP addresses and port numbers in the alerts to preserve privacy of collaborators. Worminator claims that the system topology can be either centralized or decentralized depending on the size of the network.

9.4.7   ABDIAS

Ghosh et al. proposed an Agent-Based Distributed Intrusion Alert System (ABDIAS) [18]. In the architecture design (Figure 9.5), IDSs (agents) are grouped into *communities* (neighbourhoods). Each Agent collects information inside its neighbourhood and uses a Baysian network analysis model to diagnose for possible threats. Inter-neighbourhood communication only happens if a consensus can not be reached within a neighbourhood. This system supports early warnings for pre-attack activities to gain time for administrators to respond to potential attacks. This system also supports a simple majority-based voting system to detect compromised nodes.

9.4.8   CRIM

CRIM [12] is a cooperative IDS where alerts from individual IDSs are sent to a central analyzer for clustering and correlating. A set of correlation rules are generated offline from security administrators by analyzing attack descriptions. These correlation rules are then used for analyzing alerts collected from IDSs to recognize global attack scenarios. CRIM is a semi-automatic alert correlation system since it relies on human interactions to define attack descriptions. It is also a centralized system.

9.4.9 HBCIDS

Fung et al. proposed a Host-based Collaborative Intrusion Detection System (HBCIDS) [16], where host-based IDSs are connected following a social network pattern. Each HIDS has a list of acquaintances with which they exchange alert information to gain better detection accuracy. HBCIDS uses a simple trust management system to distinguish honest nodes and dishonest nodes, therefore improve the efficiency of the collaboration. IDSs uses test messages to evaluate the trustworthiness of other nodes. In the case of intrusion detection, the weight of the diagnosis feedback from an acquaintance is proportional to its corresponding trust value. Their further work [17] proposed a Bayesian statistic model to trace the confidence level of trust estimation. The confidence level is then used to find the lowest test messages rate to guarantee trust estimation confidence. This collaboration system is a fully distributed system.

9.4.10 ALPACAS

ALPACAS [40] is a cooperative Spam filtering system that aims at preserving the privacy of emails as well as gaining scalability of the system. The system is built on top of a peer-to-peer overlay to avoid the deficiency of a centralized system. *Spam mails* and *Ham mails* are distributed on agents based on the range of their feature signatures. An email is divided into feature trunks and trunks are digested into feature finger prints to preserve content privacy of emails. Figure finger prints of an email are then sent to corresponding agents to do comparison with stored Spam emails and Ham emails by estimating the maximum signature overlap with Spam (MOS) and the maximum signature overlap with Ham (MOH). An email is labeled as Spam if the difference between MOS and MOH exceeds a certain threshold. ALPACAS is a fully distributed system.

9.4.11 CDDHT

The Cyber Desease Distributed Hash Table (CDDHT) [24] was proposed as a distributed data fusion centre system. In its architecture, each node is a local intrusion detection system which attempts to locally detect attacks and generate corresponding alerts. Each alert is assigned a disease key based on the related intrusions. The alert is then sent to a corresponding sensor fusion centre (SFC) using a DHT-based P2P system. SFCs are selected among nodes based on their capacity and resource. The purpose of this system is to avoid the bottle-neck problem by using a centralized fusion centre group and using the alert categorizing technique to load-balance the SFCs. CDDHT is a decentralized system.

9.4.12 SmartScreen Filter

SmartScreen Filter [8] is a tool in Microsoft Internet Explorer 8 which helps users avoid socially engineered malware phishing websites and online fraud when browsing the Web.

A centralized mechanism is used to maintain a list of phishing sites URLs and malicious websites URLs. Users browsing listed phishing sites or malicious websites will receive warnings to prevent them from being fraud. Users are allowed to report suspicious websites to the central server through a secure channel. Users' feedback is analyzed together with input from

SmartScreen Spam filter and input from other trusted sources to generate the URL blacklist. Other similar Phishing filters include EarthLink and eBay.

9.4.13   FFCIDN

Fast flux service networks (FFSN) are special Botnets which organize compromised nodes to form a robust Phishing domain. To detect fast-flux networks and prevent them from causing further damages, Zhou et al, [42], [41] proposed a collaborative IDN to detect FFSNs. The work is based on the observation that the number of IP addresses returned after a DNS request is larger than usual. The collaboration system collects query results from nodes of different locations and correlates them to obtain the number of unique IP addresses and the number of unique fast-flux domains. The relationship between the number of DNS queries and number of unique IP addresses and domains is traced. A corresponding DNS query threshold is derived to speed up FFSN detection. Their result shows detecting FFSN using collaboration from nodes in different name domains is more efficient than detecting from a single node. This system is a centralized system.

9.5   Open Challenges and Future Directions

Although several CIDNs have been proposed and built, there are still many challenges mainly related to efficiency. The first challenge is the privacy of exchanged information. Data privacy is a critical issue that prevents users from joining collaborations. This is especially realistic when the collaboration crosses administrative boundaries. Some existing works propose to digest sensitive information in alerts such as IP addresses and port numbers to prevent privacy leaking [26]. However, to have data privacy while keeping efficiency on collaboration is still an open challenge. The second challenge of CIDNs is the incentive design. An incentive compatible collaboration system is a key component for the long term effectiveness of all collaborations. Using a CIDN to detect Botnets is a new, but urgent, open challenge. Botnets are getting more sophisticated and stealthy with time. Recent Peer-to-peer Botnets [19] are even more difficult to detect than traditional centralized Botnets. How to detect and remove Botnets will remain important in the next few years. A few recent papers [20], [42], [41] have appeared to address how to detect fast-flux Botnet. [41] shows that collaborative detection is more efficient than single point detection. We are expecting more work to appear in the cooperative intrusion detection.

9.6   Conclusion

Collaborative Intrusion Detection Networks enable member IDSs to make use of the collected information or experience to enhance intrusion detection accuracy. In this chapter we summarized a list of existing CIDNs and categorized them using our proposed features.

There is a pressing need for the research community to develop common metrics and benchmarks for CIDN evaluation. The taxonomy will be helpful in shaping these tasks and may be enriched with the appearance of new CIDNs.

References

[1] Bro. http://www.bro-ids.org/.
[2] CAIDA: The Cooperative Association for Internet Data Analysis. http://www.caida.org/home/.
[3] myNetWatchman. http://www.mynetwatchman.com.

[4] OSSEC. http://www.ossec.net/.

[5] SANS Internet Storm Center (ISC). http://isc.sans.org/.

[6] Snort. http://www.snort.org/.

[7] TripWire. http://www.tripwire.com/.

[8] What is SmartScreen Filter? http://www.microsoft.com/security/filters/smartscreen.aspx.

[9] M. Ahamad, D. Amster, M. Barrett, T. Cross, G. Heron, D. Jackson, J. King, W. Lee, R. Naraine, G. Ollmann, et al. *Emerging Cyber Threats Report for 2009*. Georgia Institute of Technology, 2008.

[10] M. Bishop. *Computer Security: Art and Science*. Addison-Wesley, 2003.

[11] M. Cai, K. Hwang, Y. Kwok, S. Song, and Y. Chen. 'Collaborative internet worm Containment'. *IEEE Security & Privacy*, 3(3): 25–33, 2005.

[12] F. Cuppens and A. Miege. Alert correlation in a cooperative intrusion detection framework. In *2002 IEEE Symposium on Security and Privacy*, pp. 202–215, 2002.

[13] D. Dagon, X. Qin, G. Gu, W. Lee, J. Grizzard, J. Levine and H. Owen. 'Honeystat: Local worm detection using Honeypots'. *Lecture Notes in Computer Science*, pp. 39–58, 2004.

[14] D. Dash, B. Kveton, J. Agosta, E. Schooler, J. Chandrashekar, A. Bachrach and A. Newman. 'When gossip is good: Distributed probabilistic inference for detection of slow network intrusions'. In *Proceedings of the national conference on Artificial Intelligence*, vol. 21, p. 1115. Cambridge, MA; London; 2006.

[15] H. Debar and A. Wespi. 'Aggregation and correlation of intrusion-detection alerts'. In *Recent Advances in Intrusion Detection*, Lecture Notes in Computer Science, pp. 85–103. Springer, 2001.

[16] C. Fung, O. Baysal, J. Zhang, I. Aib and R. Boutaba. 'Trust management for host-based collaborative intrusion detection'. In *19th IFIP/IEEE International Workshop on Distributed Systems*, 2008.

[17] C. Fung, J. Zhang, I. Aib and R. Boutaba. 'Robust and scalable trust management for collaborative intrusion detection'. In *Proceedings of the Eleventh IFIP/IEEE International Symposium on Integrated Network Management (IM)*, 2009.

[18] Ghosh and S. Sen. 'Agent-based distributed intrusion alert system'. In *Proceedings of the 6th International Workshop on Distributed Computing (IWDC04)*. Springer, 2004.

[19] J. Grizzard, V. Sharma, C. Nunnery, B. Kang and D. Dagon. 'Peer-to-peer botnets: Overview and case study'. In *Proceedings of the First USENIX Workshop on Hot Topics in Understanding Botnets*, 2007.

[20] T. Holz, C. Gorecki, K. Rieck and F. Freiling. 'Detection and mitigation of fastflux service networks'. In *Proceedings of the 15th Annual Network and Distributed System Security Symposium (NDSS)*, San Diego, CA, USA, February 2008.

[21] G. Inc. *Gartner survey shows phishing attacks escalated in 2007; more than 3 billion lost to these attacks*. Press release, 2007. http://www.gartner.com/it/page.jsp?id = 565125

[22] R. Janakiraman and M. Zhang. 'Indra: a peer-to-peer approach to network intrusion detection and prevention'. *WET ICE 2003. Proceedings of the 12th IEEE International Workshops on Enabling Technologies*, 2003.

[23] Le, R. Boutaba and E. Al-Shaer. 'Correlation-based load balancing for network intrusion detection and prevention systems'. In *Proceedings of the 4th international conference on Security and privacy in communication network*. ACM New York, NY, USA, 2008.

[24] Z. Li, Y. Chen and A. Beach. 'Towards scalable and robust distributed intrusion alert fusion with good load balancing'. In *Proceedings of the 2006 SIGCOMM workshop on Large-scale attack defense*, pages 115–122. ACM New York, NY, USA, 2006.

[25] W. Lin, L. Xiang, D. Pao and B. Liu. 'Collaborative Distributed Intrusion Detection System'. In *Future Generation Communication and Networking, 2008. FGCN'08. Second International Conference on*, volume 1, 2008.

[26] M. Locasto, J. Parekh, A. Keromytis and S. Stolfo. 'Towards collaborative security and P2P intrusion detection'. In *Information Assurance Workshop, 2005. IAW'05. Proceedings from the Sixth Annual IEEE SMC*, pp. 333–339, 2005.

[27] B. Morin and H. Debar. 'Correlation of intrusion symptoms: An application of chronicles'. In *Lecture Notes in Computer Science*, pp. 94–112. Springer, 2003.

[28] J. Nazario and T. Holz. 'As the net churns: Fastflux botnet observations. In Malicious and Unwanted Software'. 3rd International Conference on MALWARE, pp. 24–31, 2008.

[29] Networks. http://atlas.arbor.net.

[30] P. Ning, Y. Cui and D. S. Reeves. 'Constructing attack scenarios through correlation of intrusion alerts'. In *ACM Conference on Computer and Communications Security*, pp. 245–254. ACM, 2002.

[31] T. H. Project. 'Know your enemy: Fastflux service networks', 13 July, 2007. http://www.honeynet.org/book/export/html/130.

[32] R. Richardson. 'CSI computer crime and security survey'. *Computer Security Institute*, 2007.

[33] O. Sheyner, J. W. Haines, S. Jha, R. Lippmann and J. M. Wing. 'Automated generation and analysis of attack graphs'. In *IEEE Symposium on Security and Privacy*, pp. 273–284, 2002.

[34] J. Ullrich. DShield. http://www.dshield.org/indexd.html.

[35] Valdes and K. Skinner. 'Probabilistic alert correlation'. In *Lecture Notes in Computer Science*, pp. 54–68. Springer, 2001.

[36] F. Valeur, G. Vigna, C. Krgel and R. A. Kemmerer. 'A comprehensive approach to intrusion detection alert correlation'. *IEEE Trans. Dependable Sec. Comput.*, 1(3): 146–169, 2004.

[37] R. Vogt, J. Aycock and M. Jacobson. 'Army of botnets'. In *Proc. ISOC Symp. On Network and Distributed Systems Security*, 2007.

[38] N. Weaver, V. Paxson, S. Staniford and R. Cunningham. 'A taxonomy of computer worms'. In *Proceedings of the 2003 ACM workshop on Rapid Malcode*, pp. 11–18. ACM New York, NY, USA, 2003.

[39] V. Yegneswaran, P. Barford and S. Jha. 'Global intrusion detection in the domino overlay system'. In *Proceedings of Network and Distributed System Security Symposium (NDSS)*, 2004.

[40] Z. Zhong, L. Ramaswamy and K. Li. 'Alpacas: A large-scale privacy-aware collaborative anti-spam system'. In *INFOCOM 2008. The 27th Conference on Computer Communications. IEEE*, pp. 556–564, 2008.

[41] C. V. Zhou, C. Leckie and S. Karunasekera. 'Collaborative detection of fastflux phishing domains'. *Journal of Networks*, 4: 75–84, February 2009.

[42] C. V. Zhou, C. Leckie, S. Karunasekera and T. Peng. 'A self-healing, self-protecting collaborative intrusion detection architecture to trace-back fastflux phishing domains.' In *The 2nd IEEE Workshop on Autonomic Communication and Network Management (ACNM 2008)*, April 2008.

10

Cooperation Link Level Retransmission in Wireless Networks

Mehrdad Dianati[1], Xuemin (Sherman) Shen[2] and Kshirasagar Naik[2]

[1]*Centre for Communication Systems Research (CCSR), Department of Electronic Engineering, University of Surrey, Surrey, UK*
[2]*Department of Electrical and Computer Engineering, University of Waterloo, Waterloo, Ontario, Canada*

10.1 Introduction

Wireless communications open up new challenges and potentials in the design of network protocols. Wireless channels are more vulnerable to interference and interception leading to less reliable communications. Furthermore, the quality of a wireless channel is time-variant and location-dependent. Since there is limited possibility of physical isolation of signals, there is no access to unlimited bandwidth as it has to be shared among many consumers. On the other hand, due to the open nature of wireless media, there is a significant potential of coordination among dependent and cooperation among independent communicating parties. These new constraints and dimensions of freedom necessitate redesign of many aspects of traditional communication protocols to fit the characteristics of the new environment.

The traditional perception of the communication protocols, as shown in Figure 10.1, is to split the protocol stack into: 1) application adaptation and interfacing layers; 2) networking and unification layers; and 3) link adaptation layers. Application adaptation and interfacing layers comprise a set of protocols to facilitate interfacing of different applications to network resources. Adaptation mechanisms may be implemented to smooth out variations in the quality of network services in order to improve user experience. Inter-networking and inter-operability issues usually are addressed in networking and unification layers. Since networking layers glue a variety of small networks together, clear and consistent standardization is a major design constraint for those layers. Link adaptation layers, including physical and data link layers are to provide a high level of transparency to the upper layers. They need to be optimized to fit to the characteristics of communication channel, and provide a uniform interface to the upper layers. Therefore, many studies in wireless networks naturally have been concentrated on physical and

Cooperative Networking, First Edition. Edited by Mohammad S. Obaidat and Sudip Misra.
© 2011 John Wiley & Sons, Ltd. Published 2011 by John Wiley & Sons, Ltd.

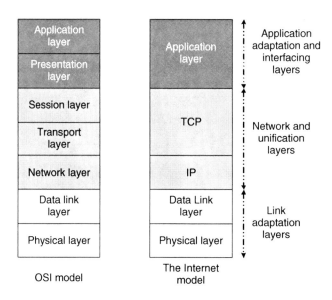

Figure 10.1   Communication protocol stack.

data link layer problems. In particular, in terms of network protocols, data link layer for wireless channels exposes a variety of challenging problems from multiple access issues to scheduling and efficient retransmission schemes. Furthermore, unlike the traditional layering model, cross layer optimization techniques are deemed to be necessary to improve the utilization of limited channel resources.

Data link layer design for wireless fading channels is a rich and wide research area, which is obviously beyond the scope of this chapter. In this chapter, we only focus on link layer retransmission. The intention is mainly to demonstrate how cooperative communications techniques can be used to improve performance of the link layer. Our approach, in this chapter, is particularly to utilize the valuable information which can be obtained from the prediction of the fading process, either implicitly or explicitly by means of the existing facilities of the traditional retransmission schemes. As we discuss throughout this chapter, using such techniques can achieve significant gain in terms of the important QoS metrics such, as throughput, average delay and delay jitter. The readers should note that we aim to demonstrate how cooperation can be exploited in upper layers of communications protocol stack by presenting an example. We hope that this will encourage further studies, particularly, on cross layer optimization link and network layers together considering the impacts of any potential cooperation of autonomous communicating parties.

The rest of this chapter is organized as follows. In Section 10.2, we discuss some of the fundamentals that will be used in the rest of the chapter. This will include the basic channel models that we will use in the performance analysis. We will also briefly discuss the basics of link level retransmission. Sections 10.3, 10.4 and 10.5 is the main part of this chapter where we introduce the system model and the proposed link level retransmission scheme, namely, Node Cooperative Stop and Wait (NCSW). This section also includes the analytical model for performance of the proposed scheme which is accompanied by simulation based performance analysis. These two alternative approaches to performance analysis are given to demonstrate validity of the analytical models and further strengthen our performance analysis. Finally, a summary of the chapter with some possible future studies are given in Section 10.10.

10.2   Background

Conventional ARQ schemes have been designed for wireline networks, where the frame errors are independent and very rare. However, due to the inherent characteristics of the fading process in wireless channels, frame errors appear in bursts, rather than independent events. When the link between two communicating nodes is experiencing frame errors, there is a high probability that the bad channel condition will continue for a considerably long period of time. The duration of a bad channel condition or channel fading may be as long as the transmission period of multiple data frames. Conventional retransmission schemes are not very effective in such an environment with bursty frame errors, and will cause significant degradation in the performance of a link layer protocol [18].

Recently, spatial diversity techniques in different forms of multiple transmit and receive antennas have been proposed to improve the quality of communication over wireless fading channels. Multiple Input and Multiple Output (MIMO) communication systems [7], [8] and the corresponding channel coding techniques, such as Space Time Coding (STC) [9], have been proposed to utilize space diversity in the next generation wireless networks. However, implementation of multiple antennas on the small mobile devices is quite difficult due to the device size and cost constraints. An alternative form of space diversity can be achieved in a multi-user environment by allowing the communicating nodes to cooperate [1], [9], [10]. In cooperative communications, each node not only transmits and receives its own data, but also can provide an alternative path for the other pairs of communicating nodes. In other words, each node acts as a relay node to facilitate better communications between other pairs of nodes at the link level.

The theoretical and implementation aspects of cooperative diversity in the physical layer have been areas of active interest among researchers [11]–[14]. However, to the best of our knowledge, the impacts of cooperative techniques on the upper layers of communication protocols have not been thoroughly studied so far, which we advocate by demonstrating the potential benefits, using an example in this chapter.

10.2.1   Modeling of Fading Channels

Design of an efficient data link layer protocol requires a good study of the characteristics of the underlying wireless channels. These characteristics depend on many aspects of the propagation environment such as the carrier frequency, surrounding environment and the mobility of users.

Typically, there is a Non-Line-of-Sight (NLoS) propagation environment in land mobile communications, with several paths from a transmitter to a receiver. Signals propagate after reflections, diffractions and scattering. At the receiver, plane waves arrive from many different directions and with different delays as shown in Figure 10.2. This property is called multi-path propagation. The multiple plane waves combine to produce a composite received signal. Given that wireless systems operate at very high frequencies, the carrier wavelength is relatively small. Hence, small differences in the propagation delays of different paths due to relative mobility of transmitter, receiver, or the surrounding obstacles will cause constructive or destructive combination of the arriving signals from different paths. The temporal variations in the amplitude and the phase of the composite received signal is called *envelope fading*. A two-dimensional isotropic scattering environment, where the plane waves arrive from all directions with equal probability, is a very common scattering model for many land mobile wireless systems. Envelope fading can be caused by large obstacles such as hills and big buildings. These variations have usually very slow dynamics and are typically known as shadowing. However, another component of envelope fading is associated with small obstacles which could be close to the receiver. This component of fading has a relatively fast variation depending on the mobility of the receiver, transmitter, or the

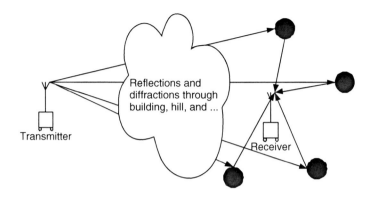

Figure 10.2   A typical wireless propagation environment.

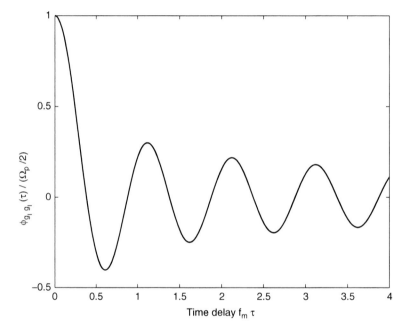

Figure 10.3   Autocorrelation of in-phase component of received band pass signal.

obstacles. Nonetheless, it has been shown that, unlike Additive White Gaussian Noise (AWGN), channel variations due to fading is a non-while random process [15]. This fact is indicated by the autocorrelation and power spectrum density of a typical fading process, as shown in Figure 10.3 and Figure 10.4. As can be seen, there is a strong correlation among the samples of fading process over a relatively large period of time, compared to the activities of link layer in typical communication systems. This means that there is a possibility of predicting the channel status in future if there is some current knowledge of the channel. This behaviour of fading channels can be modeled by some nice random processes as we discuss in the following. See Figure 10.3.

For flat fading channels, where the power spectrum density of the channel is constant over the entire frequency band of the information signal, a two-state Markov process, as illustrated in Figure 10.5, can adequately describe the process of frame success or failure [17] during a

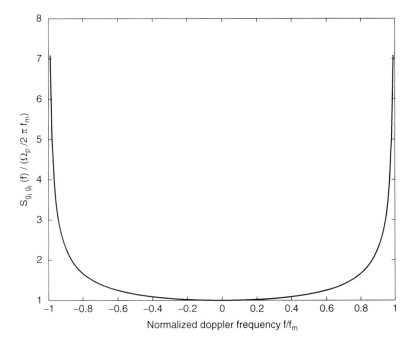

Figure 10.4 Normalized psd of the in-phase component.

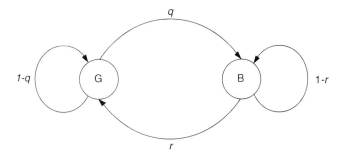

Figure 10.5 Markovian model for frame success/failure process over a fading channel.

transmission period. In this model, the channel is deemed to be in good state (G) if a transmitted frame can be correctly decoded by the receiver; otherwise, when the channel is in bad state (B), the transmitted frame cannot be decoded successfully, and needs to be retransmitted by the sender. If the channel is in state G during the transmission time of the current frame, it will either stay in state G with probability $(1 - q)$ or move to state B with probability q in the transmission time of the next frame. Similarly, starting from state B, the channel will either stay in state B with probability $(1 - r)$ or move to state G with probability r. Thus, the transition probability matrix for the two-state Markov process is

$$\begin{bmatrix} 1 - q & q \\ r & 1 - r \end{bmatrix}.$$

Let $\varepsilon(k)$ be the frame success or failure process, where the discrete time index, k, denotes the transmission time of the kth frame. Let T_f be the duration of a single frame. Assuming a non-adaptive channel coding rate and quasi-static fading process (that is, the fading level is constant for the entire frame duration), frame k cannot be decoded properly by the receiver, if the channel power gain is below a certain threshold γ at time instant k, that is,

$$\varepsilon(k) = \begin{cases} B, & \text{if} \quad |\zeta(k)|^2 \le \gamma \\ G, & \text{if} \quad |\zeta(k)|^2 > \gamma, \end{cases} \tag{10.1}$$

where $\zeta(k)$ represents the complex value of the fading envelop in the transmission time of frame k. For slow fading channels, the fading envelop is roughly constant for the entire frame duration, that is, the fading envelope can be considered as a quasi-static process. For a quasi-static Rayleigh fading channel, $\zeta(k)$ has a Rayleigh pdf. From [18] the transition parameters of the two-state Markov process in Figure 10.5 for a Rayleigh fading model can be given by

$$r = \frac{Q(\theta, \rho\theta) - Q(\rho\theta, \theta)}{e^\gamma - 1}$$

$$q = \frac{1 - e^{-\gamma}}{e^{-\gamma}} r, \tag{10.2}$$

where $Q(\cdot, \cdot)$ is the Marcum Q function,

$$\theta = \sqrt{\frac{2\gamma}{1 - \rho^2}},$$

$$\rho = J_0(2\pi f_m T_f),$$

and $J_0(\cdot)$ is the zero-order Bessel function of the first kind. Similar Markovian models for the Rician and the Nakagami flat fading channels have been given in [19] and [20], respectively.

The Markov model, described in this section, provides a useful and powerful tool for analytical modeling of communication protocols over fading channel. In the subsequent sections, after we introduce cooperative retransmission, we will use this model to study the performance of the proposed cooperative retransmission scheme.

10.2.2 Automatic Repeat Request

Data communication is always subject to random errors. The nature and frequency of transmission errors are tightly linked to the characteristics of the communication media. Dealing with transmission errors is a major functionality of communication protocols. A data packet may fail to reach its destination node due to network or link errors. Network errors usually happen if one of the forwarding routers become congested by a heavy traffic load as shown in Figure 10.6. Furthermore, packets may not be forwarded correctly over a single link due to channel errors. Transmission errors need to be addressed in different layers of communication protocol stack. Network errors are usually handled by transport layer. Link errors have to be handled in data link layer or physical layer. Error correction can be done by coding or retransmission. Data link layer and transport layer use retransmission to overcome packet errors. However, physical layer handles errors by coding and forward error correction. In wireline networks, probability of transmission errors is significantly low; therefore, data link layer retransmission is usually not necessary. Link errors can be mitigated by proper coding in physical layer. Retransmission in transport layer can also handle network errors and possible link errors that cannot be corrected

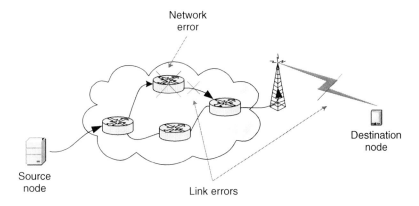

Figure 10.6   Network and link errors.

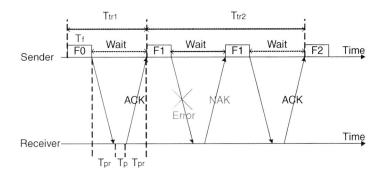

Figure 10.7   Stop and wait retransmission protocol.

by physical layer. However, in wireless networks, frequency of link errors is significantly higher than that in wireline networks. Link level retransmission is a de facto part of wireless link layer protocols [2]–[6].

In principle, link level error correction between a sender and a receiver nodes is performed in three steps: 1) error detection by the receiver; 2) feedback from the receiver to the sender, indicating successful or unsuccessful reception; and 3) retransmission of the erroneous frames. The entire process is called Automatic Repeat reQuest (ARQ) or retransmission protocol. The simplest form of retransmission protocol is *Stop-and-Wait (SW)*. The SW protocol is illustrated in Figure 10.7. The sender node transmits a single data frame. The receiver node investigates correctness of the received frame, and sends a positive acknowledgement (ACK) for a correct reception or a negative acknowledgment (NAK) for an erroneous reception. The sender node repeats the previously transmitted frame upon reception of a NAK. Otherwise, the sender node transmits the next frame in its transmission buffer.

Performance of a retransmission protocol is usually measured by its throughput, average delay and delay jitter. The throughput of a retransmission protocol is defined as the ratio of a single frame duration to the expected value of the total time spent for the transmission of a single frame. The average delay is the average time that is spent for transmission of a single frame. Delay jitter is the variance of delays, experienced by data frames. Let T_f be the frame duration, T_p be the frame processing time in the receiver node, T_{pr} be the propagation delay. The sender node spends $T_{tr} = N(T_f + T_p + 2T_{pr})$ for successful transmission of a single frame, where N

is the number of transmission trials. The number of transmission trials is a random variable. Hence, the throughput of an SW protocol is given by

$$\eta_{SW} = \frac{T_f}{\mathrm{E}[N](T_f + T_p + 2T_{pr})}. \tag{10.3}$$

The average delay is

$$\bar{T}_f = \mathrm{E}[N](T_f + T_p + 2T_{pr}). \tag{10.4}$$

Delay jitter depends on the error profile of the communication channel between a sender and receiver nodes. We will discuss delay jitter in Section 10.7.

If the propagation delay plus processing delay is relatively large, SW protocol demonstrates a poor throughput. For instance, SW protocol is not appropriate for satellite links. For those scenarios, there are more efficient retransmission protocols such as *Go Back N* and *Selective Repeat* protocols. However, SW protocol is preferred for short range land mobile communications due to its simplicity.

In the next section, we will extend this simple and effective link level retransmission to exploit inherent cooperative diversity in a network where there are multiple pairs of senders and receivers sharing the same radio channel for communications.

10.3   System Model

We consider a wireless network without a central authority as shown in Figure 10.8. Although there is no fixed infrastructure, the model can generalize other types of wireless networks with infrastructures, such as cellular networks and wireless LANs. A single node in this model can be a mobile device, a base station, or an access point. A cooperation group is a subset of nodes

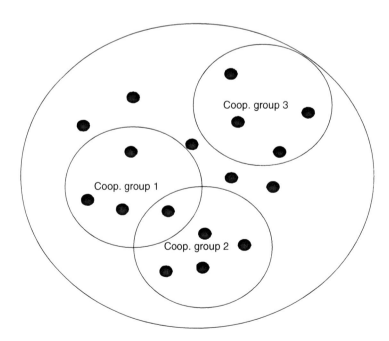

Figure 10.8   An ad-hoc wireless network model.

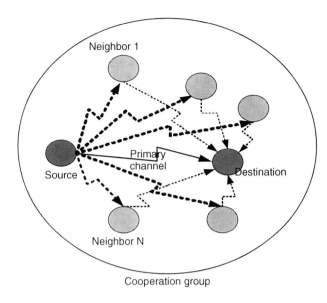

Figure 10.9 A single cooperation group.

that can reach each other with a single hop. In other words, nodes in a cooperation group are in the radio coverage area of each other.

Cooperation groups may be set up during the connection setup stage (for example, call setup in cellular networks or association in WLAN) or packet level handshaking. There is no need for extra signaling in order to form and manage cooperation groups. Cooperation groups are dynamic groups without tight membership requirement. Each neighbour node may join several cooperation groups depending on its position, capability, and willingness to cooperate. The overall structure of cooperation groups can be different in the transmission time of different packets. Thus, mobility of nodes do not disrupt the operation. As shown in Figure 10.9, each of those cooperation groups can be modeled as a single hop wireless network. At any time instant, one sender node captures the shared media to send a burst of frames to its intended destination node. During that time period the neighbour nodes in the group monitor the shared channel to assist the sender and the receiver nodes if error happens. The whole process is transparent from a sender node's point of view. Thus, a sender node performs its normal operation. A receiver node only needs to know the number and identity of the neighbour nodes during the current packet transmission time. Intuitively, node cooperation can improve the probability of successful retransmission if there is extra resource available. Otherwise, all nodes can perform their normal operations.

10.4   Protocol Model

In typical land mobile wireless networks, propagation delays among nodes are much smaller than data frame duration. For a link with short propagation delay, a frame-by-frame acknowledgment mechanism such as SW retransmission scheme, is preferred over complex schemes, such as Go Back N (GBN) and Selective Repeat (SR) retransmission schemes. For example, the IEEE 802.11 compatible products adopt an SW-like retransmission scheme. In this chapter we adopt SW as the core of the proposed ARQ scheme. For SW ARQ scheme, a sender node does not transmit the next frame until correct reception of the previous frame is confirmed by an explicit or implicit Acknowledgment (ACK). We assume that the feedback channels, used for transmitting Acknowledgment (ACK) and Negative Acknowledgement (NAK), are error free. Thus, the ACK/NAK frames can be received immediately and correctly by all the nodes in a cooperation group. Given these assumptions and a two-state Markovian frame success/failure model, the throughput of SW ARQ scheme has been given by [21] as

$$\eta_{SW} = \frac{r}{r + q},$$ (10.5)

where q and r are the parameters of the Markov model, as shown in Figure 10.5.

10.5   Node Cooperative SW Scheme

Depending on the dynamic of fading process, the average duration of bad channel condition between a sender and a receiver nodes can be as long as the transmission time of several frames. For example, for a slow fading channel with a mobility factor of 5 km/h and carrier frequency of $f_c = 2.4$ GHz, the average fading in a bad channel condition is about 60 ms [15]. For a typical frame duration of 5 ms, a simple retransmission scheme, such as the conventional SW, has to retransmit an erroneous frame for an average of 12 times. The situation may be even worse in high-rate systems with a shorter frame duration.

In the network model of Figure 10.8, since the channel status between different pairs of nodes are independent of each other, node cooperation can be used to implement a variant of space diversity, which is known as cooperative diversity [10]. Cooperation among the nodes that are in the same radio coverage can significantly improve the performance of a retransmission scheme by reducing the number of unsuccessful retransmission trials. The core concept can be explained using the model shown in Figure 10.9, where a sender node is transmitting to a destination node, and a typical neighbour node is listening to the ongoing communication. In the NCSW, the error control procedures of the sender and the receiver nodes are the same as those of a conventional SW retransmission scheme in Figure 10.10. However, the neighbour nodes in a cooperation group implement an additional functionality, as shown in Figure 10.11. If a transmitted frame cannot be decoded successfully by the receiver, as part of its normal operation, the receiver node will send a NAK to the sender node asking for retransmission of the erroneous frame, and the sender node will respond to the NAK by retransmitting the frame. The NAK signal can be either explicit or implicit through a timeout mechanism. In the conventional retransmission scheme, the neighbour nodes are oblivious to the retransmissions between the sender and the receiver nodes. However, in the NCSW scheme, all the other nodes in the cooperation group monitor the ongoing communications between the sender and the receiver nodes. The neighbour nodes decode and store a copy of the last un-acknowledged transmitted frame until the reception of a corresponding ACK. When a transmitted frame is acknowledged by the destination node, all the nodes in the cooperation group will drop their copy of the corresponding frame. However, when the neighbour nodes in a cooperation group receive a NAK from the destination node, they will cooperate with the sender in the retransmission process. Obviously, every node in a cooperation group can retransmit only if it has already received a correct copy of the requested frame. Even if a neighbour node has received a correct copy of the frame, cooperation is not mandatory, and each node may avoid cooperation for internal reasons or network considerations. This guarantees backward compatibility of the NCSW protocol with the conventional SW protocol.

If proper coding and decoding schemes, such as the Distributed Space Time Coding [16], are implemented in the physical layer, the probability of successful retransmission will be significantly increased due to the diversity gain that can be achieved. Intuitively, if the channel coding scheme is capable of achieving full diversity gain and all neighbour nodes have a

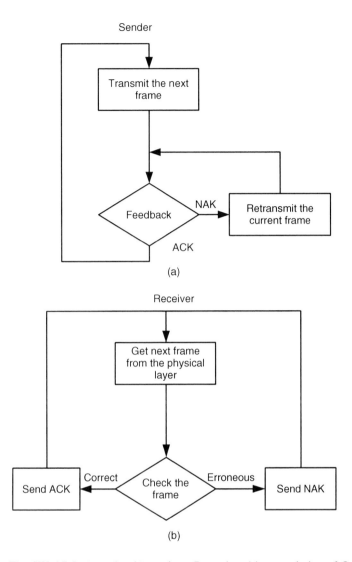

Figure 10.10   The SW ARQ a) sender; b) receiver. Reproduced by permission of © 2006 IEEE.

correct copy of a failed frame, the probability of successful retransmission for the NCSW scheme is $(1 - \prod_i^n P_{ei})$, where P_{ei} is the probability of frame error from node i in the cooperation group to the destination node, and n is the number of all the cooperating nodes plus the sender node. Theoretically, the higher the number of cooperating nodes the higher is the performance improvement. However, increasing the size of cooperation group might cause synchronization and implementation problems.

10.6   Performance Analysis

In this subsection, we develop an analytical model to describe the frame success/failure process for the proposed NCSW ARQ scheme. The model, which can be used to obtain useful perfor-mance metrics such as throughput, delay and delay jitter, is developed in three steps: 1) we

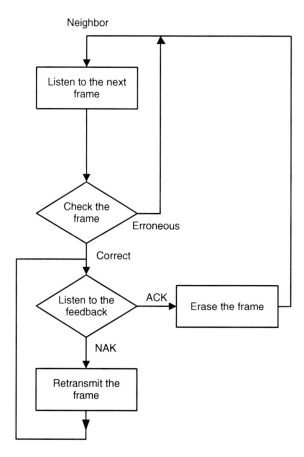

Figure 10.11   Cooperation procedure in the NCSW ARQ. Reproduced by permission of © 2006 IEEE.

derive a cooperation model for a pair of sender and receiver nodes with a single neighbour node; 2) we model the impacts of an arbitrary number of neighbour nodes in the cooperation group as an equivalent *super neighbour node*; 3) we combine the model for the super neighbour node from step 2 and the cooperation model from step 1 to obtain a model for the frame success/failure process for a pair of sender and receiver nodes and an arbitrary number of neighbour nodes.

In step 1, a sender and receiver pair with a single neighbour node, as shown in Figure 10.12, is considered. A two-state Markov model, as described in Section 10.3, is used to specify success or failure of transmission over a wireless fading channel. We use three distinct two-state Markov processes to model the *primary channel* from the sender node to the receiver node, the *interim channel* from the sender node to the neighbour node, and the *relay channel* from the neighbour node to the destination node. The corresponding transition probability matrices are denoted by

$$\begin{bmatrix} 1-q & q \\ r & 1-r \end{bmatrix}$$

for the primary channel,

$$\begin{bmatrix} 1-x & x \\ y & 1-y \end{bmatrix}$$

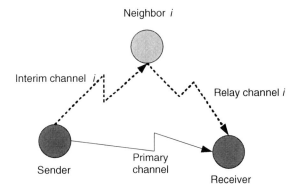

Figure 10.12 The system model for a single cooperative neighbour node. Reproduced by permission of © 2006 IEEE.

for the interim channel, and

$$\begin{bmatrix} 1-a & a \\ b & 1-b \end{bmatrix}$$

for the relay channel, where (q, r), (x, y), and (a, b) are the corresponding transition probabilities, respectively. During the transmission time of frame k, the neighbour node is in bad state (B), if it is not able to cooperate in the possible retransmission of the $(k-1)^{th}$ frame. Otherwise, the neighbour node is considered to be in good state (G). Assuming that the neighbour node is always willing to cooperate, it will be in state B at time instant k if frame $(k-1)$ cannot be decoded properly by that particular neighbour node, or the relay channel from that particular neighbour to the receiver node is in bad state at time instant k. With this argument, the status of the neighbour node at time instant k can be formulated as follows:

$$N(k) = \begin{cases} G, & \text{if } I(k-1) = G \text{ and } R(k) = G \\ B, & \text{otherwise,} \end{cases} \qquad (10.6)$$

where $I(k)$ and $R(k)$ denote the states of the interim and relay channels, respectively.

A four-state Markov model, as shown in Figure 10.13, can be used to represent the process of transition among different states of $\{I(k-1), R(k)\}$. Let π_i denote the probability of being in state S_i in Figure 10.13. The following set of linear equations can be solved to obtain π_i for $0 \le i \le 3$;

$$\begin{bmatrix} A_{00} & A_{10} & A_{20} & A_{30} \\ A_{01} & A_{11} & A_{21} & A_{31} \\ A_{02} & A_{12} & A_{22} & A_{32} \\ A_{03} & A_{13} & A_{23} & A_{33} \end{bmatrix} \begin{bmatrix} \pi_0 \\ \pi_1 \\ \pi_2 \\ \pi_3 \end{bmatrix} = \begin{bmatrix} \pi_0 \\ \pi_1 \\ \pi_2 \\ \pi_3 \end{bmatrix} \qquad (10.7)$$

$$\pi_0 + \pi_1 + \pi_2 + \pi_3 = 1,$$

where A_{ij} is the probability of transition from state S_i to state S_j. As given in (10.6), $N(k)$ can be modeled by another two-state Markov process with transition parameters of (u, v), where u and v are defined as follows:

$$u \triangleq \mathrm{P}\{N(k) = B | N(k-1) = G\}$$
$$v \triangleq \mathrm{P}\{N(k) = G | N(k-1) = B\}. \qquad (10.8)$$

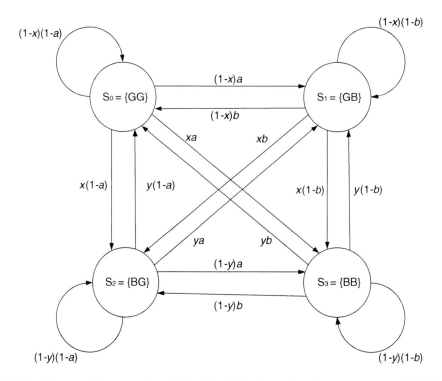

Figure 10.13 State transition model for $\{I(k-1), R(k)\}$. Reproduced by permission of © 2006 IEEE.

From the Markov process shown in Figure 10.13, it can be seen that

$$u = 1 - (1 - x)(1 - a). \tag{10.9}$$

We can rewrite v in (10.8) in a slightly different form as

$$v = \frac{\mathrm{P}\{N(k-1) = B | N(k) = G\}\mathrm{P}\{N(k) = G\}}{\mathrm{P}\{N(k-1) = B\}}. \tag{10.10}$$

The solution of equation set (10.7) can be combined with (10.10) to obtain v as

$$v = \frac{u\pi_0}{1 - \pi_0}. \tag{10.11}$$

The Markov process specified by (10.9) and (10.11) characterizes the status of a single neighbour node. In step 2, we propose an iterative algorithm to reduce a cooperation group with multiple neighbour nodes to an equivalent cooperation group with only one *super neighbour node*. Let $M \geq 2$ be the total number of neighbour nodes. In the first iteration, we combine neighbour nodes 1 and 2 into one equivalent node. Then the resulting equivalent node is combined with node 3, and so on, until all the M neighbour nodes are combined together to form a single super neighbour node.

Let $N^{(1)}(k)$ and $N^{(2)}(k)$ be the states of neighbour nodes 1 and 2 at time instant k, respectively. Since a retransmission will succeed if at least one of the neighbour nodes or the sender node can successfully deliver the frame to the receiver node, the combined cooperative node model for nodes 1 and 2, denoted by $N^{(1,2)}(k)$, can be represented by

$$N^{(1,2)}(k) = \begin{cases} G, & \text{if } N^{(1)}(k) = G \text{ or } N^{(2)}(k) = G \\ B, & \text{otherwise.} \end{cases} \tag{10.12}$$

The discrete random process $N^{(1,2)}(k)$, as specified by (10.12), can be modeled by a two-state Markov process, where its parameters are defined by

$$u^{(1,2)} \triangleq P\{N^{(1,2)}(k) = B | N^{(1,2)}(k-1) = G\}$$

$$v^{(1,2)} \triangleq P\{N^{(1,2)}(k) = G | N^{(1,2)}(k-1) = B\}. \tag{10.13}$$

Let (u_1, v_1) and (u_2, v_2) be the corresponding Markov parameters of $N^{(1)}(k)$ and $N^{(2)}(k)$, respectively. As explained in step 1, (u_1, v_1) and (u_2, v_2) can be computed by (10.9) and (10.11) for node 1 and 2, respectively. The status of $\{N^{(1)}(k), N^{(2)}(k)\}$ can be described by another four-state Markov process as shown in Figure 10.14. From Figure 10.14 and (10.13), it can be easily seen that

$$v^{(1,2)} = 1 - (1 - v_1)(1 - v_2). \tag{10.14}$$

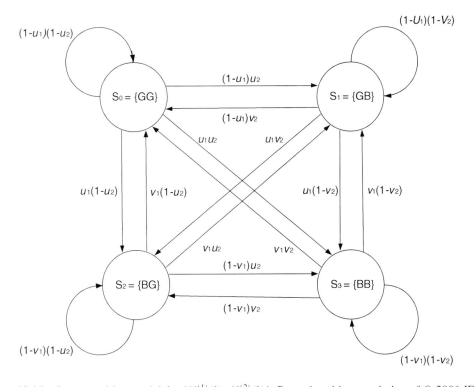

Figure 10.14 State transition model for $\{N^{(1)}(k), N^{(2)}(k)\}$. Reproduced by permission of © 2006 IEEE.

With (10.14), we can obtain $u^{(1,2)}$ as

$$u^{(1,2)} = P\{N^{(1,2)}(k-1) = G|N^{(1,2)}(k) = B\} \cdot$$
$$\frac{P\{N^{(1,2)}(k) = B\}}{P\{N^{(1,2)}(k-1) = G\}}$$
$$= \frac{v^{(1,2)}\pi_3}{1 - \pi_3}, \tag{10.15}$$

where $[\pi_0 \quad \pi_1 \quad \pi_2 \quad \pi_3]$ can be obtained by solving the equation set (10.7) in Figure 10.14.

In the next iteration, the two-state Markov model specified by $(u^{(1,2)}, v^{(1,2)})$ is combined with the two-state Markov model of neighbour node 3, which is specified by (u_3, v_3). The combination process follows the same steps as we used to combine neighbour nodes 1 and 2. This process is repeated until all the neighbour nodes in the cooperation group are considered. Finally, we obtain a two-state Markov model for the super neighbour node, which includes the impacts of all neighbour nodes. We denote the status of the super neighbour node by $N^{(1,\ldots,M)}(k)$, and use the following notations for its transition probabilities.

$$U = u^{(1,\ldots,M)}$$
$$\triangleq P\{N^{(1,\ldots,M)}(k) = B|N^{(1,\ldots,M)}(k-1) = G\}$$
$$V = v^{(1,\ldots,M)}$$
$$\triangleq P\{N^{(1,\ldots,M)}(k) = G|N^{(1,\ldots,M)}(k-1) = B\}, \tag{10.16}$$

In step 3, we model the transmission and retransmission process of the NCSW ARQ scheme with a sender and receiver pair and a single super neighbour node. Let $O(k)$ denote the state of the NCSW protocol at time instant k. $O(k)$ is either in Transmission (T) state or Retransmission (R) state according to the two-state Markov model shown in Figure 10.15. The parameters of this Markov model are defined as

$$X \triangleq P\{O(k) = R|O(k-1) = T\}$$
$$Y \triangleq P\{O(k) = T|O(k-1) = R\}. \tag{10.17}$$

In state T, the sender transmits a new frame; however, in state R, all nodes in the cooperation group retransmit the previously failed frame. Let $PC(k)$ represent the state of the primary channel at time instant k; $O(k)$ will transit between T and R states according to the logic given by Table 10.1.

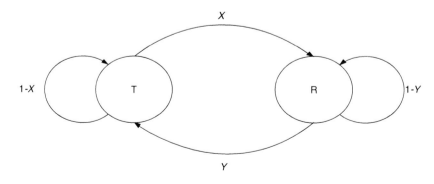

Figure 10.15 Markovian model for the NCSW ARQ. Reproduced by permission of © 2006 IEEE.

Table 10.1 State Transition Logic for $O(k)$. Reproduced by permission of © 2006 IEEE

$\{O(k-1), PC(k-1), N^{(1,\dots,M)}(k-1)\}$	$O(k)$
S_0:{T,G,G}	T
S_1:{T,G,B}	T
S_2:{T,B,G}	R
S_3:{T,B,B}	R
S_4:{R,G,G}	T
S_5:{R,G,B}	T
S_6:{R,B,G}	T
S_7:{R,B,B}	R

This table specifies the transition logic for an eight-state Markov model as shown in Figure 10.16. The corresponding transition probability matrix, $\mathbf{B}(8 \times 8)$, is given by

$$\mathbf{B} = \begin{bmatrix} 0 & 0 & 0 & 0 & \bar{r}\bar{V} & \bar{r}V & r\bar{V} & rV \\ 0 & 0 & 0 & 0 & \bar{r}U & \bar{r}\bar{U} & rU & r\bar{U} \\ q\bar{V} & qV & \bar{q}\bar{V} & \bar{q}V & 0 & 0 & 0 & 0 \\ qU & q\bar{V} & \bar{q}U & \bar{q}\bar{U} & 0 & 0 & 0 & 0 \\ 0 & 0 & 0 & 0 & \bar{r}\bar{V} & \bar{r}V & r\bar{V} & rV \\ \bar{r}U & \bar{r}\bar{U} & rU & r\bar{U} & 0 & 0 & 0 & 0 \\ q\bar{V} & qV & \bar{q}\bar{V} & \bar{q}V & 0 & 0 & 0 & 0 \\ qU & q\bar{U} & \bar{q}U & \bar{q}\bar{U} & 0 & 0 & 0 & 0 \end{bmatrix}, \tag{10.18}$$

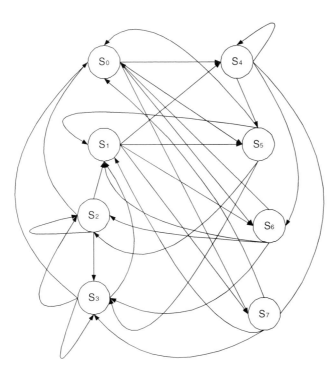

Figure 10.16 Markovian model for $\{O(k-1), PC(k-1), N^{(1,\dots,M)}(k-1)\}$. Reproduced by permission of © 2006 IEEE.

where r and q are the parameters of Markov model for the primary channel, $\bar{r} = 1 - r$, $\bar{q} = 1 - q$, $\bar{U} = 1 - U$, and $\bar{V} = 1 - V$. Let $\mathbf{P} = [p_{S_0}, \ldots, p_{S_7}]$ denote the steady state probability vector, where p_{S_i} is the steady state probability of being in state S_i in Figure 10.16. This vector can be obtained by solving a set of linear equations given by

$$\mathbf{P} \cdot \mathbf{B} = \mathbf{P}$$
$$p_{S_0} + \cdots + p_{S_7} = 1. \tag{10.19}$$

Having p_{S_i}, for $i = 0, \ldots, 7$, and Table 10.1, the parameters of the two-state Markov model for the NCSW protocol can be obtained by

$$X = \frac{p_{S_2} + p_{S_3}}{p_{S_0} + p_{S_1} + p_{S_2} + p_{S_3}}$$
$$Y = \frac{p_{S_4} + p_{S_5} + p_{S_6}}{p_{S_4} + p_{S_5} + p_{S_6} + p_{S_7}}. \tag{10.20}$$

Equation 10.20 completes the modelling of frame success/failure process as a two-state Markov model. Given this model for the NCSW ARQ scheme and (10.5), the throughput of the NCSW scheme can be obtained as

$$\eta_{NCSW} = \frac{Y}{Y + X}. \tag{10.21}$$

In the following subsection, we use (10.20) to obtain the average delay and delay jitter for the NCSW ARQ scheme.

10.7 Delay Analysis

Transmission of large frames over a wireless channel is not efficient due to high probability of frame error [22]. As shown in Figure 10.17, large data blocks from the upper layers of the communication protocol stack (that is, network layer) are usually broken down into smaller fragments to avoid retransmissions of large frames in case of transmission errors. This technique is called packet fragmentation, and has been adopted by some existing standards, such as the IEEE 802.11 MAC protocol.

Usually a sender node is allowed to capture the channel to continuously transmit all fragments of a single packet. Even in a contention based MAC protocol, the contention mechanism is designed to allow a node to transmit all the fragments of a packet as a burst before the next contention period begins. For example, in the IEEE 802.11 MAC protocol, as illustrated in Figure 10.18, the inter-frame spacing mechanism is designed to allow a sender to send a sequence of fragments without interruption. We define ARQ protocol *delay* as the time required to complete

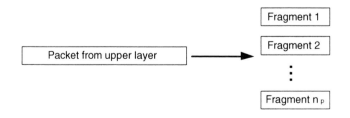

Figure 10.17 Packet fragmentation. Reproduced by permission of © 2006 IEEE.

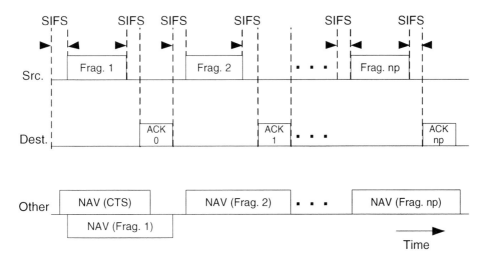

Figure 10.18 Transmission of a fragmented packet in the IEEE 802.11 MAC protocol. Reproduced by permission of © 2006 IEEE.

the transmission of all the fragments of a single packet from the network layer. Since we do not deal with any resource allocation problems in this chapter, our delay analysis does not include the queueing delay. Thus, the transmission delay of an ARQ protocol depends on the packet length and the number of retransmission trials. Without loss of generality, we assume that all packets from the network layer have identical length and are fragmented into n_p frames of identical length T_f. Thus, the analysis can be considered as a conditional one for a given value of n_p. For the case of variable packet length with a known probability distribution function (pdf), we can simply compute the expected value of the results. We also include the ACK/NAK handshake time in the frame length. In the best scenario, when there is no frame error, the entire transmission delay will be equal to $n_p T_f$. However, the average transmission delay is expected to be higher than this ideal value due to possible frame errors. Given the two-state Markov model in Figure 10.15, the average delay can be given by

$$t_p = \frac{X + Y}{Y} \cdot T_f. \tag{10.22}$$

Next, we introduce a Markov chain with absorbing state in order to obtain the delay jitter for the NCSW scheme. Let (q, r) be the parameters of the two-state Markov model in Figure 10.15 (we replace $[X, Y]$ with $[q, r]$ to simplify the notations). The transmission process of a single fragmented packet can be described by a $(2n_p + 1)$-state Markov chain with an *absorbing* state, as shown in Figure 10.19. At the beginning, the system is in $S_{n_p, T}$ where there are n_p fragments in the transmission queue. If the first transmission succeeds, the system moves to state $S_{(n_p-1), T}$; otherwise, it will move to state $S_{(n_p-1), R}$, where it retransmits the first fragment. This transition rule applies to all states, until the system moves into the absorbing state $S_{0, T}$, where all fragments of the packet in the transmission queue have been transmitted successfully. In order to make the mathematical manipulations simple, as shown in Figure 10.19, we assign integer indices from 0 to $2n_p$ to the system states. The initial and the absorbing states are given index numbers $2n_p$ and 0, respectively. All $S_{*, R}$ states are given odd numbers, and the $S_{*, T}$ states are assigned even numbers from 2 to $2n_p$. The $(2n_p + 1) \times (2n_p + 1)$ state transition probability matrix, $(\mathbf{\Pi})$, is given in (10.23), where the (i, j) element, denoted by p_{ij}, is the probability of transition from

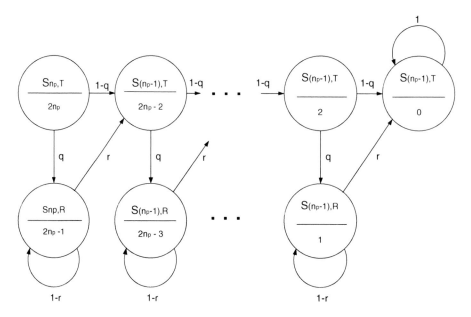

Figure 10.19 The Markovian model for the transmission process of a fragmented packet. Reproduced by permission of © 2006 IEEE.

state i to state j.

$$\Pi = \begin{bmatrix} 1 & 0 & 0 & 0 & \dots & 0 & 0 & 0 & 0 \\ r & 1-r & 0 & 0 & \dots & 0 & 0 & 0 & 0 \\ 1-q & q & 0 & 0 & \dots & 0 & 0 & 0 & 0 \\ 0 & 0 & r & 1-r & \dots & 0 & 0 & 0 & 0 \\ 0 & 0 & 1-q & q & \dots & 0 & 0 & 0 & 0 \\ \vdots & \vdots & \vdots & \vdots & \ddots & \vdots & \vdots & \vdots & \vdots \\ 0 & 0 & 0 & 0 & \dots & 0 & r & 1-r & 0 \\ 0 & 0 & 0 & 0 & \dots & 0 & 1-q & q & 0 \end{bmatrix} \tag{10.23}$$

This matrix has the structure of

$$\Pi = \left[\begin{array}{c|c} \mathbf{I} & \mathbf{0} \\ \hline \mathbf{V} & \mathbf{W} \end{array} \right]$$

where $\mathbf{I}$ is a 1×1 identity matrix, $\mathbf{W}$ is a $2n_p \times 2n_p$ matrix containing the probabilities of transition among non-absorbing states, $\mathbf{V}$ is a $2n_p \times 1$ matrix that represents the probabilities of transition from the non-absorbing states to the absorbing state $S_{0,T}$, and $\mathbf{0}$ is a $1 \times 2n_p$ zero matrix.

Let D_i be a random variable representing the number of transitions from a non-absorbing state i to the absorbing state in Figure 10.19. Delay jitter is the standard deviation of packet transmission delay, defined as

$$\sigma_p \triangleq \sqrt{\mathrm{E}[(T_p - t_p)^2]}. \tag{10.24}$$

We can rewrite (10.24) as

$$\sigma_p^2 = \mathrm{E}[(D_p - d_p)^2] \cdot T_f^2,$$

that is,

$$\sigma_p^2 = \mathrm{E}[(D_{2n_p} - d_{2n_p})^2] \cdot T_f^2. \tag{10.25}$$

Since

$$\mathrm{E}[(D_i - d_i)^2] = \mathrm{E}[D_i^2] - d_i^2, \tag{10.26}$$

we need to obtain $\delta_i^2 = \mathrm{E}[D_i^2]$.

Let d_j denote the mean value of D_j as the average number of steps required to move from a non-absorbing state j to the absorbing state in Figure 10.19. Starting from a non-absorbing state j, the system either moves to the absorbing state in one step, or gets to the absorbing state via one or more non-absorbing states. The probability of the first event is:

$$\mathbf{P}\{\text{go to the absorbing state in one step} \mid \text{current state is } j\} = p_{j0}.$$

The probability of the second event is:

$$\mathbf{P}\{\text{go to a non-absorbing state} \mid \text{current state is } j\} = \sum_{i=1}^{2n_p} p_{ji}.$$

If the latter case happens, the time of absorption is $(1 + D_i)$. Thus, we have

$$D_j = 1 \times p_{j0} + \sum_{i=1}^{2n_p} p_{ji}(1 + D_i),$$

and δ_i^2 is given by

$$\delta_i^2 = p_{j0} + \sum_{i=1}^{2n_p} \mathrm{E}[(1 + D_i)^2] p_{ji}.$$

$$= p_{j0} + \sum_{i=1}^{2n_p} (1 + 2d_i + \delta_i^2) p_{ji}$$

$$= \underbrace{p_{j0} + \sum_{i=1}^{2n_p} p_{ji}}_{\sum_{i=0}^{2n_p} p_{ji} = 1} + 2\sum_{i=1}^{2n_p} d_i p_{ji} + \sum_{i=1}^{2n_p} \delta_i^2 p_{ji}$$

$$= 1 + \sum_{i=1}^{2n_p} d_i p_{ji} + \sum_{i=1}^{2n_p} \delta_i^2 p_{ji}. \tag{10.27}$$

Organizing (10.27) into a matrix form, we have

$$
\begin{bmatrix} \delta_1^2 \\ \delta_2^2 \\ \vdots \\ \delta_{(2n_p-1)}^2 \\ \delta_{2n_p}^2 \end{bmatrix} = \begin{bmatrix} 1 \\ 1 \\ \vdots \\ 1 \\ 1 \end{bmatrix} + 2\mathbf{W} \begin{bmatrix} d_1 \\ d_2 \\ \vdots \\ d_{(2n_p-1)} \\ d_{2n_p} \end{bmatrix} +
$$

$$
+ \mathbf{W} \begin{bmatrix} \delta_1^2 \\ \delta_2^2 \\ \vdots \\ \delta_{(2n_p-1)}^2 \\ \delta_{2n_p}^2 \end{bmatrix} \tag{10.28}
$$

$$
\begin{bmatrix} \delta_1^2 \\ \delta_2^2 \\ \vdots \\ \delta_{(2n_p-1)}^2 \\ \delta_{2n_p}^2 \end{bmatrix} = (\mathbf{I} - \mathbf{W})^{-1} \begin{bmatrix} 1 \\ 1 \\ \vdots \\ 1 \\ 1 \end{bmatrix} +
$$

$$
+ 2(\mathbf{I} - \mathbf{W})^{-1}\mathbf{W} \begin{bmatrix} d_1 \\ d_2 \\ \vdots \\ d_{(2n_p-1)} \\ d_{2n_p} \end{bmatrix} \tag{10.29}
$$

Combining (10.25), (10.26) and (10.29), σ_p^2 can be obtained as

$$
\sigma_p = T_f \cdot \sqrt{(\delta_{2n_p}^2 - d_{2n_p}^2)}. \tag{10.30}
$$

10.8   Verification of Analytical Models

We simulate a single hop ad-hoc network with one pair of sender-receiver nodes and varying number of neighbor nodes, as shown in Figure 10.9. The channels among the nodes are generated by the Rayleigh fading model. The impacts of path loss and shadowing are not considered due to their very slow variations compared with the activities of link layer. The Rayleigh fading channels are simulated by low-pass Finite Impulse Response (FIR) filtering of two white Gaussian random processes. The simulation parameters for the fading channels are given in Table 10.2. The quality

Table 10.2   Simulation Parameters. Reproduced by
permission of © 2006 IEEE

Parameter	Value
Carrier frequency	2400 MHz
Relative speed of mobile nodes	5 Km/h
Sampling frequency	8000 samples per second
Frame duration	5 ms

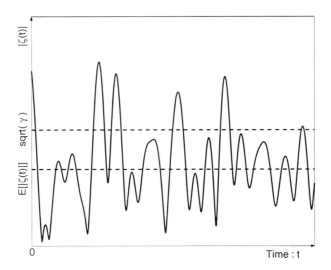

Figure 10.20   Concept of fading margin. Reproduced by permission of © 2006 IEEE.

of channels in terms of fading margin, that is, the ratio of fading threshold in (10.1) over the
mean value of fading envelop is represented by

$$L = \frac{\sqrt{\gamma}}{\mathrm{E}[|\zeta(t)|]}. \tag{10.31}$$

The mean value of the fading channels are normalized to unit; then $L = \sqrt{\gamma}$. In other words,
variations in the value of fading margin is translated to variations in the channel quality. As
illustrated in Figure 10.20, if the value of fading margin is increased, the channel quality will
be below the threshold value more frequently. This means that as the value of the fading margin
is increased, there will be more frame errors, which translates to a worse channel quality.

The frame duration is set to be 5 ms, which is reasonable for many wireless data networks.
Perfect ACK/NAK information on the feedback channels is assumed to be available for the
whole cooperation group right after a frame transmission.

We present both the simulation and analytical results for the throughput, average delay and
delay jitter of the NCSW scheme.

10.8.1 Throughput

To observe the impact of a small number of cooperative nodes on the system throughput, the SW
and the NCSW schemes with only two neighbour nodes are simulated. As shown in Figure 10.21,

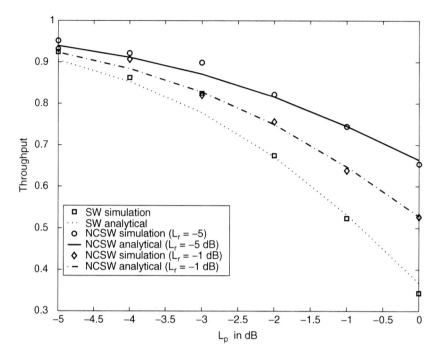

Figure 10.21 Throughput vs. fading margin of the primary channel. Reproduced by permission of © 2006 IEEE.

the results are plotted against the variations of the quality of the primary channel denoted by L_p. The quality of all the interim and the relay channels are assumed to be identical and denoted by L_r. To demonstrate the impact of the variations in the quality of the interim/relay channels, throughput of the NCSW protocol is plotted for two different values of L_r, namely $-5\,dB$ and $-1\,dB$. The results in Figure 10.21 show that with cooperation of only two neighbour nodes, throughput of the NCSW scheme can be improved up to 30%, depending on the quality of the interim/relay channels. Comparison of the analytical and the simulations result also demonstrates the accuracy of the proposed analytical model for the system throughput.

To investigate the impact of the number of neighbour nodes on the protocol throughput, the fading margin of the primary channel is set to $L_p = -1\,dB$. At this relatively low quality of the primary channel, we can clearly observe the impact of the number of neighbour nodes on the system throughput. Simulations are performed for two different fading margins for the relay/interim channels, namely $L_r = -5\,dB$ and $L_r = -1\,dB$. As shown in Figure 10.22, when the number of the cooperative nodes increases, the system throughput approaches a saturation level depending on the quality of the primary and the interim/relay channels. If the qualities of the interim/relay channels are good, having only one or two neighbour nodes can significantly improve the system performance; however, when the qualities of the interim/relay channels are poor, more neighbour nodes are required to achieve the same level of performance gain. Saturation of system throughput is also expected. In fact, regardless of the number of neighbour nodes or their channel qualities, individual frame errors cannot be avoided. However, cooperation of the neighbour nodes can reduce the durations of error bursts.

Another important issue is to separately investigate the impacts of variations in the quality of the interim and the relay channels as shown in Figure 10.23. The results of this analysis can be helpful in deciding when a node should cooperate. Such a decision can be based on the average quality of the channels from the sender and to the receiver nodes. To keep the

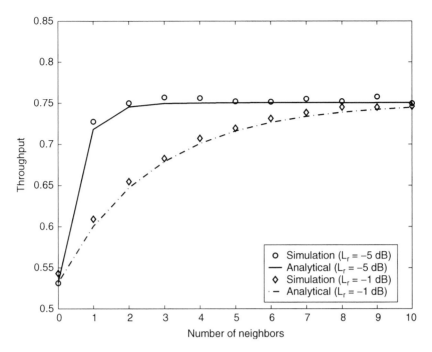

Figure 10.22 Throughput vs. number of the neighbour nodes ($L_p = -1\,dB$). Reproduced by permission of © 2006 IEEE.

system setup consistent with the previous simulations, we obtain the system throughput for only two neighbour nodes. We set the fading margin of the primary channel at a fixed level of $L_p = -1\,dB$, and perform the simulations twice. For the first run, we set the fading margin of the relay channels, denoted by L_{relay}, at a fixed level of $-5\,dB$, and let the fading margin of the interim channels, denoted by $L_{interim}$, vary from $-5\,dB$ to $0\,dB$. In the second run, for a fixed fading margin of the interim channels at $-5\,dB$ we let the fading margin of the relay channels vary from $-5\,dB$ to $0\,dB$.

It is noticed that we do not normalize the throughput to the total transmission power. Since it is very difficult to implement a distributed power control scheme, we assume that each node transmits with a fixed power. Therefore, our results are not for the study the tradeoff between the total transmission power and the achievable throughput gain. The purpose is to show that even if the total transmission power is fixed through a power control mechanism, there will be still a significant throughput gain due to the utilization of spatial diversity.

10.8.2 Average Delay and Delay Jitter

We simulate the SW and the NCSW protocols with only two cooperating neighbour nodes for our delay analysis. For the NCSW scheme, we assume moderate interim and relay channel qualities at a fading margin of $-2.5\,dB$. The other channel parameters are the same as in Table 10.2. The fading margin for the primary channel is varied from $-5\,dB$ (good channel quality) to $0\,dB$ (bad channel quality). Packets from the upper layer, that is, network layer, are fragmented into 20 frames of 5 ms duration (including ACK/NAK); therefore, if there is no transmission error, it will take 100 ms for all the fragments of a single packet to be transmitted. However, as shown in Figure 10.24, on average, it will require more than 100 ms due to frame errors. As shown in the figure, for a small number of cooperative nodes with moderate interim/relay channel qualities, the

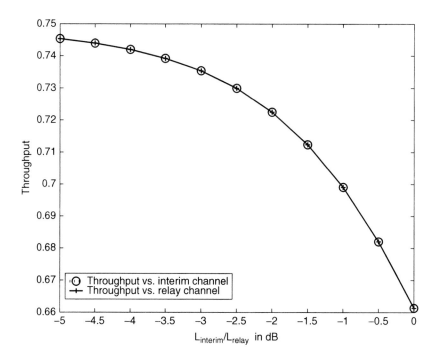

Figure 10.23 Throughput vs. the fading margin of the interim/relay channels ($L_p = -1\,dB$). Reproduced by permission of © 2006 IEEE.

NCSW protocol significantly outperforms the SW protocol in terms of transmission delay. For example, when the primary channel is poor ($L_p = 0\,dB$), the average delay for the SW scheme is about 265 ms; however, for the same channel condition, the NCSW protocol can reduce the average delay to 160 ms. Thus, due to node cooperation, the average delay is reduced by 60% when the primary link is experiencing poor condition.

With the same simulation setup, we also investigate the delay jitter caused by the transmission errors in both schemes. Once again, by comparing the results given in Figure 10.25, we observe significant improvement in the performance in terms of reduced delay jitter. For example, when the primary channel is in poor condition, the delay jitter is reduced from 140 ms to 20 ms, which is equivalent to 85% reduction. Furthermore, the exponential growth in the delay jitter is reduced to a slow linear growth. The improvement of the delay jitter is more remarkable than that of the delay and the throughput. Intuitively, since node cooperation reduces the negative impact of long and varying error bursts, this significant improvement in the jitter is expected. This is a very important advantage of the proposed scheme for real-time multimedia applications, such as voice over IP (Internet Protocol), where reducing the delay jitter is a challenging task in providing the desired quality of service.

10.9 Discussion of the Related Works

In this section, we discuss some of the very recently related publications.

In [23] and [24], a cooperative hybrid-ARQ scheme has been proposed. A system model with a sender and a receiver pair with several intermediate nodes have been considered. The intermediate nodes are geographically located between the sender and the receiver nodes. The hybrid-ARQ scheme combines Forward Error Correction (FEC) with a conventional retransmission scheme. In

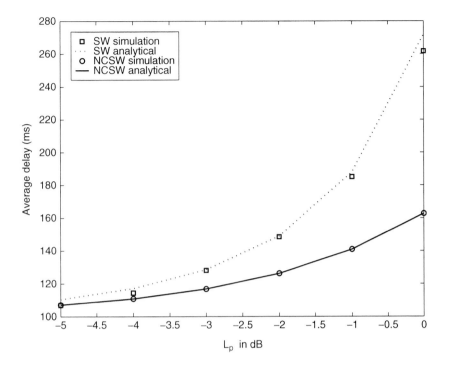

Figure 10.24   Delay vs. the fading margin of the primary channel.

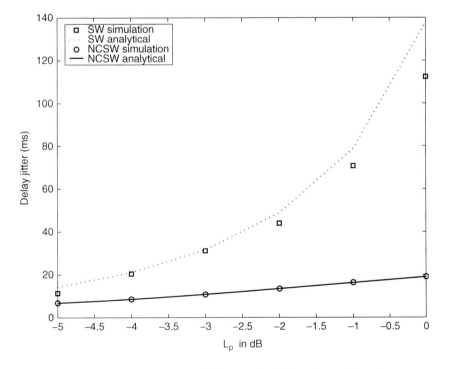

Figure 10.25   Jitter vs. the fading margin of the primary channel.

the previous hybrid-ARQ scheme, the original frame is coded into several blocks with incremental redundancy. After the sender sends the first block, if the receiver is not able to decode the frame, another block with more information is transmitted. The sender sends more blocks to increase information to the receiver node about the original frame until the receiver decodes the frame successfully. In [23]–[24], the hybrid-ARQ is extended by allowing the intermediate nodes to transmit the successive code blocks. In addition performance analysis is provided using simulations. In [25], a truncated cooperative ARQ scheme has been proposed. In the scheme, the source node and the relay nodes utilize orthogonal space-time block codes for packet retransmission. Different from our work, uncorrelated frame errors are considered. A preliminary model for throughput analysis also has been proposed. However, the model is yet to be verified. In [26], cooperation through ARQ for the uplink of an infrastructure-based network has been studied. A multiple access channel with a single antenna sources and a multiple antenna destination is considered. A system with only two nodes has been studied. For this setup, an upper-bound on the achievable diversity-multiplexing-delay tradeoff has been obtained. In [27], an ARQ scheme for a pair of sender and receiver nodes with a single relaying node has been proposed. The source is using an ARQ retransmission protocol to send data to the relay and the destination. If the relay is able to decode, both the relay and the source send the same data to the destination providing additional gains. The achievable diversity, multiplexing gain and delay tradeoff for a high signal to noise ratio regime has been obtained.

10.10 Summary

In this chapter, we discussed potential benefits of exploiting cooperative diversity in the upper layers of communication protocol stack. The work presented in this chapter is certainly far from being a comprehensive discussion of all possibilities. However, by means of an example, cooperative retransmission, we highlighted the importance and feasibility of cooperative communication protocols. The combination of our analytical and simulation based study of the behaviour of cooperative retransmission has demonstrated that throughput and delay performance of the link level retransmission can be significantly improved when cooperation among autonomous nodes is exploited. In particular, cooperation, fundamentally enhances the delay performance of link level retransmission. This is very important for the future real-time multimedia applications where delay jitter, in particular, will be a decisive factor.

For future work, based on what we have learned from the current work, we believe that there is a good possibility of exploiting cooperative techniques in more advanced mechanisms in link, network, even possibly the application layer protocols. In addition, there are some interesting and important issues regarding the integration of the cooperation with the standard systems such as those used in ad-hoc, cellular and wireless local area networks. This will involve the study of the integration of cooperative protocols with contention based MAC schemes such as the IEEE 802.11x standard series.

10.11 Acknowledgement

This work has been partially supported by a research grant from the Engineering and Physical Research Council (EPSRC) of United Kingdom under India UK Advanced Technology Centre (IU-ATC) Project. The work done in the University of Waterloo has been supported by the Natural Sciences and Engineering Research Council (NSERC) of Canada.

References

[1] E. V. D. Meulen, "Three-terminal communication channels," *Advanced Applications of Probability*, vol. 3, 1971, pp. 120–154.

[2] O. Gurbuz and E. Ayanoglu, "A transparent ARQ scheme for broadband wireless access," *IEEE Wireless Communications and Networking Conference (WCNC 2004)*, vol. 1, March 2004, pp. 423–429.

[3] S. Bhandarkar, N. E. Sadry, A. L. N. Reddy and N. H. Vaidya, "TCP-DCR: a novel protocol for tolerating wireless channel errors," *IEEE Transactions on Mobile Computing*, vol. 4, no. 5, Sept.-Oct. 2005, pp. 517–529.

[4] J. A. Afonso and J. E. Neves, "Fast retransmission of real-time traffic in HIPERLAN/2 systems," in *Proc. Advanced Industrial Conference on Telecommunications 2005*, July 2005, pp. 34–38.

[5] L. Fang, K. Jian, W. Wenbo and L. S. Yuan'an Liu, "Fast and reliable two-layer retransmission scheme for CDMA systems," *in Proc. IEEE Eighth International Symposium on pread Spectrum Techniques and Applications 2004*, Sept. 2004, pp. 769–773.

[6] C. Min and W. Gang, "Multi-stages hybrid ARQ with conditional frame skipping and reference frame selecting scheme for real-time video transport over wireless LAN," *IEEE Transactions on Consumer Electronics*, vol. 50, no. 1, Feb 2004, pp. 158–167.

[7] J. Winters, "On capacity of radio communication systems with diversity in a Rayliegh fading environment," *IEEE Journal on Selected Areas*, vol. 5, June 1987, pp. 871–878.

[8] I. E. Telatar, "Capacity of multi-antenna Gaussian channels," *European Transaction of Telecommunications*, vol. 10, Nov. 1999, pp. 585–595.

[9] V. Tarokh, N. Seshadri and A. R. Calderbank, "Space-time codes for high data rates wireless communications: Performance criterion and code construction," *IEEE Transactions on Information Theory*, vol. 44, Feb. 1998, pp. 744–765.

[10] A. Sendonaris, E. Erkip and B. Aazhang, "Increasing up-link capacity via user cooperation diversity," *IEEE International Sympousium on Information Theory*, Cambridge, MA, Aug. 1998.

[11] A. Sendonaris, E. Erkip and B. Aazhang, "User cooperation diversity, Part I: System description Sendonaris," *IEEE Transactions on Communications*, vol. 51, no. 11, Nov. 2003, pp. 1927–1938.

[12] A. Sendonaris, E. Erkip and B. Aazhang, "User cooperation diversity, Part II: Implementation aspects and performance analysis," *IEEE Transactions on Communications*, vol. 51, no. 11, Nov. 2003, pp. 1939–1948.

[13] S. N. Diggavi, N. Al-Dhahir, A. Stamoulis and A. R. Calderbank, "Great Expectations: the value of patial diversity in wireless networks," *Proc. IEEE*, vol. 92, no. 2, Feb. 2004, pp. 219–270.

[14] M. Janani, A. Hedayat, T. E. Hunter and A. Nosratinia, "Coded cooperation in wireless communications: space-time transmission and iterative decoding," *IEEE Transactions on Signal Processing*, vol. 52, no. 2, Feb. 2004, pp. 362–371.

[15] G. L. Stuber, "Principles of Mobile Communication," Kluwer Academic Publishers, 2001.

[16] J. N. Laneman and G. W. Wornell, "Distributed space-time-coded protocols for exploiting cooperative diversity in wireless networks," *IEEE Transactions on Information Theory*, vol. 49, no. 10, Oct. 2003, pp. 2415–2425.

[17] H. S. Wang, "On verifying the first-order Markovian assumption for a Rayleigh fading channel model," *IEEE Transactions on Vehicular Technology*, vol. 45, no. 2, May 1996, pp. 353–357.

[18] M. Zorzi, R. R. Rao and L. B. Milstein, "ARQ error control for fading mobile radio channels," *IEEE Transactions on Vehicular Technology*, vol. 46, no. 2, May 1997, pp. 445–455.

[19] C. Pimentel, T. H. Falk and L. Lisboa, "Finite-state Markov modeling of correlated Rician fading channels," *IEEE Transactions on Vehicular Technology*, vol. 53, no. 5, Sept. 2004, pp. 1491–1501.

[20] H. Kong and E. Shwedyk, "A hidden Markov model (HMM)-based MAP receiver for Nakagami fading channels," in *Proc. IEEE International Symposium on Information Theory*, Sept. 1995, pp. 210.

[21] C. H. C. Leung, Y. Kikumoto and S. A. Sorensen, "The throughput efficiency of the Go-Back-N ARQ scheme under Markov and related error structure," *IEEE Transactions on Communications*, vol. 3, no. 2, Feb. 1988, pp. 231–233.

[22] A. R. Parsad, Y. Shinohara and K. Seki, "Performance of hybrid ARQ for IP packet transmission on fading channel," *IEEE Transactions on Vehicular Technology*, vol. 48, no. 3, May 1999, pp. 900–910.

[23] B. Zhao and M. C. Valenti, "Practical relay networks: a generalization of hybrid-ARQ," *IEEE Journal on Selected Areas in Communications*, vol. 23, no. 1, Jan. 2005, pp. 7–18.

[24] M. C. Valenti and Bin Zhao, "Hybrid-ARQ based intra-cluster geographic relaying," *Military Communications Conference MILCOM 2004*, Nov. 2004, vol. 2, pp. 805–811.

[25] L. Dai and K. B. Letaief, "Cross-layer design for combining cooperative diversity with truncated ARQ in ad-hoc wireless networks," in Proc. GLOBECOM '05, Dec. 2005, vol. 6, pp. 3175–3179.

[26] Y. H. Nam, K. Azarian, H. El Gamal and P. Schniter, "Cooperation through ARQ," *in Proc. IEEE Workshop on Signal Processing Advances in Wireless Communications 2005*, pp. 1023–1027.

[27] T. Tabet, S. Dusad and R. Knopp, "Achievable diversity-multiplexing-delay tradeoff in half-duplex ARQ relay channels," *in Proc. International Symposium on Information Theory 2005*, June 2005, pp. 1828–1832.

11

Cooperative Inter-Node and Inter-Layer Optimization of Network Protocols

D. Kliazovich[1], F. Granelli[2] and N. L. S. da Fonseca[3]

[1]*University of Luxembourg, Luxembourg*
[2]*Department of Information Engineering and Computer Science, University of Trento*
Trento, Italy
[3]*Institute of Computing, State University of Campinas, Brazil*

11.1 Introduction

End-to-end Quality of Service (QoS) provisioning and network performance optimization under changing network environments imply challenges with solutions demanding coordination of protocols and nodes and joint optimization of the protocol parameters. The need for such coordination is a direct consequence of the designs of the TCP/IP and ISO/OSI architectures which rely on the separation of scope and objectives of the various layers. Such design philosophy implies the absence of coordination among the protocols operating at different layers.

Moreover, the widespread diffusion of the TCP/IP reference model has only strengthened interaction among protocols allowing only evolutionary rather than a revolutionary approach.

In this scenario, dynamic adjustment and optimization of the parameters of the protocol stack through inter-node and inter-layer cooperation represents a feasible option to support fine-tuning of parameters and enhancement of network performance.

The introduction of cooperation among network nodes to enable flexible adaptation of operating parameters was envisaged by J. Mitola III [1] with the introduction of the concept of *cognitive radio* – aimed at providing efficient spectrum sharing by cooperative and adaptive access to the available transmission resources. Actually, cognitive radio is concerned with the tuning of parameters at physical and link layers as well as optimization goals at these layers. The broader concept of cognitive network was introduced to cope with system-wide goals and cross-layer design and can be considered a generalization of the cognitive radio concept. A cognitive network involves cognitive algorithms, cooperative networking and cross-layer design in order to provide dynamic configuration and real-time optimization of communication systems.

Cooperative Networking, First Edition. Edited by Mohammad S. Obaidat and Sudip Misra.
© 2011 John Wiley & Sons, Ltd. Published 2011 by John Wiley & Sons, Ltd.

The main contribution of the chapter is an analysis of cooperative inter-node and inter-layer networking issues and solutions from an architectural point of view. Moreover, a framework for cooperative configuration and optimization of communication protocols performance is introduced.

The proposed architecture is concerned not only with the initial setup of protocol parameters, but also with their timing, reconfiguration and optimization during the network runtime. The architecture requires the introduction of a cognitive plane operating 'in parallel' with the protocol layers and which is capable of monitoring each protocol layer parameter as well as controlling them by issuing configuration commands. For the duration of the optimization process, the cognitive plane monitors the feedback from all the protocol layers, which includes reports on the values of target parameters. For example, the metric at the physical layer can be the obtained data rate, while at the application layer the feedback metric can be the perceived quality of real-time multimedia applications.

11.2   A Framework for Cooperative Configuration and Optimization

11.2.1   Tuning TCP/IP Parameters[1]

The TCP/IP reference model [2] is the 'de facto' standard for communication on the Internet. It contains a large variety of protocols, whose parameters need to be adequately set for proper functioning. Table 11.1 presents a snapshot of the most widely used protocols, their main configuration parameters and corresponding performance metrics the parameters' affect.

Application layer provides the environment for supporting and running user applications. Configurable parameters and quality metrics at this layer depend on the nature of applications. For File Transfer (FTP) applications [3], a configurable parameter could be the number of parallel connections and the main quality metric is the file transfer goodput. For Voice over IP (VoIP) applications controlling coding rate, coding interval and Forward Error Correction (FEC) [4] impact the voice quality commonly expressed using Mean Opinion Score (MOS) metric [5].

For video streaming applications, streaming bitrate, framerate and keyframe interval determine the quality of video flow perception. High bitrate values allow video transmissions with high resolutions; high framerate values improve perception of the video samples involving high motions, while shorter keyframe intervals improve decoding capabilities in the presence of frame losses or transmission errors.

Transport layer is generally represented by Transmission Control Protocol (TCP) [6] and User Datagram Protocol (UDP) [7] protocols. While UDP is lightweight, with the main task of providing differentiation of IP datagrams between different port numbers, TCP implements complex mechanisms to achieve reliable data transfer, including ARQ, flow control and congestion control.

Indeed, congestion window of TCP connection is the main mechanism controlling outgoing rate and is a key to window evolution strategy. Controlling the aggressiveness of congestion window increase factor (α) and decrease factor (β) allows adjusting the tradeoff between network utilization, protocol fairness and the level of network congestion improving data goodput performance.

Network layer is responsible for routing packets across interconnected networks or network domains. Hence the main performance metric is the quality of the selected route expressed as the number of hops or end-to-end delay.

---

[1] Excerpt taken from D. Kliazovich, N. Malheiros, Nelson L. S. da Fonseca, F. Granelli, and E. Madeira, "CogProt: A Framework for Cognitive Configuration and Optimization of Communication Protocols," 2nd International Conference on Mobile Lightweight Wireless Systems. Reproduced by permission of © 2010 ICST.

Table 11.1 TCP/IP Protocols and Parameters. Reproduced by permission of © 2009 IEEE

Protocol Layer	Protocol Name	Parameter	Metric
Application	File Transfer (FTP)	Number of parallel transfers	FTP Goodput
	VoIP	• Coding rate • Coding interval • FEC strength	• Mean Opinion Score (MOS)
	Video Streaming	• Streaming bitrate • Frame Rate • Keyframe interval	• Peak signal-to-noise ratio (PSNR) • Structural Similarity (SSIM) • Video Quality Measurement (VQM)
Transport	Transmission Control Protocol (TCP)	• Congestion window (w) • Slow start threshold (ssthreshold) • Aggressiveness of window increase (α) and decrease (β) • Protocol version	• Flow Goodput
Network	Routing	• Routing type	• Route setup delay • End-to-end delay
Link	MAC	• Contention window (cwnd) • Fragmentation threshold (MTU) • Retransmission scheme	• Data rate provided to higher layers • Medium access delay
Physical	PHY	• Transmit rate, power • Type of modulation, coding • Frequency channel	• Transmission rate • Bit error rate

Link layer serves network layer requests and controls the access to the actual transmission medium. Most of the controllable parameters in this layer are determined by the communication technology in use. In Career Sense Multiple Access (CSMA) protocols, such as WiFi IEEE 802.11 [8] or Ethernet IEEE 802.3 [9], the main tunable parameters correspond to the size and evolution of the contention window, as well as the retry limit, which corresponds to the maximum number of retransmission attempts taken at the link layer before a packet is discarded.

Physical layer parameters are defined by the nature of the transmission medium. Wireless access technologies can provide control over the power level of the transmitted signal, choice of the type of modulation and frequency channel allocation. Physical layer performance can be defined in terms of the data rate achievable, as well as Bit Error Rate (BER) achieved for the transmitted bit stream.

As one can see, several parameters exist in the TCP/IP stack that can be tuned, and, even more important, some of them can impact overall applications performance. For instance, the parameters governing the TCP transmission window and the initial value of the window have great influence on the performance, especially in networks with high bandwidth delay product [10].

11.2.2 Cooperative Optimization Architecture

Figure 11.1 outlines the main functional blocks of an architecture enabling cooperative optimization of the values of protocol parameters. To some extent, the proposed architecture is a

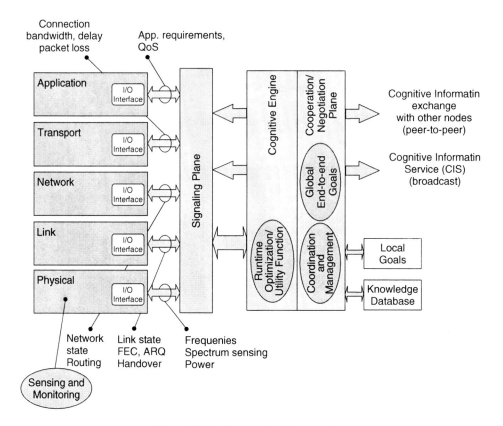

Figure 11.1   Cooperative framework.

potential instantiation of the concept of cognitive network under the constraint of enforcing given by interoperability with existing TCP/IP stack.

To operate with a standard protocol stack, each protocol layer is enhanced with a small software module which should be able to both obtain information internal to the specific layer (observation) or to tune its internal parameters (action). The information sensed at different protocol layers is delivered to the cognitive plane implemented at the cognitive node. The cognitive plane performs data analysis and decision making processes.

Results of data analysis could lead to information classified as knowledge which is storable in the local knowledge database.

The main task of the *cognitive engine* at every node is the optimization of different protocol stack parameters in order to converge to an optimal operational point given the network condition. The operational point can be expressed by a utility function that combines reports from running applications as well as other layers of the protocol stack. For that, cognitive adaptation algorithms include phases such as observation, data analysis, decision-making and action.

The decisions made by the cognitive engine at the node aim to optimize the protocol stack performance and are driven by the goals specified in the local database. The scope of these goals is local (at node level). Most of them are generated by the demands and the QoS requirements of user applications running at a given cognitive node.

Global optimization goals are defined on an end-to-end basis. The achievement of these goals requires cooperative actions from different network nodes which are implemented using the *cooperation/negotiation plane* operating closely with the cognitive engine to achieve the target goals.

While goals and knowledge databases are directly connected to the cognitive plane of the node and allow instant information exchange, the cognitive plane communication with the protocol stack is performed by the *signaling plane*. The signaling plane is responsible for providing a proper way for the delivery of signaling information delivery. Depending on the signaling type required, for instance, indication of parameter values, signaling threshold violation, or a callback-like indication, different signaling methods are required.

The signaling plane allows information exchange not only between the cognitive engine and different protocol layers of a single node but also provides two interfaces for communication with other network nodes, for enabling the exchange of parameters' values or targeted end-to-end optimization goals. One interface operates on a peer-to-peer basis which allows information exchange between any two nodes of the network in a distributed manner. An alternate (or complementary) one, called *Cognitive Information Service (CIS)*, corresponds to a network broadcast channel where information inserted by a given node is heard by all the nodes of the network segment. CIS signaling has obvious scalability limitations. Because of that, it is mainly used in well-defined parts of the network with limited number of nodes, such as a WiFi cell.

The cooperation and negotiation plane is responsible for harvesting cognitive information available at other network nodes, filtering and managing them in a distributed manner. Information harvesting can either be scheduled or be pursued by using instant requests or interrupts. Moreover, information could be node-specific or specific to a particular data flow.

The analysis of information gathered from cognitive nodes helps the cognitive engine to construct global knowledge and goals. Upon every adjustment, such information is reported back to cognitive nodes, so that they can adjust their appropriate local databases and, as a result, their behaviour.

A main characteristic of the cognitive network architecture is scalability, assured by the use of a combination of centralized (at the node level) and distributed (at the network level) techniques. In particular, at node level, the core cognitive techniques (such as data analysis, decision making and learning) are concentrated in the cognitive planes of the nodes and implemented in a centralized manner. Furthermore, observation and action software add-ons to the protocol layers serve only as instruments and cognitive planes are typically 'non-intelligent'. Distributing cognitive process among protocol layers (especially the learning and decision making functions) would require complex algorithms for synchronization and coordination between intra-layer cognitive processes. Alternatively, it seems that a single centralized cognitive process at node level brings a simpler solution, while implementation of cognitive process at network layer must be distributed or clustered implemented.

11.3 Cooperative Optimization Design

11.3.1 Inter-Layer Cooperative Optimization[2]

The cooperative optimization framework presented in this section aims at supporting dynamic configuration and optimization of communication protocols. It provides a way for network elements to adapt their configuration and protocol stack parameters in order to constantly adapt the values of protocol parameters to changing network conditions. The process of search for optimal

---

[2] Excerpt taken from D. Kliazovich, N. Malheiros, Nelson L. S. da Fonseca, F. Granelli, and E. Madeira, "CogProt: A Framework for Cognitive Configuration and Optimization of Communication Protocols," 2nd International Conference on Mobile Lightweight Wireless Systems. Reproduced by permission of © 2010 ICST.

setup of protocol parameters is performed by using cognitive algorithms [11] and by sharing information among network nodes.

The proposed approach is based on the cooperative architecture presented in the previous section and it extensively relies on quality feedback loops as well as on commands allowing the control of internal to the protocol parameters. The core idea is to enable each node to randomly select minor variations of some parameters, test them and use the information to identify the best parameter setting given the operating context.

The main task of the cognitive plane is the adaptation of different protocol stack parameters in order to converge to an optimal operational point given the network state. This way, the cognitive adaptation algorithms include phases such as data analysis, decision-making and action, as illustrated in Figure 11.2.

One of the main design requirements for the presented cooperative framework is to provide cognitive adaptation with minimal changes in the protocol stack. In the proposed approach, each protocol parameter P is expressed in terms of its default value P_{def} and its operation range $[P_{min}, P_{max}]$. The operation of the protocol is initiated with parameter P set to its default value. Then, the cognitive mechanism begins searching for optimal P values.

At the end of a given interval I, the cognitive mechanism measures using a defined quality metric and stores the obtained performance from the current value of P accordingly. Then, the mechanism selects a value of P that provides the best performance. That value is assigned to the mean of a normally distributed random number generator. Finally, a new value for P is chosen in the range $[P_{min}, P_{max}]$ using the random number generator. The initial mean for the number generator is P_{def}. This loop continuously adjusts the mean of the normal distribution to the value of P that provides the best performance under the current network scenario. The mean of the normal distribution converges to the best P value for the current network state. The standard deviation assigned to the normal distribution affects the aggressiveness of the mechanism in trying new values of P at each interval I.

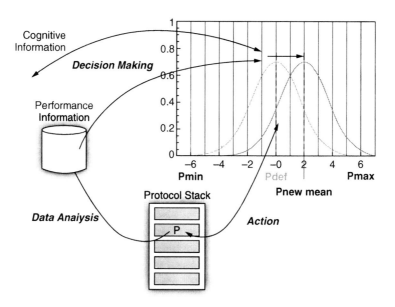

Figure 11.2 Cognitive adaptation algorithm. Reproduced by permission of © 2010 ICST.

11.3.2   Inter-Node Cooperative Optimization

This section extends the previous section by adding the dimension of inter-node cooperation which allows network nodes to exchange available cognitive information as well as to provide the means for making joint decisions.

Similar to the signaling plane, the Cognitive Information Service (CIS) provides the means for cognitive information exchange and its aggregation among network nodes.

Figure 11.3 shows the CIS implementation in a corporate network segment. CIS servers may become a bottleneck if overloaded since they provide centralized solutions for cognitive information management and aggregation. Consequently, they should be used in well-defined network segments with a limited number of nodes.

Taking into account that the cognitive communications in the direction from the CIS server to the cognitive nodes are often point-to-multipoint, they can be implemented either at the IP level using broadcast protocols or at the link layer. The layer-2 implementation will bring efficiency in cognitive information exchange and lead to the reduction of signaling overhead.

The configuration offered by CIS to a new node will not necessarily be optimal since the ideal configuration was inferred without information on running applications, traffic demands and the peers the node is willing to communicate with. However, such configuration can potentially offer better startup performance than using fixed default values for the protocols, as is currently done for the protocols of the TCP/IP stack.

Depending on the nature and requirements for information exchange, three different signaling methods can be used: in-band signaling, on-demand signaling and broadcast signaling.

In-band signaling is the most effective signaling method from the point of view of overhead reduction. Cognitive information can be encapsulated into ongoing traffic flows, for example into optional packet header fields [9], and delivered without waste of bandwidth resources.

Due to low overhead, in-band implementation of CIS is best suited for networks with wireless technologies used for access networks such as WiFi, WiMAX and cellular, where the bottleneck of the end-to-end connection is typically at the wireless link.

Another advantage of in-band signaling is that cognitive information can be associated with the portion of application data it is delivered with.

However, the main shortcoming of in-band signaling is the limitation of signaling to the direction of the packet flow, making it not suitable for cognitive schemes requiring instant communication between nodes having no ongoing data exchange.

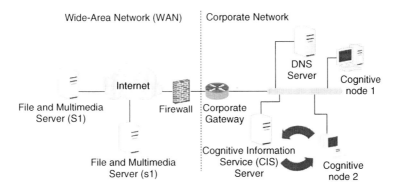

Figure 11.3   Cognitive Information Service in cooperative network segment. Reproduced by permission of © 2009 IEEE.

On-demand signaling method operates on a request-response basis and can be complementary to in-band signaling. It is designed for cases requiring instant cognitive information delivery between network nodes. Cognitive information becomes available at the requesting node following round-trip time delay, which makes it well suited for Wired/Wireless LAN scenario.

One of the core signaling protocols considered in on-demand signaling is the Internet Control Message Protocol (ICMP). Generation of ICMP messages is not constrained by a specific protocol layer and can be performed at any layer of the protocol stack. However, signaling with ICMP messages involves operation with heavy protocol headers (IP and ICMP), checksum calculation, and other procedures which increase processing overhead.

Broadcast signaling method allows point-to-multipoint cognitive information delivery from CIS server to the network nodes located in the same segment, while keeping low overhead. Broadcasting is especially suited for wireless networks following cellular organization.

Cognitive information is encapsulated into a beacon periodically broadcast by wireless gateways (access points or base stations), and thus fits scenarios where cognitive information delivery is tolerant to delays and can be performed at regular intervals.

11.4   A Test Case: TCP Optimization Using a Cooperative Framework

11.4.1   Implementation

In order to present the benefits of the proposed cooperative optimization framework we extended the Network Simulator (ns2) [12] with the required functionalities.

Simulated network topology is presented in Figure 11.4. It consists of four cognitive nodes S running intra-layer cognitive engine and CIS server performing inter-node cognitive operations all connected using 100 Mb/s, 0.1 ms links in a star topology centred at router R1. Such connectivity aims at mimicking operation of Ethernet network segment. Similar configuration is followed by the destination nodes D, which do not implement any cognitive functionality.

Routers R1 and R2 are connected with four links with propagation delays of 10 ms, 50 ms, 100 ms and 200 ms, and Packet Error Rates (PERs) of 0.0, 0.05, 0.1 and 0.15.

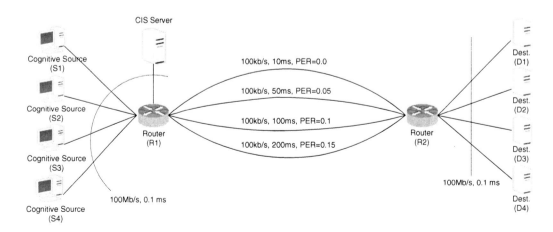

Figure 11.4   Simulated topology. Reproduced by permission of © 2009 IEEE.

A maximum of four flows can be started in the simulated topology between nodes corresponding S and D. For example, the first flow is initiated between S1 and D1 and flows through L1 link.

Different values of link PERs will require different window increase strategies, controlled by the parameter α, so that optimal performance can be achieved. High values of α are expected to bring better throughput for high PERs. However, in the case of no errors, high values of α will lead to multiple congestion-related losses and throughput degradation due to retransmission.

Different RTTs are designed to influence the choice of both α and β parameters through different time required for S nodes to react to congestion- or link-related losses.

The Network Simulator (ns2) module that implemented the cooperative framework focused on the cognitive adaptation of the main parameters controlling TCP protocol steady state behaviour, that is, speed of window increase α and aggressiveness of multiplicative decrease β. The main performance feedback metric is the end-to-end TCP goodput which is the amount of data successfully delivered to the receiver.

At the beginning of each flow, α and β are set to their default values 1 and 0.5, respectively. Then, right after the end of the TCP slow start phase, a timer is started for guiding the cognitive engine implemented inside each network node which corresponds to intra-layer cognitive functionality. Whenever the timer expires, intra-layer cognitive engine performs the steps defined by Algorithm 1. First, it computes, analyzes and stores the throughput value. Then, it selects α and β values corresponding that lead to optimal throughput values and initializes α and β with the mean values given by the normally distributed random number. Finally, samples taken from these generators are used to obtain α and β values that will be used in the next sampling interval.

The measured performance in terms of TCP goodput is averaged using an exponentially weighted moving average as follows:

$$T_a = T_a * (w) + T_m * (1 - w),$$

where w is the weight assigned to the average goodput (Ta) computed for the corresponding value of α. Our experiments showed that $w = 0.5$ provides a good tradeoff between current and past values of the goodput.

The inter-node level implementation of the cognitive engine is presented in Algorithm 2 and it follows a similar approach. At regular intervals, each cognitive node communicates the chosen α and β values and the throughput in the immediate past interval. Communication is pursued using the ICMP protocol. Moreover, typically CIS service is implemented in the same network segment of cognitive nodes and should not lead to performance degradation of the data flows. Nevertheless, the interval used for cognitive nodes to report to CIS should be chosen carefully considering possible overhead issues.

Algorithm 1: Intra-layer Cognitive Engine

Analyze Performance Metrics

Get the number of bytes received since last time interrupt nBytes

Get the time elapsed from the last timer interrupt smapleInterval

Calculate the average throughput R_i as nBytes/sampleInterval

Calculate the weighted throughput $R_i = R_{i-1} \times (p) + R_i \times (1 - p)$, where p is the weight value given to the past history

Store the weighted Ri value in two-dimensional array on a position defined by the current α and β values $[\alpha, \beta]$

Get parameters corresponding to optimal performance

Set Max Throughput to zero
For each element of the two-dimensional array with Ri **do**
    If current throughput Ri is greater than Max Throughput **then**
        Set Max Alpha equal to α
        Set Max Beta equal to β
    Endif
Endfor

Choose parameters for next sampling interval

Set Normal Distribution Mean to Max Alpha
Get new α value from the random number
Set Normal Distribution Mean to Max Beta
Get new β from the distribution

Algorithm 2: Inter-node Cognitive Engine

Receive feedback and obtain cumulative throughput maximum

Set the Total Throughput R_{TOT} equal to zero
For each cognitive node **do**
Receive α, β values, and the measured Throughput Ri
Increase R_{TOT} to Ri
Endfor
Store R_{TOT} along with α and β parameters to each node
Get max R_{TOT} and α_{max} and β_{max} parameters corresponding for every cognitive node involved into R_{TOT} calculation

Configure each cognitive node parameters

For each cognitive node **do**
    If current α is not equal to α_{max} **then**
        Set α equal to α_{max}
    Endif
    If current β is not equal to β_{max} then
        Set β equal to β_{max}
    Endif
Endfor

11.4.2 Inter-Layer Cognitive Optimization[3]

In order to obtain the details of inter-layer cognitive optimization, we first limited our scenario to only two TCP connections simulated in separate experiments: flow F_1 between S_1 and D_1

---

[3] D. Kliazovich, F. Granelli, Nelson L. S. da Fonseca, R. Piesiewicz, "Cognitive Information Service: Basic Principles and Implementation of A Cognitive Inter-Node Protocol Optimization Scheme," IEEE Global Communications Conference. Reproduced by permission of © 2009 IEEE.

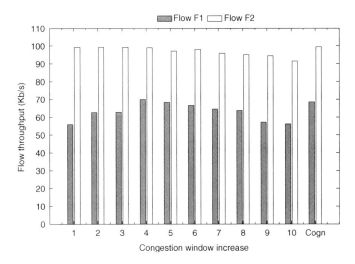

Figure 11.5   Average throughput performance of a single TCP flow with different fixed congestion window increase (α) parameters and cognitive adaptation approach. Reproduced by permission of © 2009 IEEE.

involving L_1 link and flow F_3 between S_3 and D_3 involving L_3. The average flow throughput is chosen as the main performance metric, which is measured for different values of α with β fixed at 0.5. We used 1s as sampling interval for the cognitive engine and 0.5 for standard deviation of normal distributed random numbers.

Results presented in Figure 11.5 show the ability of the cognitive engine to converge to the optimal value of parameter α. For F_1, the optimal value of α is equal to 1 due to the absence of link errors – since every packet loss is caused by congestion; while for F_3 the optimal value of α is 4, which corresponds to a balance between increase rate after loss related to link error and the amount of retransmissions performed due to multiple congestion-related losses at the bottleneck buffer.

Figure 11.6 presents the distribution of the α values chosen by intra-layer cognitive engine during a TCP flow lifetime. Each value is obtained by dividing the time at which TCP flow congestion control used a given value of alpha divided by the total simulation time. As expected for the flow F_1 most of the chosen α values are gathered around $\alpha = 1$, while for flow F_3 $\alpha = 4$, most of the time.

Results confirm that the proposed cognitive adaptation leads to significant improvements in a dynamic network environment by performing both intra-layer and inter-layer optimizations. It was shown that fixing the α value performance degradation under specific scenarios can happen. The proposed cognitive mechanism avoids that problem. If there were no global optimal values for a protocol parameter, certainly the presented cognitive adaptation can provide the best average performance by adapting the protocol behaviour to current network conditions.

11.4.3   Inter-Node Cognitive Optimization[4]

The previous section showed the benefits of using an intra-layer cognitive engine for tuning the performance of a single TCP flow. In this section, we focus on joint optimization of the

---

[4] D. Kliazovich, F. Granelli, Nelson L. S. da Fonseca, R. Piesiewicz, "Cognitive Information Service: Basic Principles and Implementation of A Cognitive Inter-Node Protocol Optimization Scheme," IEEE Global Communications Conference. Reproduced by permission of © 2009 IEEE.

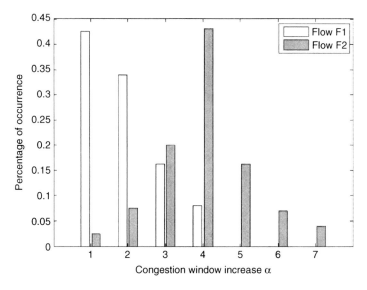

Figure 11.6   Density of congestion window increase (α) parameters chosen by intra-layer cognitive engine. Reproduced by permission of © 2009 IEEE.

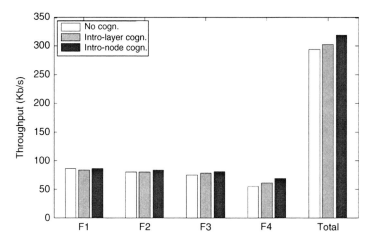

Figure 11.7   Multi-flow TCP throughput performance for case with no cognitive adaptation, intra-layer cognitive adaptation, and inter-node cognitive adaptation. Reproduced by permission of © 2009 IEEE.

performance of multiple TCP flows. With this aim in mind, we simulated four TCP NewReno flows between specific S and D pair of nodes as illustrated in Figure 11.4.

Figure 11.7 presents the throughput results for each individual flow as well as the cumulative throughput for the following three cases: i) no cognitive adaptation, ii) with intra-layer cognitive adaptation, and iii) with inter-node cognitive adaptation (CIS server). In both cases when cognitive adaptation is used, α and β TCP flow control parameters are tuned as outlined in the section above.

As expected, the throughput decreases for long links with high error rates. The main reasons for such throughput reductions are link errors and well-known RTT unfairness for flows with different RTTs competing for the same buffer resources.

However, it can be observed that intra-layer cognitive engine can easily solve the problem of link losses by adapting α and β parameter, converging to the optimal throughput value. However, it cannot cope with the problem of RTT unfairness, which requires coordination between flows. Such coordination is performed at the inter-node level by the CIS server, which leads to higher performance.

11.5   Conclusions

This chapter introduces the novel concept of cooperative network optimization that is carried at inter-layer and inter-node basis. Based on the proposed concept, protocols from the TCP/IP can be extended to dynamically tune their configuration parameter values based on the past performance.

Results demonstrated that cooperation between the protocol layers of a protocol stack can significant enhance the performance when compared to a fixed non-cooperative approach. Moreover, the exchange of information on configuration and performance among network nodes can further improve the performance of data transfer.

One of the main points behind the design of the cooperative optimization framework is the ability to tune configuration parameters in runtime and in a distributed manner. This ensures fast convergence and optimal protocol stack performance adaptation in dynamically changing network environments.

References

[1] J. Mitola III, 'Cognitive radio for flexible mobile multimedia communications,' *Mobile Networks and Applications*, vol. 6, no. 5, September 2001.

[2] M. W. Murhammer and E. Murphy, *TCP/IP: Tutorial and Technical. Overview*,' Upper Saddle River, NJ: Prentice-Hall, 1998.

[3] J. Postel and J. Reynolds, 'File Transfer Protocol (FTP),' RFC 959, IETF, October 1985.

[4] Clark, George C., Jr., and J. Bibb Cain, *Error-Correction Coding for Digital Communications*, New York: Plenum Press, 1981, ISBN 0-306-40615-2.

[5] ITU-T, P.800, 'Methods for Subjective Determination of Transmission Quality,' Aug. 1996.

[6] J. Postel, 'Transmission Control Protocol,' RFC 783, September 1981.

[7] J. Postel, 'User Datagram Protocol,' RFC 768, Aug. 1980.

[8] IEEE 802.11 Wireless Local Area Networks. Available from: http://grouper.ieee.org/groups/802/11/

[9] ANWIEEE Std 802.3, 'Carrier Sense Multiple Access with Collision Detection,' 1985.

[10] T. Kelly, 'Scalable TCP Improving performance in highspeed wide area networks', *Computer Communication Review*, vol. 32, no. 2, April 2003.

[11] K. Machova and J. Paralic, *Basic Principles of Cognitive Algorithms Design. Proceedings of the ICCC International Conference Computational Cybernetics*, Siofok, Hungary, 2003.

[12] The network simulator ns2. Available from: http://www.isi.edu/nsnam/ns.

12

Cooperative Network Coding

H. Rashvand[1], C. Khirallah[2], V. Stankovic[3] and L. Stankovic[3]

[1]*School of Engineering, University of Warwick, Coventry, U.K.*
[2]*Institute for Digital Communications, School of Engineering, University of Edinburgh, Edinburgh, U.K.*
[3]*Department of Electronic and Electrical Engineering, University of Strathclyde, Glasgow, U.K.*

12.1 Introduction

Recent exploitation of the wireless channel as a flexible but constrained, hostile and extremely adverse medium has enabled the rapid growth of pioneering technological development for vast new emerging applications such as distributed sensor networks that demand the use of innovative techniques as cooperative communications to control and make better use of the components constructing the media. Cooperative systems are now playing important roles in wireless communications networks boosting up the system performance in terms of reduced power consumption, increased system capacity and greater resilience. For example, using physical layer cooperation among wireless nodes resemble virtual multiple-input multiple-output (MIMO) antenna configurations that provide receivers with spatial diversity gain or rate multiplexing gain, depending on receiving multiple replicas of the same information or different information over various independent fading channels, to mitigate multipath fading or equally increase the network throughput.

Cooperative communications at the network level play critical roles in combating the outage probability, increasing channel capacity, providing diversity-multiplexing trade-off, extending coverage area, and many other measures which improve the system's performance. Cooperative communication is also a building block for relaying messages, for example, by making use of classic cooperative schemes to relay messages in the form of incremental redundancy or repetition coding. One of the most interesting application areas of cooperative communication at the network level are enabled by using network coding (NC).

Most dynamic and unstructured networks with distributed sources and destinations are wireless and due to distributed variable interference conditions they suffer from heavy outage and extensive loss of data such that recent developments in cooperative communications find the use of diversity extremely useful. Though the diversity is an inherent part of the broadcasting nature of wireless media and abundantly available, this however comes at the cost of additional complexity to the network routing process, networking optimization and effective node capacity as well as increasing waste of bandwidth resources due to additional overheads. Network coding, which is a promising evolution of simple routing protocols allows mixing messages from different nodes before later sending this mixture on shared links, instead of separate links for all messages, hence, network coding will eventually increase the network's throughput by reducing the overall number of required links in the network. In other words, NC increases links' bandwidth efficiency. Further enhancement can come from the integration of cooperative strategies and network coding that offers some degrees of diversity.

Following an introduction to network coding, and cooperative strategies in both high signal-to-noise ratio (SNR) and low SNR channel conditions, this chapter provides a detailed description of using network coding in cooperative relay networks as well as a practical network spread coding scheme for cooperative communications in multipath fading channels.

12.2   Network Coding Concept

Network coding, introduced by Ahlswede et al. [1] some 10 years ago, originally set out to enable a better use of links in upcoming complex, multi-hop relay and mesh networks for their multiple source and multiple sink transmission requirements. At its lowest level, network coding allows routers to mix the information content inside packets, received from other nodes, before forwarding the mixture to destination nodes in the network. NC can be regarded as an extension to the traditional network routing for solving the problem of 'networks with distributed sources and sinks' for scarce satellite resources. The concept of NC, however, differs from simple network routing due to the fact that NC aims at whole network optimization rather than individual classes of users or applications being the objective. A better understanding of NC lies in its use in new complex and unstructured networks where multiple diversity links can be easily made available and often sources and sinks are potentially distributed, randomly or clustered. Example networks are overgrown networks such as Internet, the new generation of distributed sensor systems (DSS), such as the new wireless sensor networks (WSN), and ever growing mobile ad hoc networks (MANET).

NC has been rapidly growing into a potentially powerful tool in the design of communication networks. Unlike traditional use of error control coding in the networks where coding is performed at the edges (end-to-end) to detect and/or correct the errors on individual packets for a given link, NC seeks to combine different diversified routes in a multipath network routing fashion at the network level for the purpose of better usage of network resources. NC simply facilitates intermediate network nodes to combine user data packets received over different incoming links.

12.2.1   Example

The classic butterfly network example is shown in Figure 12.1 [1], where two sets of information, A and B, originating from two independent sources X1 and X2 are delivered to two destinations, Y1 and Y2, by using a simple coding A+B on the single shared link, J-K, in association with the dedicated side links of X1-Y1 and X2-Y2. A+B in a binary information system would represent a simple exclusive-OR (XOR) operation. This straightforward but fundamental example shows the concept of NC in how sharing a common link can save the need of having a fourth link for sending A and B from X1 and X2 independently to Y1 and Y2 resulting in 25% save in network resources.

In order to elaborate the properties of NC in further detail we can examine this example. For the sake of simplicity we consider the total cost for overall transmission of information as the

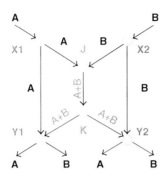

Figure 12.1   The butterfly network [1]. Reproduced by permission of © 2010 IEEE.

Direct	Y1	Y2
X1	c_{11}	c_{12}
X2	c_{21}	c_{22}

NC	Y1	Y2
X1	c_{11}	ρcs
X2	$(1-\rho)cs$	c_{22}

(a)   (b)

Figure 12.2   The cost matrix for connecting two sources to two sinks.

main objective for our network design. Under this requirement we base our calculations on the total cost of transferring two independent but equal in size information from two independent sources to two sinks, simultaneously.

Without the use of NC four independent links are needed to carry four pieces of information: two links for transferring information A from X1 to Y1 and Y2; two links for transferring information B from X2 to Y1 and Y2. In general, the total cost for direct links (that is, without NC) sums up to four different entries in a 2-by-2 cost matrix $[C]$ as shown in Figure 12.2(a). The cost for providing the service is given by:

$$C_{\text{direct}} = (c_{11} + c_{12}) \cdot I_1 + (c_{21} + c_{22}) \cdot I_2, \tag{12.1}$$

where, $I = [I_1, I_2]$ is the information rate vector (in bits per second) representing A and B, c_{ij} is the cost per bit then C_{direct} in (12.1) gives the total cost (in rate, per second).

With NC, however, a simultaneous delivery of information from multiple-source to multiple-sink can benefit from sharing links for transferring some of the information. In the butterfly network example of Figure 12.1, we use the matrix of Figure 12.2(b) to get a smaller overall cost, as

$$C_{NC} = (c_{11} + \rho \cdot c_s) \cdot I_1 + [(1 - \rho) \cdot c_s + c_{22}] \cdot I_2, \tag{12.2}$$

where, c_s indicates the total cost of the shared link, where ρ ($0 \le \rho \le 1$) portion of the shared link is utilized for transmission of information I_1 and remaining $(1-\rho)$ portion of link for transmission of I_2.[1]

Assuming the same cost for all links, $c_{11} = c_{12} = c_{21} = c_{22} = c_s = c_0$, and also all information rates are fixed at the equal rates, $I_1 = I_2 = I_0$ and $\rho = 0.5$, we then have the following total costs for these two network topologies.

---

[1] All overheads such as cost of local connections and coding etc are neglected.

For the direct connection topology of Figure 12.2(a), the cost is:

$$C_{\text{direct}} = \sum_{i=1}^{2} \sum_{k=1}^{2} c_{ik} \cdot I_i = 4c_0 I_0, \tag{12.3}$$

with network coding, using entries in Figure 12.2(b), the cost is:

$$C_{NC} = \sum_{i=1}^{2} \sum_{k=1}^{2} c_{ik} \cdot I_i = 3c_0 I_0. \tag{12.4}$$

It is easy to see a '3/4 cost ratio' from (12.3) and (12.4), indicating the 25% reduction in the cost due to the use of NC.

In the above example the effectiveness of NC is limited to 25% due to the limited number of sources (and sinks), that is, $n = 2$. Practical multi-source (and multi-sink) applications are extensively represented in new applications such as distributed sensor networks with n, in general, much larger than 2 which enable NC to offer greater savings. For example, for $n = 3$ or 4, NC reduces the cost by one third (33.3%) or 3/8 (37.5%), respectively. Hence, it is easy to show that the NC saving can be generalized as:

$$Saving = \frac{n^2 - n}{2n^2}. \tag{12.5}$$

All the above mentioned NC cases use two dimensional matrix models, however, as we will discuss further in this chapter NC comes with a great flexibility to be combined with cooperative relaying strategies for superior performance enhancement in many unstructured and complex networks such as multi-hop relay and mesh networks that make extensive use of multipath diversity in order to:

- increase quality of service in highly volatile media;
- reduce probability of outage in high mobility applications;
- increase bandwidth with very poor SNRs.

Although originally NC is known for its cost or equivalently classic throughput enhancement, it has also shown its potential for other improvements as well. NC can, for example, reduce energy consumption required to multicast a packet by reducing the number of transmissions per packet. This in effect can reduce the overall delay in a network when measured in number of hops for a packet to reach its destination. Besides its classic uses, physical-layer NC (PNC) [2] is proposed in order to make a better use of NC over wireless broadcast channels. PNC can be combined with channel coding to be used in various relay-based configurations such as two-way communication, multiple access, multi-hop transmission and multicasting. PNC, in effect, can exploit the natural features of the wireless channel to provide much higher capacity than the classic packet-level NC mentioned above. However, due to their strict synchronization requirements PNC signals suffer from variable delays, which may impose some limits to PNC offered applications in wireless networks.

Analog network coding (ANC) [3] is another form of NC that is effectively a PNC but exploits signal interference at intermediate nodes to increase the throughput in the network while relaxing the synchronization between mixed signals under severe conditions. ANC, however, generally, works well only in a channel with a high SNR and without fading in the communication channel.

Network spread coding (NSC) [4] is an additional approach to PNC that brings features of NC and spread spectrum technique together for exploiting advantages of these two for better bandwidth efficiency offered by NC with higher robustness of spread spectrum against interference and noise due to use of linearly independent complete complementary (CC) sequences. CC is a class of spreading codes that combines the local and global encoding vectors and maintains their orthogonality over asynchronous communication channels under adverse and high interference channel conditions. Cooperation with network coding is discussed in [5], [6].

12.3   Cooperative Relay

The information-theoretic properties of classic three node relay networking with a source that transmits information, a destination that receives information and a relay that both receives and transmits can be traced back to the seminal work of Cover and El Gamal [7]. This work analyzes the relaying capacity under an additive white Gaussian noise (AWGN) relayed channel, and comes up with several optimum relaying strategies.

Inspired by the recent development in wireless networks, and based on information-theoretical concepts of relay channel, cooperative communications aims at increasing diversity gain and reducing the outage probability for many practical wireless fading channels [8], [9]. Most works on wireless cooperation strategies involve two phases. The coordination phase and the cooperation phase. In the first phase, the strategy is to decide on the best source node broadcasting method to adopt for its signals being sent to both destination and relay. In the second phase the decisions that have to be made involve further processes on the overhead signals and the method of forwarding them to the destination. The terminology full-duplex is used for the relay nodes to transmit and receive simultaneously [7] compared to the half-duplex setup [10], [11] where relays cannot transmit and receive simultaneously in the same band, that is, relays cannot use the same frequency band in frequency division multiple access (FDMA) and orthogonal frequencies (OFDMA) or the same time-slots in time division multiple access (TDMA) systems. Further details for relaying strategies using full-duplex setup achieve capacities closer to the upper bound and full-duplex practical relaying designs are proposed in [12]. The half-duplex relaying remains as the most commonly used for cooperation communications due to its simplicity. It is shown that the sub-optimality of cooperative strategies using the half-duplex relaying compared to those using the full-duplex relaying is not due to the half-duplex limitation itself but rather due to their use of the orthogonal FDMA and TDMA relaying and that use of non-orthogonal relaying strategies such as non-orthogonal amplify-and-forward (NAF) and dynamic decode-and-forward (DDF) should improve the achieved rates and diversity gains for the source transmitting in the first phase [13].

The three-node relay channel is shown in Figure 12.3, where h_{SR}, h_{RD} and h_{SD} are the channel coefficients for the Source-Relay (S-R), Relay-Destination (R-D), and Source-Destination (S-D), respectively, which are modeled as zero-mean, complex Gaussian random variables with

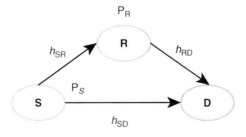

Figure 12.3   Three-node wireless relay channel.

variances $\sigma_{SR}^2, \sigma_{RD}^2$, and σ_{SD}^2. P_R and P_S denote the power transmitted by the relay and the source, respectively.

In the first phase, the source node broadcasts its information signal x_{S1} towards the relay and destination nodes at transmission power P_{S1}, embedded in zero-mean complex additive Gaussian channel noise, denoted by z_{SR} and z_{SD1}. The received signals at the relay y_{SR} and the destination y_{D1} are given by:

$$y_{SR} = h_{SR}x_{S1} + z_{SR}, \text{ and} \tag{12.6}$$

$$y_{D1} = h_{SD}x_{S1} + z_{SD1}. \tag{12.7}$$

In the second cooperation phase, the source node transmits its signal x_{S2} towards the destination at transmission power P_{S2}, while the relay forwards a processed version of the source's signal (that is received in the first phase) to the destination at transmission power P_R. The received signal at the destination is $y_{D2} = y_{RD} + y_{SD2}$ where:

$$y_{RD} = h_{RD}f(y_{SR}) + z_{RD}, \text{ and} \tag{12.8}$$

$$y_{SD2} = h_{SD}x_{S2} + z_{SD2}. \tag{12.9}$$

where $f(y_{SR})$ in (12.8) is a specifically defined function of the signal received at the relay, y_{SR}, and assuming that the source and relay nodes transmit at equal power $P_{S1} = P_{S2} = P_R = 1$. This function depends on the cooperation strategy and the way, in which, the relay process signals received from the source. The following section provides a summary of the cooperative strategies used in both high SNR and low SNR channel conditions.

12.4   Cooperation Strategies

In this section we provide an overview of some commonly used cooperation strategies and then in the following section discuss their performance for high and low SNR conditions.

In order to achieve optimum relay cooperation, signals received by relays can be processed differently. Therefore several cooperative relaying strategies are proposed [7]–[10]. Cooperation strategies can be classified into two categories of amplify-and-forward (AF) and decode-and-forward (DF). AF is one of the simplest relaying strategies with a low implementation cost [9] where upon receiving a copy of a source signal, embedded in noise, and sometimes interference, each relay whilst satisfying specific power constraints simply amplifies the noisy signal before forwarding it towards the destination without further processing. DF, however, regenerates new signals by demodulating and decoding the received signals, that is, usually processing received signals, then encoding and modulating them before forwarding to the destination [8].

The comparisons between AF and DF strategies in respect of their capacity bounds and outage probability indicate that AF and DF can outperform each other under different channels and application scenarios. One condition differentiating these two is the quality of channels in use for different SNR operating regions and distance or signal loss between the source and relay, and the position of the relay relative to the source and destination. For example, AF provides a better performance when the relay is located half way between the source and destination [14] no matter what the channel condition. On the other hand, DF outperforms AF when the source-relay channel ensures error-free detection of the received signal at the relay [10], [11].

Janani et al. [15] claim a sophisticated cooperation strategy called 'coded cooperation' that combines channel coding with a DF cooperative based operating strategy for a superior performance under sufficiently reliable source-relay channel condition.

For coded-cooperation, the main cooperating partners, source and relay, fundamentally encode their data into two codewords. Each codeword consists of two parts. The first part of a codeword contains the source or relay information plus some parity bits generated from a cyclic redundancy check (CRC) code, whilst the second only contains parity bits derived from the first part. The different parts of each partner codeword are transmitted via two statistically independent fading paths over two time slots. In the first cooperation phase, each node transmits the first part of its codeword and based on the success of its detection (determined by CRC) by the partner, the partner will either generate and forward parity bits of the received part, or it will discard the received part and transmit additional parity bits for its own data. Comparison between the three cooperation strategies AF, DF and coded-cooperation reveals that in general the coded-cooperation demonstrates a better block error rate than basic AF and DF over all SNRs, whilst AF approaches the DF performance for SNR region from mid to high [15].

An interesting cooperation strategy that is known to outperform DF but is restricted to the condition of the relay being placed closer to the destination is compress-and-forward (CF). In CF, originally suggested in 1979 in the pioneering work of Cover and El Gamal [7], the relay exploits redundancies present within the correlated signals received by the relay and the destination to compress the received signal with certain distortion rates. The CF relay also employs source coding with side information (Wyner-Ziv coding (WZC) [16]) to compress its received signal for use at the destination as side information. A significant problem with CF is experienced as it has a high implementation complexity. In comparison with DF, CF approaches the max-flow min-cut upper bound [17] on capacity if the relay-destination channel is more reliable than the source-relay channel. In contrast, DF achieves higher rates when the relay is closer to the source [11], [18]. Madsen and Zhang [10] present the upper bounds and lower bounds on the ergodic capacity and the outage capacity for three-node wireless relay networks that operate under Rayleigh fading channels and employ half-duplex or full-duplex relaying. Also, the effect of the power allocation, between the source and relay, on the performance is analyzed. Further analysis of the CF capacity bounds is also presented in [10].

Avestimehr and Tse [19] investigate a frequency division cooperative system called bursty amplify-and-forward (BAF) for the low SNR and low outage probability fading channel conditions.

12.4.1 Performance Measures

In this section we examine some cooperative strategies that are at their optimum under high SNR or low SNR channel conditions. Then, in order to be able to make fair judgements, compare different cooperative systems using the following performance measurement metrics:

- outage capacity;
- diversity multiplexing tradeoffs (DMT);
- energy requirement for half-duplex wireless relay channel.

Shannon in his classic work defines the channel capacity as the maximum data rate that can be sent over the channel while ensuring a reliable communication [20]. However, as capacity derivations of Shannon assumes stationary random variations for the channel but with deep fades due to poor fading channel condition, the obtained values are normally far below those of Shannon. This assumption therefore results in a significant reduction in the transmission rates from those expected from Shannon capacity under the AWGN channel conditions and we need some more accurate measures.

In order to simplify the capacity measurements of time-varying channels for their fading speed and the availability of the channel state information (CSI) at the transmitter and receiver we consider two associated channel capacities of ergodic capacity and outage capacity as the performance metrics. Ergodic or 'average capacity' is defined as the probabilistic average of the

Shannon capacity for AWGN channel over all instantaneous channel fading realization states. In other words, in theory achieving the ergodic capacity requires use of sufficiently long codewords that can span all possible states of the fading channel [21]. For example, in fast fading channels with the channel states being varied significantly from one symbol to the next, the ergodic capacity coincides with the Shannon capacity obtained in AWGN channels. This leads to the knowledge of the performance metrics of cooperation strategies, initially derived for the AWGN channels, in fast fading channels.

Madsen and Zhang [10] derive the capacity upper and lower bounds for fast fading channels. However, many applications suffer from severe delay constraints that force the practical systems to use finite length codewords. In flat fading channels where each codeword is affected by a single channel state, the ergodic capacity is not a valid measure since it only represents the upper bound of the transmission achieved for all codewords. To this end, Ozarow et al. [22] introduces a more meaningful capacity called 'outage capacity' defined as the maximum data rate that can be sent over a channel for a given outage probability. An outage event occurs when the signals are transmitted at higher rates than the mutual information of the fading channel, as happens within a poor state of the channel.

In wireless networks, several techniques exploit the random nature of multipath fading channels to either increase data rates for a given outage probability (for example, multiplexing gain) or reduce the outage probability for a given data rate (for example, diversity gain). The multiplexing gain is achieved by multiplexing different data streams onto independent fading channels, while the diversity gain mitigates the effects of fading by transmitting the same data stream over independent fading channels. Several forms of time, frequency, polarization and spatial diversity have implementations in practical modern communications [23], [24]. Similarly, relaying among single-antenna nodes in a wireless network creates a virtual multiple antenna system and spatial diversity.

Since the capacity of a general relay channel has remained an open issue for over 40 years, research interests are diverted towards the design and analysis of optimum cooperation strategies that efficiently exploit the resources of the relay channels such as diversity and degree of freedom. Due to the difficulty in finding cooperation strategies that are optimum in all SNR channels, research instead is focused on two SNR channel conditions: (i) the high SNR channel condition where optimum cooperation strategies achieve high diversity gains or high degrees of freedom and very simple yet powerful performance measure is the diversity-multiplexing tradeoffs analysis [25]; (ii) the low SNR channel with more attention on the energy efficient strategies [19].

12.4.1.1   High Signal-to-Noise Ratio Regime

Several wireless communication designs are proposed to either increase the diversity gain or increase the multiplexing gain [26]–[28]. In [28], the authors propose the use of CC sequences for increasing spectral efficiency for MIMO code division multiple access (CDMA) system operating under selective fading channels. The results show that MIMO CC-CDMA demonstrates a superior performance over the traditional Walsh spreading sequences.[2]

In their seminal work Zheng and Tse [25] show that the outage capacity for the system is equivalent to the DMT in the high SNR channel condition. Therefore they propose to use DMT as a performance measure for various MIMO schemes. The results for the DMT demonstrate that although it is not possible to achieve full diversity and full multiplexing gains simultaneously,

---

[2] The traditional Walsh spreading sequences suffer from high level of intersymbol interference and multiple access interference levels.

it is possible to use part of available antennas to increase the data rate and then use the remaining antennas to increase the error reliability which indicates tradeoffs between these two gains. It is shown in [25] and [29] that, in order for a transmission scheme to achieve a mutual multiplexing gain r and diversity gain d at high SNR condition, it should be able to send data at rate $R(\text{SNR})$ and an average error probability $P_e(\text{SNR})$, where both are functions of channel SNR and satisfy:

$$\lim_{\text{SNR}\to\infty} \frac{R(\text{SNR})}{\log \text{SNR}} \geq r, \text{ and} \tag{12.10}$$

$$\lim_{\text{SNR}\to\infty} \frac{\log P_e(\text{SNR})}{\log \text{SNR}} \leq -d. \tag{12.11}$$

The optimal tradeoff curve for any scheme operating in Rayleigh fading channels, and with fading block length l exceeding the total number of transmit antennas M_T and the receive antennas M_R (that is, $l \geq M_T + M_R - 1$), becomes a piecewise linear function connecting the points $(r, d_{\text{optimal}}(r))$. $d_{\text{optimal}}(r)$ is defined as the best achievable diversity gain at a given spatial multiplexing gain r, where [25]:

$$d_{\text{optimal}}(r) = (M_T - r)(M_R - r), \quad 0 \leq d_{\text{optimal}} \leq M_T M_R, \text{ and} \tag{12.12}$$

$$r = \frac{R}{\log \text{SNR}}, \quad 0 \leq r \leq \min\{M_T, M_R\}. \tag{12.13}$$

For example, let us consider a MIMO scheme with $M_T = M_R = 6$ that satisfies (12.10)–(12.13) transmitting at $R = 20$ bits per second (bits/sec), and the channel SNR = 20 and 30 dB, according to (12.12) and (12.13) the maximum diversity gains achieved with $r \approx 3$ and 2 are $d_{\text{optimal}} = 9$ and 16 for SNR = 20 and 30 dB, respectively. Hence, increasing the diversity gain comes at the price of reducing the multiplexing gain and vice versa. Several adaptive schemes that switch between the two gains are investigated in [30].

Figure 12.4 shows the optimal diversity multiplexing tradeoff curve $d_{\text{optimal}}(r)$ for the general number of $m = M_T, n = M_R$ and $l \geq M_T + M_R - 1$, and the two extreme cases of no

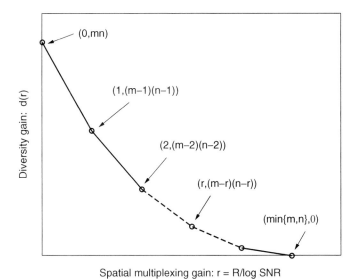

Figure 12.4 The optimal diversity multiplexing tradeoffs curve $d_{\text{optimal}}(r)$ for the general number of $m = M_T, n = M_R$ and $l \geq M_T + M_R - 1$, [25]. Reproduced by permission of © 2010 IEEE.

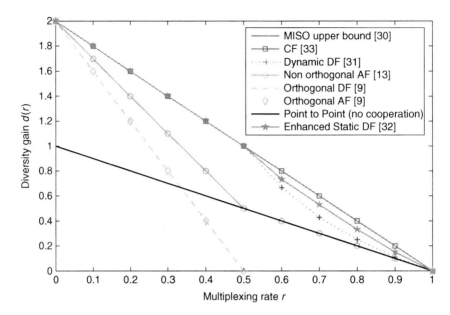

Figure 12.5 Diversity-multiplexing tradeoffs for some orthogonal and non-orthogonal cooperative strategies, half-duplex relay channels [31]. Reproduced by permission of © 2010 IEEE.

DMT: (i) maximum diversity gain $d_{max} = mn$, with no spatial multiplexing gain, hence, $r_{min} = 0$ (that is, point $(0, d_{max})$), (ii) and the maximum spatial multiplexing gain $r_{max} = \min\{m, n\}$ with $d_{min} = 0$ (that is, with no diversity gain achieved, point $(r_{max}, 0)$).

Pawar et al. [31] propose the DMT of 2-by-1 multiple-input single-output (MISO) channel as the DMT upper bound for cooperative schemes in which the relay employs full-duplex and fully cooperate with the source to transmit to the destination:

$$d_{MISO}(r) = 2(1 - r), \quad 0 \le r \le 1. \tag{12.14}$$

Figure 12.5 shows the achieved DMT for several cooperative strategies: orthogonal AF and orthogonal DF [9], the dynamic DF [13], non-orthogonal AF [14], and the partial DF [31], [32], and CF [33]. Due to the fact that the time division restriction is forced onto orthogonal AF and DF, and that relay and source never transmit at the same time, the maximal achievable diversity and multiplexing gain are $d = 2$ and $r = 1/2$. On the other hand, allowing the relay and source to transmit simultaneously in the non-orthogonal strategies, may result in an improved performance for the DMT. For example, the non-orthogonal AF, though not achieving the MISO upper bound still exhibits a better performance than the orthogonal AF.

Another interesting work is from Azarian et al. [13], where a strategy called the dynamic decode and forward (DDF) is proposed in which the relay listens to the source and as soon as it is able to decode the received data correctly, it starts encoding and relaying to the destination. Figure 12.5 shows that DDF is optimal and achieves the upper bound as long as the multiplexing gain $r \le 1/2$ (that is, low multiplexing gain), then the relay requires a long time to decode the received messages and it will be too late for the relay to join the cooperation. To overcome this delay, Prasad and Varanasi [32] propose an equally interesting partial DF (PDF) strategy in which the relay is not forced to decode the entire received message; instead the source message is split and transmitted to the relay in parts. The cooperation is performed over two phases, in the first phase the relay decodes the part of the source message then it joins in the cooperation.

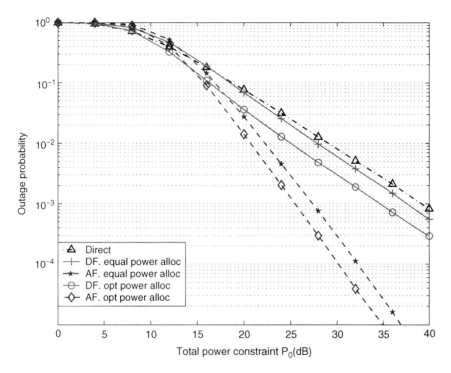

Figure 12.6 Outage probability comparison between AF and DF cooperation strategies in the three-node relaying network for the case of equal power allocation and optimal power allocation [34]. Reproduced by permission of © 2010 IEEE.

In the second phase, the relay forwards the decoded first part to the destination while ignoring the transmitted second part of the source message. This way, the destination combines the first part of the source message received from the source and relay. Then in the second phase the destination only decodes the message part received from the source.

Yuksel and Erkip [33] prove that the CF cooperation strategy achieves the DMT upper bound for any multiplexing gain r and for any number of antennas used at the node. However, this scheme comes at the expense of knowing the CSI at both relay and destination.

Laneman et al. [9] discuss that an equal power allocation for the source and relay (that is, $P_S = P_R$ in Figure 12.3) is not always optimum. They argue that an optimal power allocation is highly dependent on the channel link quality (that is, h_{SR}, h_{RD} and h_{SD}). Figure 12.6 shows an outage probability comparison between AF and DF cooperation strategies in the three-node relaying network for the case of optimal and equal power allocation with $P_R + P_S \leq P_0$, where P_0 is the total available power.

As shown in Figure 12.6 for optimal power allocation both AF and DF gain about 3 dB as compared with the case of equal power allocation ($P_R = P_S$). On the other hand, as expected DF outperforms the AF strategy at the low SNR condition.[3] However, at high SNR channel condition AF outperforms the DF and achieves much higher diversity gains.[4]

Hong et al. [34] analyze the optimal power allocation, subject to a total power constraint $P_R + P_S \leq P_0$, that maximizes the achieved channel capacity using the standard max-flow min-cut

---

[3] This is due to noise constraints imposed by AF at high SNR condition.
[4] DF performance is highly dependent on the quality of the S-R link.

formula [17], which is generally used to obtain upper bounds on achieved capacity in a network, that is,

$$C_{DF} = \max_{\{P_S, P_R\}} \min \left\{ \frac{1}{2} \log(1 + |h_{SR}|^2 P_S), \frac{1}{2} \log(1 + |h_{SD}|^2 P_S + |h_{RD}|^2 P_R) \right\}, \text{ and} \qquad (12.15)$$

where, C_{DF} is the channel capacity for DF in a three-node network. Then, optimal power allocations for P_R and P_S are given by:

$$P_S = P_0 \frac{|h_{RD}|^2}{|h_{SR}|^2 + |h_{RD}|^2 - |h_{SD}|^2}, \text{ and} \qquad (12.16)$$

$$P_R = P_0 \frac{|h_{SR}|^2 - |h_{SD}|^2}{|h_{SR}|^2 + |h_{RD}|^2 - |h_{SD}|^2}. \qquad (12.17)$$

This shows that more power should be allocated to the source than to the relay (that is, $P_S > P_R$) since the former needs to transmit over two links S-R and S-D.

Zhang et al. [35] offer a review of power allocation methods for the three-node relay scenario using DF and AF strategies. A further work containing extensive knowledge of CSI at all three nodes comes from Hong et al. [34], whilst [36] provides interesting work for partial CSI knowledge. Using the CSI, they develop an optimal power allocation scheme for high SNR channel condition for both DF and AF cooperation strategies. These schemes tend to minimize the symbol error rate (SER) performance for the systems using M-ary-phase-shift-keying (M-PSK) and M-ary-quadrature-amplitude-modulation (QAM). In the case of DF, the asymptotic symbol-error-rate (SER) and the optimum power allocation for P_S and P_R are given by:

$$P_{SER} \approx \frac{N_0^2}{b^2} \cdot \frac{1}{P_S \sigma_{SD}^2} \left(\frac{A^2}{P_S \sigma_{SR}^2} + \frac{B}{P_R \sigma_{RD}^2} \right), \qquad (12.18)$$

$$P_S = P_0 \frac{\sigma_{SR} + \sqrt{\sigma_{SR}^2 + 8(A^2/B)\sigma_{RD}^2}}{3\sigma_{SR} + \sqrt{\sigma_{SR}^2 + 8(A^2/B)\sigma_{RD}^2}}, \text{ and} \qquad (12.19)$$

$$P_R = P_0 \frac{2\sigma_{SR}}{3\sigma_{SR} + \sqrt{\sigma_{SR}^2 + (8A^2/B)\sigma_{RD}^2}} \qquad (12.20)$$

where, b = binary-phase-shift-keying (BPSK), N_0 is the noise variance, A and B are constants with values depending on the constellation size of the modulation used. For example, for M-PSK [36]:

$$A = \frac{M-1}{2M} + \frac{\sin \frac{2\pi}{M}}{4\pi}, \qquad (12.21)$$

$$B = \frac{3(M-1)}{8M} + \frac{\sin \frac{2\pi}{M}}{4\pi} - \frac{\sin \frac{4\pi}{M}}{32\pi}. \qquad (12.22)$$

For Quadrature-phase-shift-keying (QPSK) modulation, $A = 3/8 + 1/(4\pi) = 0.4546$ and $B = 9/32 + 1/(4\pi) = 0.3608$.

Similar to (12.16) and (12.17), equations (12.19) and (12.20) show that the optimal power allocation depends on the quality of the channels S-R and R-D, and the modulation in use. However, surprisingly, the optimal power allocation does not depend on the quality of the channel S-D, assuming the availability of all channels S-R, R-D, and S-D and high SNR. From (12.19) and (12.20) the approximate optimum power ranges are: $0 < P_R < P_0/2$ while $P_0/2 < P_S < P_0$.

In the case of AF, the asymptotic SER and the optimum power allocation for P_S and P_R are given by:

$$P_{SER} \approx \frac{BN_0^2}{b^2} \cdot \frac{1}{P_S \sigma_{SD}^2} \left(\frac{1}{P_S \sigma_{SR}^2} + \frac{1}{P_R \sigma_{RD}^2} \right), \tag{12.23}$$

$$P_S = P_0 \frac{\sigma_{SR} + \sqrt{\sigma_{SR}^2 + 8\sigma_{RD}^2}}{3\sigma_{SR} + \sqrt{\sigma_{SR}^2 + 8\sigma_{RD}^2}}, \quad \text{and} \tag{12.24}$$

$$P_R = P_0 \frac{2\sigma_{SR}}{3\sigma_{SR} + \sqrt{\sigma_{SR}^2 + 8\sigma_{RD}^2}}, \tag{12.25}$$

where, optimum power allocation is independent of the modulation type and depends on the link quality of the channels S-R and R-D. This is due to the AF relaying only amplifying and forwarding the received messages without decoding them. From (12.24) and (12.25) the approximate optimum power ranges are: $0 < P_R < P_0/2$ while $P_0/2 < P_S < P_0$. Figures 12.7(a) and (b) show simulation results for a comparison between the AF and DF cooperation strategies assuming that $N_0 = 1, \sigma_{SD}^2 = 1$ and QPSK modulation. In Figure 12.7(a) the variances of the channel links S-R and R-D are $\sigma_{SR}^2 = \sigma_{RD}^2 = 1$, hence, from (12.24) and (12.25) the optimum power ratios for the AF cooperation strategy, which are independent of the modulation type, are $P_S/P_0 = 2/3$ and $P_R/P_0 = 1/3$, where P_S and P_0 are named P_1 and P, respectively, in Figures 12.7(a) and (b). Similarly from (12.19)–(12.22) the optimum power ratios for the DF strategy are $P_S/P_0 = 0.6270$ and $P_R/P_0 = 0.3730$. In this case, the AF strategy has slightly worse SER performance (≈ 0.5 dB) than the DF strategy. On the other hand, in Figure 12.7(b) with $\sigma_{SR}^2 = 1$ and $\sigma_{RD}^2 = 10$, $P_S/P_0 = 0.7968$ and $P_R/P_0 = 0.20321$, for the DF strategy, and $P_S/P_0 = 0.8333$ and $P_R/P_0 = 0.1667$, for the AF strategy, the performance gain of the DF is almost 1 dB. This shows that the better the channel link quality of R-D, compared to that of the S-R (that is, $\frac{\sigma_{RD}^2}{\sigma_{SR}^2} \geq 1$), the higher the performance gain of the DF compared to that of the AF. However, the DF performance gain is bounded by the ratio of the cooperation gains of DF and AF:

$$\gamma = \frac{\Delta_{DF}}{\Delta_{AF}} \rightarrow \frac{\sqrt{B}}{A}. \tag{12.26}$$

Table 12.1 shows that DF significantly outperforms the AF strategy in the case of relay closer to the destination, and low constellation size (for example, BPSK), otherwise, AF remains the most favourable cooperation strategy for networks operating at high data rates (for example, 16QAM) due to its low complexity and almost similar performance to that of the DF strategy.

12.4.1.2   Low Signal-to-Noise Ratio Regime

In our previous calculation of the upper bound for high SNR channel conditions such as the DMT curves we have made many assumptions which are not valid for applications such as wireless sensor networks that operate within limited bandwidth and energy resources. Hence, other types of cooperation strategies that can ensure efficient energy transfer, through fading networks under low SNR channel condition, are proposed. Avestimehr and Tse [19] show that in a simple point to point network, operating at low SNR and low outage probability ε, the loss in the achieved capacity given as a ratio between the outage capacity C_ε and the AWGN capacity C_{AWGN}, is significantly higher than that observed at high SNR with $C_\varepsilon \approx \varepsilon C_{AWGN}$, as shown in Figure 12.8.

One solution to improve the outage capacity at low SNR is to employ diversity. Figure 12.9 shows that the loss in the outage capacity, of a network operating in low SNR channels, is

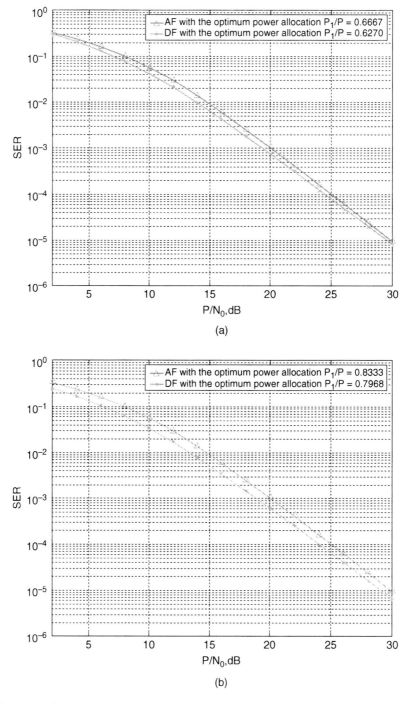

(a)

(b)

Figure 12.7 (a) Performance comparison of the cooperation systems with either AF or DF cooperation protocol with QPSK signals for $\sigma_{SR}^2 = \sigma_{RD}^2 = 1$ [36]. Reproduced by permission of © 2008 Springer. (b) Performance comparison of the cooperation systems with either AF or DF cooperation protocol with QPSK signals for $\sigma_{SR}^2 = 1$ and $\sigma_{RD}^2 = 10$ [36]. Reproduced by permission of © 2008 Springer.

Table 12.1   The Performance Gain γ for $\dfrac{\sigma^2_{RD}}{\sigma^2_{SR}} \geq 1$ and $\dfrac{\sigma^2_{RD}}{\sigma^2_{SR}} \leq 1$

Cases Different Modulation Schemes

	Modulation	Cooperation Gain Ratio $\gamma = \dfrac{\Delta_{DF}}{\Delta_{AF}}$
$\dfrac{\sigma^2_{RD}}{\sigma^2_{SR}} \geq 1$	BPSK	$\sqrt{3}$
	QPSK	1.3214
	16QAM	1.1245
$\dfrac{\sigma^2_{RD}}{\sigma^2_{SR}} \leq 1$	BPSK	1.1514
	QPSK	1.0851
	16QAM	1.0378

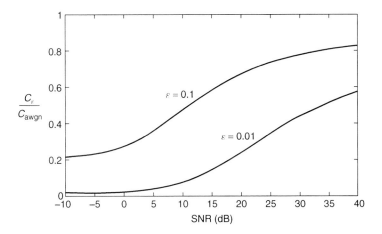

Figure 12.8   The loss in the outage capacity C_ε in the fading channel to AWGN capacity under Rayleigh fading, for $\varepsilon = 0.1$ and $\varepsilon = 0.01$ [19]. Reproduced by permission of © 2010 IEEE.

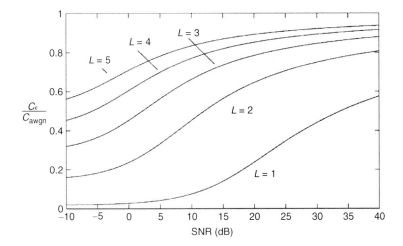

Figure 12.9   The loss in the outage capacity C_ε in the fading channel to AWGN capacity under Rayleigh fading, for $\varepsilon = 0.01$ and different L-fold receive diversity [19]. Reproduced by permission of © 2010 IEEE.

by a factor of $\varepsilon^{1/L}$ instead of ε (that is, no diversity case), where L is diversity order. This improvement in the outage capacity is higher at low outage probability.

In [19] authors assume a three-node relay scenario with slow Rayleigh fading channels, half-duplex relaying constraints implemented using frequency division (FD) and CSI available at the receiver only. Then, upon the max-flow min-cut formula they obtain an upper bound on outage capacity in the relay channel as:

$$C_{\text{Relay}} \approx \sqrt{\frac{2\text{SNR}_{\text{SR}}\text{SNR}_{\text{RD}}\text{SNR}_{\text{SD}}}{\text{SNR}_{\text{RD}} + \text{SNR}_{\text{SR}}}} \varepsilon, \tag{12.27}$$

where, SNR_{SR} is the average SNR received at the relay over the Rayleigh fading channel between the source and relay (h_{SR}), SNR_{RD} and SNR_{SD} are the received SNRs at the destination over both channels h_{RD} and h_{SD}. Then the outage rate achieved using cooperation strategies AF, DF and BAF:

$$R_{\text{DF}} \approx \sqrt{\frac{2\text{SNR}_{\text{SR}}\text{SNR}_{\text{RD}}\text{SNR}_{\text{SD}}}{2\text{SNR}_{\text{RD}} + \text{SNR}_{\text{SR}}}} \varepsilon, \tag{12.28}$$

$$R_{\text{AF}} \approx \varepsilon\text{SNR}_{\text{SD}}, \tag{12.29}$$

$$R_{\text{BAF}} \approx \sqrt{\frac{2\text{SNR}_{\text{SR}}\text{SNR}_{\text{RD}}\text{SNR}_{\text{SD}}}{\text{SNR}_{\text{RD}} + \text{SNR}_{\text{SR}}}} \varepsilon. \tag{12.30}$$

From (12.27)–(12.30) and assuming that the relay is equidistant to the source and destination and that $\text{SNR}_{\text{SD}} = \sigma_{\text{SD}}^2\text{SNR}$, $\text{SNR}_{\text{SR}} = \sigma_{\text{SR}}^2\text{SNR}$, $\text{SNR}_{\text{RD}} = \sigma_{\text{RD}}^2\text{SNR}$, and that the variances of the channel gains $\sigma_{\text{SD}}^2 = \sigma_{\text{SR}}^2 = \sigma_{\text{RD}}^2 = 1$, Figure 12.10 shows the achieved outage rate, given in NATS per second (NATS/S), for DF, AF and BAF at two outage probabilities $\varepsilon = 1\%$ and 10%. As expected, both AF and DF fail to achieve the max-flow min-cut upper bound, since AF relays have the lowest outage rate, equal to that of the point-to-point link, due to the high level of noise at low SNR conditions; whereas for DF relays operating in practical channels, by failing to decode some of the received data, will be unsuccessful in ensuring 'continuous' energy transfer, the condition for max-flow min-cut upper bound, to the destination regardless of the source-relay channel condition [23].

On the other hand, BAF achieves the relay channel upper bound thanks to its transmission scheme that is based on transmitting at high power and very low effective transmission rates. This way the relay amplifies the data received at high SNR (that is, low noise level) and forwards the data to the destination without any interruption in the energy transfer between the source and destination [19].

12.5   Cooperative Network Coding

As mentioned earlier NC, pioneered by the work of Ahlswede et al. [1], suggests that nodes of new complex data communication networks should go far beyond traditional storage and forwarding of received messages, to take on more functionality, carry local intelligence, and be involved in other processes to ease up the application complexity for the users. For example, nodes in distributed systems can become intelligent members that can participate in routing, coding and other light signal processing. The possibilities and advantages over the traditional routing schemes are numerous: such as increased throughput, power efficiency, reduced retransmissions per packet, reduced delays, reduced hop-counts per packet and robustness against link failures [37].

In the original work [1] authors introduce the upper bound on the achievable rates using the max-flow min-cut theorem, and prove that networks using NC can always achieve the multicast

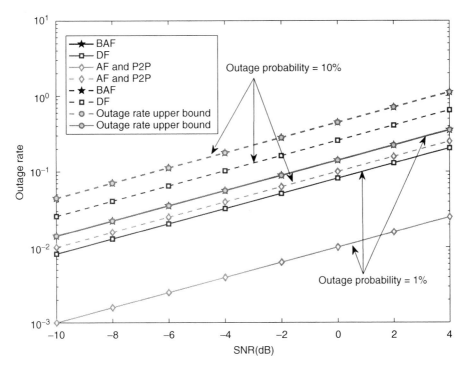

Figure 12.10 Approximate outage rate (NATS/S) at low SNR as a function of the SNR for different cooperation strategies BAF, DF, AF, compared with the outage probability upper bound and the lowest outage capacity of the point-to-point (P2P) link. Results are obtained for $\sigma_{SD}^2 = \sigma_{SR}^2 = \sigma_{RD}^2 = 1$ and at two outage probabilities $\varepsilon = 1\%$ and 10%.

capacity, while traditional routing techniques are not optimal and cannot achieve any of these upper rates. To this end, Li et al. [38] study linear network coding, where the intermediate nodes perform linear operation on the received messages, to prove that for a finite field with alphabets sufficiently large the linear NC can actually achieve the multicast capacity.

Although the original work does not clearly mention the uses of NC for wireless networks, it is easy to apply the concept to any complex and unstructured network being an Internet with many duplicated sites as a multiple source and many users of the same sites as multiple sinks or being a wireless DSS with sensors scattered all over place for the same sort of information being detected for further processing such as data fusion. Strong potentials of NC extension to wireless networks are further studied in [37], [38], [39], with detailed discussion of information exchange in wireless networks offered in [40]. However, existing packet-level NC designs offer limited features for the current classic signal processing capabilities due to: (i) the increased complexity and cost of encoding and decoding nodes, which present a major obstacle for NC applications such as wireless sensor networks, where, the NC scheme is probably needed most to reduce the number of transmissions and preserve the scarce energy supply; (ii) wireless networks are broadcast channels in nature, where, a signal transmitted by one node may reach several other nodes at the same time, and a destination node may receive signals from multiple nodes simultaneously, that can result in excessive interference and therefore reduction in overall network throughput.

Zhang et al. [2] show that PNC can make use of the additive nature of the wireless channels to seek higher capacity than the packet-level NC. However, strict synchronization conditions of the PNC limits its use in practical wireless networks, where signals suffer from variable delays. On the other hand, ANC from Katti et al. [3], exploits signal interference at the intermediate nodes to increase throughput, whilst relaxing the condition of synchronization between the mixed signals. Indeed, in simpler network topologies, ANC scheme outperforms packet-level NC, when a high SNR is assumed with no fading in the communication channel, and the mixed signals have similar power levels. Otherwise, severe degradation in performance can be experienced.

As mentioned earlier, cooperative relay strategies are analyzed under some ideal relaying conditions such as the interference-free links, the ideal knowledge of the CSI at the receivers and ignoring loss in resources due to the relay's inability to listen and transmit simultaneously or the relay being in a non-cooperative state due to failure in decoding messages received over high outage probability links. However, in practical networks ignoring interference effect and resource loss may reduce the expected outage rates. To overcome such problems, several solutions based on combining PNC with cooperative communications for the two-way or multi-way traffic are proposed in [40]–[44]. An example of a two-way traffic is two nodes A and B communicating their messages via a relay node R. In [41] authors introduced a two-way relaying scheme based on PNC termed as denoise-and-forward (DNF), which consists of two stages. A multiple access (MA) stage, in which both nodes A and B simultaneously transmit their messages to R and a broadcast stage of the de-noised (but not decoded) received combined messages to both A and B that obtain each others messages but cancel out their own. Popovski and Yomo [43] group two-way relaying schemes into a generic three-step DF scheme, and two-step AF [41] and DNF [42] schemes, then investigate how to maximize the achieved rate for those schemes using PNC. Results obtained demonstrate the ability of DNF to achieve much higher two-way throughputs than those offered by DF and AF.

More recently Koike-Akino et al. extend these studies of relaying systems using DNF for more complex constellations and the use of diversity gain for further throughput enhancements [44].

Xiao et al. study the complex case of combining cooperative strategies with NC and error control codes [6] introducing the concept of cooperative network coding as a way to address inefficient resource usage of a network. In their proposed scheme the transmitting nodes perform some algebraic superposition of locally generated information prior to that of the partners encoded using a convolutional code generator matrix. This system outperforms classic cooperative strategies for its time sharing and simple message superposition, for example, using simple XOR operation for locally generated and relayed bits. With the channel coding involved, however, this scheme comes at the expense of increased complexity.

Yu et al. combine NC with cooperative communication to reduce overall inter-user interference [5]. This improvement is achieved by increasing the diversity gain in multiuser fading channels compared to the traditional time-sharing relaying strategies. For this they consider a scheme where each relay transmits codewords that contain three parts:

- its own message;
- parts of its partner's previously transmitted codeword (parity bits) or codewords;
- parity bits generated from joint encoding of the two parts.

As mentioned earlier DMT is used extensively for measurement based comparison of various strategies in this chapter. Two directly related works are from [45] and [46] where authors analyze the outage capacity and provide some DMT curves for cooperative network coding (CNC) and show that CNC outperforms traditional DF and it can also improve selecting DF for both the outage capacity and the DMT whilst AF can only perform in a similar way to the CNC for three-node cooperation networks. Additionally, CNC has the ability to increase the coverage area of wireless networks using the conventional cooperative relaying provided that the relay-destination link is of a better quality than the source-relay link. In other words, the relay is closer to the destination.

In [4], the authors propose the use of the network spread coding (NSC) scheme as a novel PNC based on spread spectrum using the mutually orthogonal complete complementary (CC) sequences [47]. NSC uses the linearly independent CC sequences to generate local and global encoding vectors [38] that maintain their orthogonality over asynchronous communications, high interference and adverse channel conditions. Similar to ANC [3], signal-mixing occurs within a channel at the physical layer. However, in contrast to ANC and PNC [2], the proposed NSC scheme can operate at different SNR levels and under a high level of interference caused by the de-synchronization mixing of signals in mobile fading channels. The proposed NSC scheme can use two operation modes, at intermediate nodes to facilitate various complexity/cost/performance trade-offs: (i) the low-complexity 'forward' mode (NSC-F) where each intermediate node simply forwards the received mixed signal to its destination without further processing; (ii) the more error-robust 'despread-spread-forward' mode (NSC-DSF) where each intermediate node despreads the incoming mixed signals to recover the transmitted signals and then re-spreads and forwards them to their destination nodes. The proposed NSC system brings together NC and spread spectrum techniques, exploiting the advantages of both, that is, bandwidth efficiency of NC and interference and noise robustness of spread spectrum.

Figure 12.11(a)–(c) shows a single-session multicast wireless network with two-source nodes sending signals to two-destination nodes through one intermediate node (relay) using the: (i) the general NC scheme [1], which can be either packet-level NC or PNC, shown in Figure 12.11(a), (ii) the traditional spreading scheme using CC sequences but without NC, shown in Figure 12.11(b), and (iii) the proposed NSC scheme as in Figure 12.11(c).

Assume a wireless network with a set of N source nodes S and a set of destination nodes T. Suppose first that source nodes are connected to the destination nodes via intermediate nodes, which only forward received signals. Each node is equipped with an omni-directional antenna, and communication links are half-duplex and experience independent fading and additive white Gaussian noise (AWGN).

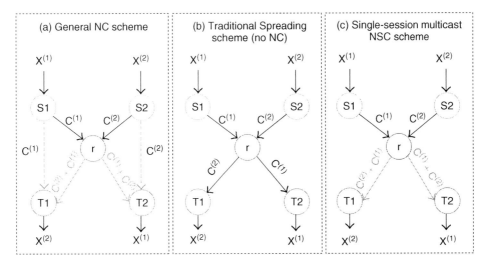

Figure 12.11 The example of the standard Butterfly-like wireless network [1]. (a) A general NC; (b) a traditional spreading scheme without NC; (c) the single-session multicast NSC (NC + CC). Reproduced by permission of © 2010 IEEE.

To ensure mutual orthogonality among transmitted signals, each source node uses a unique set of N CC sequences to encode its signals and transmits resulting spread signals over N orthogonal carrier frequencies. Note that this poses a limitation that the total number of source nodes available in a network must be equal to or less than the number of the sets of CC sequences N, while the total number of destination nodes P is arbitrary.

Let $\mathbf{x}^{(i)} = [x_1^{(i)}, x_2^{(i)}, \ldots, x_n^{(i)}, \ldots, x_K^{(i)}]^T$ be a stream of K source symbols to be encoded at the source S_i based on the offset stacked (OS) spreading technique [47]. K symbols are spread using N CC sequences into N sequences of length $M = \tau \times (2K - \tau)$, where $\tau \in \{1, 2\}$ is the number of chip-shifts during the OS spreading. Following the NC encoding technique used in [37], the local encoding matrix $\mathbf{G}^{(i)}$ for the i-th source node, where $1 \le i \le K$, of dimension $MN \times K$ can be defined as:

$$\mathbf{G}^{(i)} = [\mathbf{g}_1^{(i)}, \ \mathbf{g}_2^{(i)}, \ldots, \mathbf{g}_n^{(i)}, \ldots, \mathbf{g}_N^{(i)}],^T \tag{12.31}$$

where, $\mathbf{g}_n^{(i)}$ is of dimension $M \times K$ and is generated using the n-th CC sequence $\mathbf{w}_{i,n}$ of a matrix $\mathbf{W}$ and is given by:

$$\mathbf{g}_n^{(i)} = \begin{bmatrix} g_{n1}^{(i)} & 0 & \cdots & 0 \\ g_{n2}^{(i)} & g_{n1}^{(i)} & \cdots & 0 \\ \vdots & \cdots & & 0 \\ g_{nK}^{(i)} & g_{n(K-1)}^{(i)} & \cdots & g_{n1}^{(i)} \\ 0 & g_{nK}^{(i)} & \cdots & g_{n2}^{(i)} \\ \vdots & 0 & \cdots & \vdots \\ 0 & \cdots & 0 & g_{nK}^{(i)} \end{bmatrix} \tag{12.32}$$

$g_{nj}^{(i)}$ is the j-th chip of the n-th CC sequence $\mathbf{w}_{i,n}$ assigned to node S_i. Hence, the encoded data stream at source node S_i is $\{\mathbf{c}_1^{(i)}, \mathbf{c}_2^{(i)}, \ldots, \mathbf{c}_n^{(i)}, \ldots, \mathbf{c}_N^{(i)}\}$ with $\mathbf{c}_n^{(i)} = \mathbf{g}_n^{(i)}\mathbf{x}^{(i)}$ being the n-th encoded stream of length M chips, which is transmitted on a carrier frequency f_n at transmission rate $R_n = \tau \times (2K - \tau)/K$. R_n is defined as the number of spread-chips per source symbol (c/s) transmitted on f_n (for example, $R_n = 7/4$ c/s for $K = 4$ chips and $\tau = 1$). Note that the larger the τ the lower the transmission rate or the larger the required bandwidth.

For simplicity, assume a simple network with a single intermediate node (relay) r that is receiving all source nodes' signals. Let r be only one hop away from other source nodes. All signal components sharing the same carrier frequency are simply summed together at r. Thus, the signal received at node r over the n-th carrier frequency f_n is of length M chips and given by:

$$\mathbf{y}_n = \sum_{i=1}^{N} \alpha_i \mathbf{c}_n^{(i)} = \sum_{i=1}^{N} \alpha_i \mathbf{g}_n^{(i)}\mathbf{x}^{(i)} \tag{12.33}$$

here, α_i is a complex Rayleigh fading channel coefficient for the wireless link/path between source node S_i and intermediate node r. Then, the combined signal received at node r over all N carrier frequencies is:

$$\mathbf{Y} = [\mathbf{y}_1, \ \mathbf{y}_2, \ldots, \mathbf{y}_n, \ldots, \mathbf{y}_N] + \mathbf{z}_r. \tag{12.34}$$

$\mathbf{z}_r$ is a vector of N AWGN components, each with zero mean and unit variance.

Node r broadcasts the combined signal y_n over the carrier frequency f_n, thus the signal received at destination node T_p is:

$$\mathbf{Q}_p = \beta_p \mathbf{y}_n + \mathbf{z}_p, \qquad (12.35)$$

where, β_p is the complex Rayleigh fading channel coefficient for the wireless link/path (which can consist of one or more hops) between node r and destination node T_p. z_p is the AWGN with zero mean and unit variance at node T_p.

In the absence of noise and knowledge of the fading channel profile, the i-th destination node can decode received signals by finding the inverse of the local encoding matrix $\mathbf{G}^{(i)}$ using Gaussian elimination as in NC [38]. However, the mutual orthogonality of CC sequences enables the receivers to decode signals simply by multiplying signals received over different carrier frequencies with the transposition of the suitable local encoding vector $\mathbf{g}_n^{(i)}$ and then summing the results for all N frequencies and finally performing normalization by KN [28]. For example, the decoded signal at node T_p is given by:

$$\tilde{\mathbf{x}}^{(i)} = \frac{1}{KN} \left(\sum_{n=1}^{N} [\mathbf{g}_n^{(i)}]^T \mathbf{y}_n \right). \qquad (12.36)$$

The above scheme can easily be extended to the case with multiple hops between the source and relay nodes. Moreover, the intermediate nodes can at the same time play the role of the source and relay node. However, the total number of active sources must not exceed the total number of CC sequences N.

In the following we further explain the operation of the proposed NSC-F scheme by means of a simple example shown in Figure 12.11(c). Source nodes S1 and S2 attempt to multicast two binary streams $\mathbf{x}^{(1)}$ and $\mathbf{x}^{(2)}$, respectively, of length four bits each to destination nodes T1 and T2. $\mathbf{x}^{(1)}$ and $\mathbf{x}^{(2)}$ are encoded using two independent linear encoding matrices $\mathbf{G}^{(1)}$ and $\mathbf{G}^{(2)}$ into $\mathbf{C}^{(1)}$ and $\mathbf{C}^{(2)}$, respectively, that get combined in the shared physical multiple access channel (MAC) as $\mathbf{C}^{(1)} + \mathbf{C}^{(2)}$ (with added noise and fading) before the intermediate node r. For $\tau = 1$, each source node sends information at the rate of 14 chips per 4 symbols on both carrier frequencies ($R_1 = R_2 = 14/4$ c/s), and the effective combined transmission rate R at node r is 14 chips per 8 source symbols on two carrier frequencies. Next, node r transmits the combined streams $\mathbf{C}^{(1)} + \mathbf{C}^{(2)}$ to the destination nodes by simply forwarding $\mathbf{C}^{(1)} + \mathbf{C}^{(2)}$ (together with the added noise) to both destination nodes T1 and T2, which then perform OS despreading to reconstruct $\mathbf{x}^{(1)}$ and $\mathbf{x}^{(2)}$.

Results show considerable bandwidth saving as much as 50% compared to the traditional spreading scheme operating without NC while achieving comparable BER performance [4]. This bandwidth saving is expected to increase with the increase of multicast sessions in the wireless networks.

Figure 12.12 shows the Diversity multiplexing tradeoffs comparison between the network coded cooperation (NCC), conventional cooperation, repetition coding, non cooperation and the ideal cooperation curve. NCC has higher DMT than the conventional cooperation because the NCC achieves full diversity gain of 2 in the high SNR channel condition while still offering higher spatial multiplexing gain than the conventional cooperation. Thus networks using NCC promise increased throughput.

Ding et al. [49] study the extension of the PNC to the multipath fading channels, in which path-loss and phase-shift hinders the PNC usage, and proposed combining network coding and diversity (NCD), shown in Figure 12.13. The NCD scheme is performed in two steps. In the first step, the nodes broadcast their messages with no relaying at this stage. Then the NCD requests the help of the upper medium access layer (MAC) to sort the available relays according to their

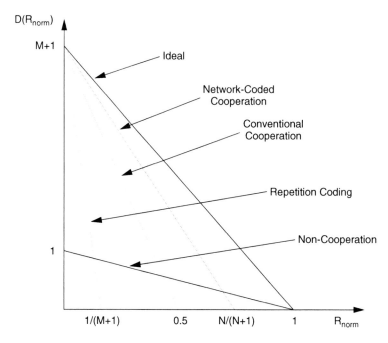

Figure 12.12 Diversity multiplexing tradeoffs comparison between the network coded cooperation, conventional cooperation, repetition coding, non cooperation and the ideal cooperation curve [48]. Reproduced by permission of © 2010 IEEE.

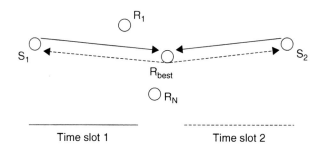

Figure 12.13 A diagram of the proposed NCD transmission strategy [49]. Reproduced by permission of © 2010 IEEE.

local channels qualities. This ensures better PNC performance. The chosen best relay employs AF strategy and broadcasts the mixture. [49] provides detailed simulation results comparing the outage capacity and ergodic capacity of the proposed NCD, conventional PNC scheme, and the direct transmission.

Figure 12.14 shows the outage capacity of the three transmission schemes at different SNR, where the proposed NCD achieves larger outage and ergodic capacities than those of the conventional PNC in the high SNR channel condition, while PNC still outperforms NCD for larger distances between the two source and destination, and operation in the low SNR channels, the latter due to the inevitable high noise effect on the performance of the AF strategy. For example, at SNR $=$ 10 dB, and at 10% outage capacity of PNC offers only 2.5 bits/s/Hz,

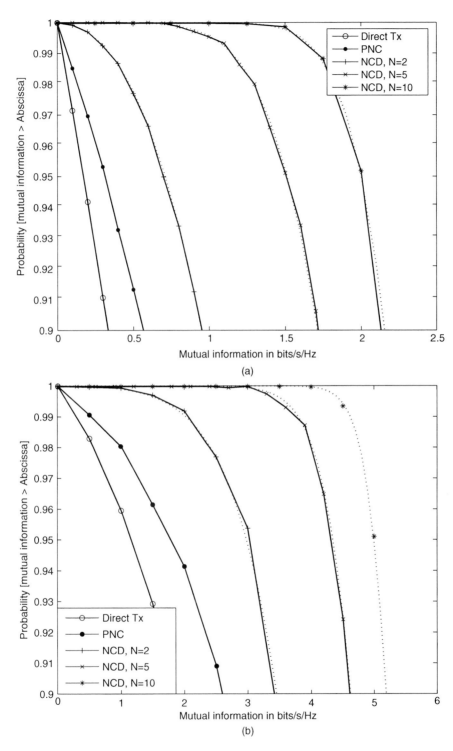

Figure 12.14 (a) and (b) Mutual information complementary cumulative distribution functions for (a) SNR = 10 dB, and (b) SNR = 20 dB. The distance of the two sources is 2 m. Solid line represents the results obtained by using the Monte-Carlo simulations, and the dotted line represents the results calculated by using the proposed analytical formulations [49]. Reproduced by permission of © 2010 IEEE.

compared to the 1.9 bit/s/Hz in the case of the no cooperation, while the proposed NCD achieves 3.5 bits/s/Hz [49].

12.6 Conclusions

This chapter provides an up-to-date summary of the trends in using the cooperative strategies in conjunction with network coding with some results for both high SNR and low SNR channel conditions. In doing this we have looked at uses of PNC in cooperative relay networks, current problems of cooperative communications, and examined the new concept of combining network coding and cooperative communication in order to improve systems performance and spectral efficiency.

Additionally, the chapter presents a detailed description of some practical network spread coding schemes for cooperative communications over multipath fading channels. The proposed scheme promises a saving of extra bandwidths compared to traditional spreading schemes operating without the NC. This bandwidth saving is expected to increase with a higher number of multicast sessions. NSC shares the same key idea of NC for combining the incoming packets and providing destinations with just enough combinations of input symbols as required to decode.

References

[1] R. Ahlswede, N. Cai, S.-Y. R. Li and R. W. Yeung, 'Network information flow,' *IEEE Trans. on Info. Theory*, vol. 46, issue 4, pp. 1204–1216, 2000.
[2] S. Zhang, C. S. Liew, and P. P. Lam, 'Hot topic: Physical layer network coding,' Proc. of the 12th annual international conference on Mobile computing and networking (MobiCom), LA, CA, USA, Sept. 2006.
[3] S. Katti, S. Gollakota and D. Katabi, 'Embracing wireless interference: Analog network coding,' ACM SIGCOMM, Kyoto, Japan, Aug. 2007.
[4] C. Khirallah, V. Stankovic, L. Stankovic and D. Poutouris, 'Network spread coding,' Proc. of the 4th Workshop on Network Coding, Theory and applications (Netcode2008), Hong Kong, Jan. 2008.
[5] M. Yu, J. T. Li and R. S. Blum, 'User cooperation through network coding,' Proc. of the IEEE International conference on communications (ICC), Glasgow, June 24-27, 2007.
[6] L. Xiao, E. T. Fuja, J. Kliewer and J. D. Costello, 'A network coding approach to cooperative diversity,' *IEEE Trans. Info. Theory*, vol. 53, issue 10, Oct. 2007, pp. 3714–3722.
[7] T. M. Cover and A. El Gamal, 'Capacity theorems for the relay channel,' *IEEE IT*, vol. 25, no. 5, p. 572, Sept. 1979.
[8] A. Sendonaris, E. Erkip and B. Aazhang, 'User cooperation diversity-Part I: System description and user cooperation diversity-Part II: Implementation aspects and performance analysis,' *IEEE Trans. Commun.*, vol. 51, no. 11, pp. 1927–1948, Nov. 2003.
[9] J. Laneman, D. Tse, and G. Wornell, 'Cooperative diversity in wireless networks: efficient protocols and outage behaviour,' *IEEE Trans. Info. Theory*, vol. 50, no. 12, pp. 3062–3080, Dec. 2004.
[10] A. Høst-Madsen and J. Zhang, 'Capacity Bounds and Power Allocation for Wireless Relay Channels,' *IEEE Trans. Info. Theory*, vol. 51, no. 6, pp. 2020–2040, Jun. 2005.
[11] G. Kramer, M. Gastpar and P. Gupta, 'Capacity strategies and capacity theorems for relay networks,' *IEEE Trans. Info. Theory*, vol. 51, no. 9, pp. 3037–3063, Sept. 2005.
[12] M. Khojastepour, N. Ahmed and B. Aazhang, 'Code design for the relay channel and factor graph decoding,' Proc. of the 38th Asilomar Conference, vol. 2, California. pp. 2000–2004, 2004.
[13] K. Azarian, H. El-Gamal and P. Schniter, 'On the achievable diversity-multiplexing trade-off in half-duplex cooperative channels,' *IEEE IT*, vol. 51, no. 12, p. 4152, Dec. 2005.
[14] R. U. Nabar, H. Bolcskei and F. W. Kneubuhler, 'Fading relay channels: performance limits and space-time signal design,' IEEE JSAC, Aug. 2004.
[15] M. Janani, A. Hedayat, T. E. Hunter and A. Nosratinia, 'Coded cooperation in wireless communications: Space-time transmission and iterative decoding,' *IEEE Trans. Sig. Proc*, vol. 52, no. 2, pp. 362–37, Feb. 2004.
[16] A. D. Wyner and J. Ziv, 'The rate-distortion function for source coding with side information at the decoder,' *IEEE Trans. Info. Theory*, vol. 22, pp. 1–10, Jan. 1976.
[17] R. L. Ford and R. D. Fulkerson, *Flows in Networks*, Princeton: Princeton University Press, 1962.

[18] V. Stankovic, A. Host-Madsen, Z. Xiong, 'Cooperative diversity for wireless ad hoc networks: capacity bounds and code designs,' *IEEE Sig. Proc. Mag.* 22, pp. 37–49, 2006.

[19] A. S. Avestimehr and D. Tse, 'Outage Capacity of the Fading Relay Channel in the Low SNR Regime,' *IEEE Trans. Info. Theory*, vol. 53, no. 4, pp. 1401–1415, April 2007.

[20] C. E. Shannon Communications in the presence of noise. Proc. IRE, pp. 10–21, 1949.

[21] A. Goldsmith, *Wireless Communications*, Cambridge University Press, 2005.

[22] L. H. Ozarow, S. Shamai and A. D. Wyner, 'Information theoretic considerations for cellular mobile radio,' *IEEE Trans. Vehicular Technology*, 43(2), 359–378, May 1994.

[23] Y. Zhang, H. H. Chen, M. Guizani, *Cooperative Wireless Communications*, Auerbach Publications, Taylor & Francis Group, USA, 2008.

[24] T. S. Rappaport, *Wireless Communications: Principles and Practice*, Prentice-Hall, Inc., Upper Saddle River, New Jersey, Second edition.

[25] L. Zheng and D. N. C. Tse, 'Diversity and multiplexing: a fundamental tradeoff in multiple-antenna channels,' *IEEE Trans. Info. Theory*, vol. 49, pp. 1073–1096, May 2003.

[26] G. Foschini, G. Golden, R. Valenzuela, and P. Wolniansky, 'Simplified processing for high spectral efficiency wireless communication employing multi-element arrays,' IEEE JSAC, vol. 17, Nov. 1999.

[27] V. Tarokh, N. Seshadri and A. Calderbank, 'Space-time codes for high data rates wireless communications: Performance criterion and code construction,' *IEEE Trans. Info. Theory*, vol. 44, pp. 744–65, 1998.

[28] C. Khirallah, P. Coulton, H. Rashvand and N. Zein, 'Multi-user MIMO CDMA systems using complete complementary sequences,' *IET Proc. Commun.*, vol. 153, no. 4, pp. 533–540, Aug. 2006.

[29] D. N. C. Tse, P. Viswanath and L. Zheng, 'Diversity-multiplexing tradeoff in multiple-access channels,' *IEEE Trans. Info. Theory*, vol. 50, pp. 1859–1874, Sept. 2004.

[30] R. W. Heath, Jr. and A. J. Paulraj, 'Switching between multiplexing and diversity based on constellation distance,' Proc. of the 38th Allerton Conference on communication, control, and computing, Monticello, IL, USA, Sept. 30–Oct. 2, 2000.

[31] S. Pawar, A. S. Avestimehr and D. N. C. Tse, 'Diversity-multiplexing tradeoff of the half-duplex relay channel,' Proc. of 46th Allerton conference on communication, control, and computing, Monticello, IL, USA, Sept. 2008.

[32] N. Prasad, M. K. Varanasi, 'High Performance Static and Dynamic Cooperative Communication Protocols for the Half Duplex Fading Relay Channel,' Proc. of the IEEE GLOBECOM, San Francisco, CA, USA, 27 Nov. – 1 Dec. 2006.

[33] M. Yuksel and E. Erkip, 'Multi-antenna cooperative wireless systems: A diversity multiplexing tradeoff perspective,' *IEEE Trans. Info. Theory*, Special Issue on Models, Theory, and Codes for Relaying and Cooperation in Communication Networks, vol. 53, no. 10, pp. 3371–3393, Oct. 2007.

[34] Y.-W. Hong, W.-J. Huang, F.-H. Chiu and C.-C. J. Kuo, 'Cooperative communications in resource-constrained wireless networks,' *IEEE Sig. Proc. Mag.*, vol. 24, pp. 47–57, May 2007.

[35] Q. Zhang, J. Zhang, C. Shao, Y. Wang, P. Zhang and R. Hu, 'Power allocation for regenerative relay channel with Rayleigh fading,' Proc. of the IEEE Vehicular Technology Conference (VTC-spring), vol. 2, pp. 1167–1171, Milan, Italy, May 2004.

[36] W. Su, A. K. Sadek, K. J. R. Liu, 'Cooperative communication protocols in wireless networks: Performance Analysis and Optimum Power Allocation,' *Wireless Personal Communications*, vol. 44, no. 2, 181–217, DOI: 10.1007/s11277-007-9359-z.

[37] P. A. Chou, and Y. Wu, 'Network Coding for the Internet and wireless networks,' *IEEE Sig. Proc. Mag.* vol. 24, issue 5, 2007, pp. 77–85.

[38] S.-Y. R. Li, R. W. Yeung and N. Cai, 'Linear network coding,' *IEEE Trans. Info. Theory*, vol. 49, pp. 371–381, 2003.

[39] Y. Wu, P.A Chou, Q. Zhang, K. Jain, W. Zhu and S.-Y. Kung, 'Network planning in wireless ad hoc networks: a cross-layer approach,' *IEEE JSAC*, 2005, vol. 23, pp. 136–150.

[40] Y. Wu, P. A. Chou and S.-Y. Kung, 'Information exchange in wireless networks with network coding and physical-layer broadcast,' Proc. of the 39th Annual Conference on Information Sciences and Systems (CISS), Baltimore, MD., USA, Mar. 2005.

[41] P. Popovski and H. Yomo, 'Bi-directional amplification of throughput in a wireless multi-hop network,' Proc. of the IEEE 63rd Vehicular Technology Conference (VTC), Melbourne, Australia, May 2006.

[42] ——, 'The anti-packets can increase the achievable throughput of a wireless multi-hop network,' Proc. of the IEEE International Conference on Communication (ICC), Istanbul, Turkey, Jun. 2006.

[43] ——, 'Physical network coding in two–way wireless relay channels', Proc. IEEE International Conference on Communications (ICC), Glasgow, June 24-27, 2007.

[44] T. Koike-Akino, P. Popovski and V. Tarokh, 'Optimized constellations for two-way wireless relaying with physical network coding,' *IEEE JSAC*, vol. 27, no. 5, June 2009. pp. 773–787.

[45] D. H. Woldegebreal and H. Karl, 'Network-coding-based cooperative transmission in wireless sensor networks: diversity-multiplexing tradeoff and coverage area extension,' In Proc. of the 5th European conference on Wireless Sensor Networks (EWSN), Bologna, Italy, Jan. 2008.

[46] Y. Wang, C. Hu, H. Liu, M. Peng and W. Wang, 'Network coding in cooperative relay networks,' Proc. of the 19th IEEE International Symposium on Personal, Indoor and Mobile Radio Communications (PIMRC), Cannes, French Riviera, France, Sept. 2008.

[47] H.-H. Chen, J. Yeh and N. Suehiro, 'A multicarrier CDMA architecture based on orthogonal complementary codes for new generations of wideband wireless communication', *IEEE Commun. Mag.*, vol. 39, 2001, pp. 126–135.

[48] C. Peng, Q. Zhang, M. Zhao and Y. Yao, 'On the Performance Analysis of Network-Coded Cooperation in Wireless Networks,' Proc. of the 26th IEEE International Conference on Computer Communications (INFOCOM), Anchorage, Alaska, USA, May 2007.

[49] Z. Ding, K. Leung, D. L. Goeckel and D. Towsley, 'On the study of network coding with diversity', *IEEE Trans. on Wireless Communications*, vol. 8, no. 3, pp. 1247–1259, Mar. 2009.

13

Cooperative Caching for Chip Multiprocessors

J. Chang[1], E. Herrero[2], R. Canal[2] and G. Sohi[3]
[1]*Intelligent Infrastructure Lab, HP Labs, Palo Alto, CA*
[2]*Department of Computer Architecture, Universitat Politècnica de Catalunya Barcelona, Catalunya*
[3]*Department of Computer Sciences, University of Wisconsin-Madison, Madison, WI*

13.1 Caching and Chip Multiprocessors

13.1.1 Caching Background

Caching is widely used in computing and communication systems to improve data access efficiency. By transparently keeping data copies needed for future accesses in local storage, caches can reduce the number of expensive operations of fetching data from their original store and therefore save time, energy and network bandwidth. The motivation for caching comes from several fundamental technology and application observations. First, the growth of data volume has been observed to outpace the Moore's law of component scaling, causing the number of components to increase in a system and hence the unavoidable need for accessing remote data. While Internet is a classic example of this trend, the same observation can be made for microprocessors: today's multicore CPUs are actually distributed systems each with multiple processor cores and caches interconnected via the on-chip network. Second, the disparate scaling trends of compute vs. storage/communication devices lead to an ever-increasing performance penalty for remote data access, necessitating techniques to cost-effectively reduce such accesses. For example, processor access of main memory data nowadays takes the same amount of time to complete 100s of arithmetic instructions, while off-chip memory bandwidth and energy efficiency scale poorly with current technology. Caching frequently used data on-chip to avoid main memory accesses is crucially important for both performance and power reasons. Finally, the locality of reference as a fundamental computing principle [29] allows caches to capture frequently used data and be effectively used in many scenarios, including virtual memory, processor caches, storage and database caches and web proxy caches, and so on.

Cooperative Networking, First Edition. Edited by Mohammad S. Obaidat and Sudip Misra.
© 2011 John Wiley & Sons, Ltd. Published 2011 by John Wiley & Sons, Ltd.

13.1.2   CMP (Chip Multiprocessor)

CMPs (Chip Multiprocessors) or multicores have been widely adopted and commercially available [3, 45, 61, 75, 128] as the building blocks for future computer systems. This is a dramatic shift from the historical trend of single-processor performance scaling. Up until around year 2004, high-performance microprocessors are usually single-core based, and the performance improvement mainly comes from two sources: (1) increased processor frequency as well as (2) parallel execution of multiple instructions (so called instruction-level parallelism). Both sources are unsustainable as they require processor designs that significantly increase design complexity and power consumption with diminishing performance benefits.

Instead of building highly complex, power-hungry, single-threaded processors, CMP designers integrate multiple, potentially simpler, processor cores on a single chip to improve the overall throughput while reducing power and complexity. This move has caused a scaling trend 'right-turn', where per-chip frequency, power and single-thread performance take a visual right turn from the past exponential growth curve and remain mostly flat. On the other hand, the number of processor cores on a single chip grows exponentially to match Moore's law. As the number of processor cores increases [62], a key aspect of CMP design is to provide fast data accesses for on-chip computation resources. Although caching has been one of the first and most widely used techniques to improve memory access speed in single-core chips, it faces several challenges when used in multi-core environments.

13.1.3   CMP Caching Challenges[1]

Unlike conventional designs where caches are dedicated to a single processor, CMP caches serve multiple threads running concurrently on physically distributed processor cores. This change of execution paradigm both aggravates caching demands and introduces new challenges that cannot be sufficiently addressed by prior caching proposals.

13.1.3.1   Limited Off-Chip Bandwidth

The main purpose of on-chip cache memory is to streamline processor operation by reducing the number of long-latency off-chip accesses. Based on the von Neumann architecture, processor computation involves frequent accesses of the memory system to fetch/store instructions and data. Historically, the performance gap between processor and DRAM has been increasing exponentially for more than two decades [101], which makes memory operations very expensive (costing hundreds of processor cycles). Even with large on-chip caches, high-performance processors often spend more than 50% of the time waiting for memory operations to complete [94]. Recently, improvement of single processor performance has slowed down as frequency scaling approaches its limit, but the 'memory wall' problem is likely to persist for CMPs due to limited off-chip bandwidth. Technology trends [42] indicate that off-chip pin bandwidth will grow at a much lower rate than the number of processor cores (and thus their aggregate memory bandwidth requirement) on a CMP chip. Without disruptive technology (for example, proximity communication [30]), the increasing bandwidth gap has to be bridged by efficient organization and use of available CMP cache resources. Emerging software such as 'Recognition, Mining and Synthesis' (RMS) workloads [60] can also stress the memory bandwidth requirement. Many such programs have poor data locality, either due to inherent streaming/scanning behaviours in

[1] Reproduced by permission from Doctoral Dissertation 'Cooperative Caching for Chip Multiprocessors'. University of Wisconsin at Madison, Madison, WI, USA © 2007.

the workload, or because better locality is only possible when their large working sets can be simultaneously satisfied by the last-level cache [64]. Therefore, both technology and software trends demand the CMP cache resources to be well utilized to reduce off-chip accesses.

13.1.3.2 Growing On-Chip Wire Delay and Power Consumption

In future technology, on-chip wire delay [54] will increase to a point where cross-chip cache accesses are far more expensive than local cache accesses. Without careful data placement, such non-uniform latencies reduce the benefit of on-chip caches because on average only a small fraction of blocks are located in cache banks that are close to their consuming processors. To reduce on-chip cache access latency, single-core designs exploit non-uniform cache architecture (NUCA) [19, 72] by migrating frequently used data into closer and faster cache banks. However, with multiple distributed cores accessing shared data, migration may be ineffective because the competition between different cores often leaves shared data in banks that are farther from all requesters [11]. In contrast, private cache organization reduces on-chip access latency as it locally replicates frequently accessed data, but such replicas can waste capacity and incur more expensive off-chip misses. CMP caching thus faces the conflicting requirements of saving off-chip bandwidth and reducing on-chip latency, and has to trade off between techniques that reduce off-chip vs. cross-chip misses [9]. With power comes a first-order metric, CMP caches should also facilitate per-core and per-cache power managements. On-chip caches should also be co-designed with on-chip network as it has significant power footprint and performance impact.

13.1.3.3 Destructive Inter-Thread Interference

The ineffectiveness of shared data migration among CMP cores demonstrates that competition of shared resources can lead to destructive inter-thread interference. For multiprogrammed workloads, threads also compete for cache capacity and associativity which can cause lowered performance (for example, due to thrashing [27]), unexpected performance (for example, due to unfair resource allocation [73]), lack of performance QoS [63] (that is, no guarantee in providing certain baseline performance [131]) and lack of control over the per-thread and overall performance (for example, no priority support). Without hardware solutions, their remedy can complicate the task of operating systems (for example, to improve fairness and avoid priority inversion) and server administration (for example, to maintain QoS for consolidated server workloads). Because fairness, QoS and priority support are important requirements for CMP users, and are often assumed by software running on CMPs, CMP caching schemes have to attack the problem of destructive interference and answer the challenge of simultaneously satisfying multiple, potentially conflicting, requirements such as throughput and fairness improvement.

13.1.3.4 Diverse Workload Characteristics

With multiple execution contexts available, CMPs can support both single-threaded and multithreaded workloads as well as their multiprogramming combinations. These workloads demonstrate different caching characteristics and therefore prefer different cache organizations or caching policies. For example, multithreaded workloads with large working sets prefer a shared cache for better effective capacity [64] while smaller workloads prefer a private cache organization for better on-chip latency [9]. Workloads with little sharing can benefit from dynamic migration [11], but aggressive sharing requires careful tradeoff between replication and migration. Furthermore, LRU-based cache replacement performs well for workloads with good temporal locality, while frequency-based policies (for example, LFU) are more suitable for workloads with poor locality [122]. Diverse workload preferences suggest that using a single caching scheme with fixed policies is unlikely to provide robust performance for a wide range of workloads. An ideal CMP caching scheme should be able to combine the strengths of different cache organizations and policies, and dynamically adapt to suitable behaviours to accommodate individual workload's caching requirements.

13.2 Cooperative Caching and CMP Caching

13.2.1 Motivation for Cooperative Caching

Multiple layers of caches can be deployed in a system to form a hierarchy to gradually bridge the steep speed gap between the central processing unit (for example, CPU) and the data store (for example, main memory). For example, in a multicore environment, each processor usually has small level one (L1) caches with sub-nanosecond latency followed by larger but slower L2 caches that are about 10 nanoseconds away, backed by even slower off-chip main memory whose latency is around 50–100 nanoseconds as well as microsecond or millisecond level non-volatile storage devices. Connected by a high-speed on-chip network, the ensemble of on-chip L2 caches can be collectively viewed as a single entity that is much faster than off-chip memory. The speed gap between processor core and off-chip memory, combined with the relatively low-latency of remote L2 cache accesses, motivates us to coordinate the data contents across multiple L2 caches to maximize its overall effectiveness for all the processor cores.

Similar scenarios occurred in the contexts of network file caches and web proxy caches. In the early 1990s, the improvement of cluster network speed over hard disk seek latency allowed a server to access a file cached in its peer's main memory buffer cache faster than fetching from its local hard disk, making it practical to globally manage the networked filed caches as an ensemble. Dahlin et al. [25] proposed cooperative caching schemes by exploiting this trend and demonstrated significant performance benefits. In the late 1990s, when the web became popular, the same observation was made because fetching a web page from another proxy cache was much faster than from the remote website and reduces both network bandwidth and remote server loads. Fan et al. [37] designed a web cache sharing scheme called summary cache to exploit this idea. The common theme across the above examples is to exploit the efficiency benefits of scale-out commodity building blocks (for example, processor cores, cluster servers or distributed web caches) against the cost and scalability limits of scale-up high-end hardware, and the availability of high-speed local interconnect that is much faster than fetching data from their original store, in order to achieve coordinated management of cache contents across distributed building blocks for various optimization goals.

13.2.2 The Unique Aspects of Cooperative Caching[2]

The idea of caching was first documented in the IBM System/360 implementation [83] which used a high-speed buffer to bridge the processor-memory speed gap. For a given cache size (which is determined by engineering tradeoffs), the cache's efficacy is largely determined by its data placement and replacement policies. Theoretically, Belady's MIN replacement algorithm [12] is optimal by providing a provable upper limit for hit ratio. However, because this algorithm

---

[2] Reproduced by permission from Doctoral Dissertation 'Cooperative Caching for Chip Multiprocessors'. University of Wisconsin at Madison, Madison, WI, USA © 2007.

evicts data with farthest reuse distance, it is limited in: (1) being an offline algorithm that requires both future knowledge and unbounded look-ahead to make replacement decisions, (2) assuming uniform cache hit latencies and miss latencies and (3) not considering conflicts between multiple cache resource consumers. Below we will use multicore and cooperative caching examples to illustrate how cooperative caching makes unique extensions in these aspects. Notice we will also use CC and cooperative caching interchangeably.

13.2.2.1 Online Cache Replacement and Placement

To attack the first limitation of the MIN algorithm, many practical, online replacement policies have been proposed, among which the least recently used (LRU) and least frequently used (LFU) policies are two typical examples. The two policies are near optimal, respectively, for programs with strong and weak temporal locality [34].[3] LRU is arguably the most widely used policy because its implementation is simpler than LFU and it can quickly adapt to working set changes. To provide good caching performance for a wide range of workloads, many software policies (for example, [66, 91]) and hardware designs (for example, [33, 122]) are proposed to combine the benefits of both LRU and LFU. Cooperative caching achieves the same goal, but instead uses cache partitioning to isolate workloads with weak locality from those with good locality and by integrating LRU replacement with cache partitioning.

In the context of CMP caching, flexible data placement is also needed to exploit the benefits of advanced cache replacement policies [48], reduce inter-thread conflict misses [139], provide QoS support via cache partitioning [70], and keep frequently accessed data close to the processor [19]. Highly associative caches (used by most CMP caching proposals) are thus needed to allow flexible data placement at the cost of extra area, latency, power and complexity overhead. Page coloring and remapping [112] can reduce conflict misses with low associativity, but require profile-based software optimization. Across cache banks, distance-aware placement attempts to keep frequently used data in the closest (and thus fastest) banks. For example, D-NUCA [72] dynamically can migrate 'hot' data towards the processor core, and NuRapid [19] further decouples data placement from tag placement using forward pointers.

CC improves cache associativity through cooperation among private caches, each with lower associativity, without extra hardware and software overhead. The aggregate cache is managed by an approximation of global LRU via a combination of local LRU and global placement history, therefore can support fine-grained sharing for both multithreaded and multiprogrammed workloads. To achieve distance-aware data placement, CC relaxes the inclusion requirement between L1 and L2 caches, and improves data locality using private caches that keep frequently used data close to the requesting processors.

13.2.2.2 Non-Uniform Latencies

In web caches that buffer files from multiple servers, cache misses have non-uniform latencies. Because Belady's MIN algorithm can only support uniform cache latency, new cache replacement policies are needed to prioritize data with variable costs and provide the best average access latency [138]. Similarly, processor caches can also exploit the cost difference between misses

---

[3] The notions of strong and weak temporal locality were formally treated by Coffman and Denning [34] and were later explained by Megiddo and Modha in the context of paging [91]: LRU is the optimal policy if the request stream is drawn from an LRU Stack Depth Distribution (SDD) which is "useful for treating the clustering effect of locality but not the non-uniform of page referencing." LFU is optimal if the workload can be characterized by the Independent Reference Model (IRM), which assumes that each reference is drawn in an independent fashion from a fixed distribution over the set of all pages in the auxiliary storage.

having non-uniform memory-level parallelism [43], as shown by proposals for cost-sensitive and MLP-aware uniprocessor caches [65, 105]. Distributed web and file caches can also cooperate to form a logically shared, global cache with non-uniform hit latencies [25, 37, 41]. CMP cooperative caching research is directly inspired by the above observation, and borrows from existing cooperative file caching policies.

The most related research to CMP caching, in terms of handling non-uniform access latencies, focuses on the memory organization of shared memory multiprocessors. To improve scalability of small-scale Uniform Memory Access (UMA) machines (SMP systems [18]), Non-uniform Memory Access (NUMA) systems statically partition the memory space among processor/memory nodes [79, 81]. Because local memory accesses can be several times faster than remote accesses in a NUMA system, a significant amount of research was done to improve data locality via a Cache-Only Memory Architecture (COMA) that uses local memory to attract frequently used data [47, 111], NUMA augmented with remote data caches or victim caches [92, 141], adaptation between COMA and CC-NUMA [36, 76] according to perceived memory pressure, and software techniques for page migration and replication [118, 130].

Due to similar latency/capacity tradeoffs, techniques to improve the average memory latency of distributed shared memory (DSM) systems can also be used to improve CMP caching. To illustrate this, Table 13.1 lists some of the corresponding proposals in DSM memory organization and CMP caching and their common features. CMP caching faces similar issues as their counterparts in DSM page caching such as cache coherence, replacement policies, control of replication/migration and scalability, although different tradeoffs and implementations are needed for DSM vs. CMP architectures.

13.2.2.3   Shared Resource Management

The third limitation of the MIN algorithm, its inability to deal with competition among multiple cache consumers, has been addressed by paging techniques (that is, caching at the memory level). Sharing physical memory among competing threads, and specifically avoiding destructive interference such as thrashing [27], is a classic problem for virtual memory management [28]. This problem is solved in software by proactively co-scheduling programs whose aggregate working set can be contained by available resources (for example, balanced-set schedulers [39]), or reactively reducing the level of multiprogramming (that is, the number of in-memory programs) to fit resource constraints. Similarly, Fedorova studied CMP-aware operating system schedulers [38] to avoid cache thrashing [40] and maintain fair cache usage by adjusting the CPU time quantum allocated to different threads.

Table 13.1   Corresponding Proposals for Organizing DSM Memory and CMP Caches. Reproduced by permission from Doctoral Dissertation 'Cooperative Caching for Chip Multiprocessors'. University of Wisconsin at Madison, Madison, WI, USA © 2007

DSM Memory	CMP Cache	Common Features
UMA [18]	Shared cache (UCA)	High capacity, uniformly long latency
NUMA [72, 79, 81]	Shared, banked, cache (S-NUCA)	High capacity, non-uniform latencies
COMA [47, 111]	Private caches [56]	Lowered capacity, low latency
RC-NUMA [141]	Adaptive private/shared NUCA [31]	Partition between private/shared space
VC-NUMA [92]	VC-CMP [98], Victim replication [140]	Victim caching
R-NUMA [36]	CMP-NuRapid [20]	Counter-based hints for relocation
AS-COMA [76]	CMP cooperative caching [16]	Biased replacement
OS support [130]	ASR [10]	Selective replication

Memory partitioning is another mechanism to avoid interference. As an example, local memory management policies (such as the WS policy used in Windows operation systems) ensure that each thread has enough pages to hold its working set [26] and only replace pages from a thread's local page pool. The concept of working set and local replacement is also exploited by CMP caching partitioning schemes (for example, [31, 106, 124, 137]). On the other hand, global replacement policies allow different threads to share space in a fine-grained manner, but are prone to inter-thread pollution. WSClock [15] achieves the benefits of both schemes by integrating working set based partitioning with global LRU replacement. In a similar vein, we integrate cache partitioning and LRU-based latency reduction policies to combine their strengths. Verghese et al. [131] recognized the need for performance isolation in a central server environment, and proposed mechanisms to provide isolation under heavy load while allowing sharing under light load. Later, the notion of performance isolation was extended to provide flexible QoS [125], including QoS guarantee, fairness and differentiated services. Waldspurger [132] introduced several novel policies for virtual machine servers, which: (1) identify and reclaim least valuable pages, (2) eliminate redundancy overhead and (3) support performance isolation. CMP cooperative caching attacks the same problems for processor caching via: (1) global replacement of inactive blocks, (2) replication-aware replacement and (3) cache partitioning support. To simultaneously improve throughput, fairness and QoS, CMP cooperative caching can also extend spatial partitioning with timesharing, which has been well studied and implemented by operating systems [50, 133].

13.2.2.4   CMP Cache Schemes

Besides studies in CMP cache organizations [56, 96, 126], many hybrid caching schemes have been proposed to reduce the average memory access latency for CMPs [10, 11, 20, 49, 57, 82, 98, 119, 139, 140]. Due to the importance of CMP caching, the list of proposals is long and keeps growing. Instead of an exhaustively comparison, below we will provide an overview and taxonomy of prior schemes.

13.2.2.5   Shared Cache Based Proposals

Oi and Ranganathan [98] made the analogy between CMP caching and CC-NUMA memory organization, and evaluated the benefit of using part of the shared L2 cache as fix-sized victim caches. They discovered that aggressive replication (using large victim caches) can hurt performance, and for SPLASH2 benchmarks [136], suggested using 1/8 of total L2 capacity for victim caching. Zhang and Asanovic [140] proposed a dynamically adaptable form of victim caching, called Victim Replication (VR), in a tile-based CMP. Their scheme allows an L1 victim to be cached in the local L2 bank, potentially evicting data without replicas. A random replacement policy and a directory-based coherence protocol are required by their implementation to simplify replication control and replica identification. Because VR keeps replicas in both the home node cache and all consumer caches, it can waste capacity when little data sharing exists (for example, in multiprogrammed workloads). This issue was resolved by victim migration [139] via home block migration, implemented with extra shadow tags to keep track of the migrated data.

Beckmann and Wood [11] studied CMP-NUCA schemes to mitigate the impacts of increasing on-chip wire delay. Their results showed that, different from single-core caching, dynamic migration (D-NUCA) in a CMP is ineffective for widely shared data. They also identified the issues with CMP-NUCA implementation and power consumption, and proposed using LC transmission lines to reduce wire delay. Li et al. [82] extended their work with 3D die-stacking and network-in-memory, and demonstrated better performance. CMP cooperative caching techniques are orthogonal to these new technologies and can potentially exploit fast wires and 3D caches to improve latency and capacity. Chishti et al. proposed CMP-NuRapid [20] to optimize replication and capacity in CMP caches. Their design, like CMP-NUCA, uses individual algorithms to optimize special sharing patterns (that is, controlled replication for read-only sharing, in-situ communication for read-write sharing, and capacity stealing for non-sharing). In contrast, CC aims to achieve these optimizations through a unified technique: cooperative cache placement/replacement, which can be implemented in either a centralized or distributed manner.

CMP-NuRapid implements a distributed directory/routing service by maintaining forward and reverse pointers between the private tag arrays and the shared data arrays. This implementation requires extra tag entries that may limit its scalability, and increase the complexity of the coherence protocol (for example, the protocol has to avoid creating dangling pointers). CC tries to avoid such issues by using a simple, centralized directory engine with less space overhead. Based on NUCA, Huh et al. [57] introduced a cache organization to support a spectrum of sharing degrees, which denote the number of processors sharing a pool of their local L2 banks. The average access latency can be optimized by partitioning the aggregate on-chip cache into disjoint pools, to fit the running application's capacity requirement and sharing patterns. Their study showed that static mappings with a sharing degree of 2 or 4 can provide the best latency, and dynamic mapping can improve performance at the cost of complexity and power consumption. CC is similar in trying to support a spectrum of sharing points, but achieves it through cooperation among private caches and adaptive cooperation throttling. Cho and Jin [21] recently proposed an OS-level page allocation approach to address CMP caching's locality, capacity and isolation issues. Based on shared cache organization, their proposal maps physical pages into cache slices to exploit locality and uses 'virtual multi-cores' to provide isolation between multi-programmed threads. However, this approach requires both hardware page mapping support [67], and significant modifications in the operating systems (for example, location-aware page allocation, locality and capacity aware OS scheduling). Comparatively, CC answers CMP caching challenges with simple hardware extensions, while being amenable to software-based solutions by providing capacity sharing and isolation mechanisms.

13.2.2.6 Private Cache Based Proposals

Harris proposed synergistic caching [49] for large scale CMPs, in which neighboring cores and their private L1 caches are grouped into clusters to allow fast access of shared data. Synergistic caching uses three duplication modes (that is, beg, borrow and steal), each corresponding to replication, use without replication and migration. Because no single duplication mode performs best across all benchmarks, reconfiguration was suggested, although not evaluated, to statically or dynamically choose the best mode. Speight et al. [119] studied adaptive mechanisms in private cache based designs to reduce off-chip traffic. They used an L2 snarf table to identify locally evicted blocks that might be reused soon. Upon local eviction, such blocks are kept on-chip via write-backs to peer caches. The host cache will replace either invalidated or shared clean blocks to make room for them, potentially reducing expensive off-chip misses when they are reused later. This scheme is limited by only supporting multithreaded workloads, and differs from CC in its write-back and global replacement policies. Motivated by the need for dynamic adaptation, Beckmann et al. [10] proposed Adaptive Selective Replication (ASR) for multithreaded workloads. ASR uses private caches and dynamically seeks an optimal degree of replication for each individual thread. The cost and benefit of current and future replication levels are estimated using on-chip counters, which are considered when determining whether a thread can benefit from more aggressive replication. Their cost/benefit estimation mechanisms can be used to throttle CC's replacement-aware cache replacement. Although ASR was proposed to control replication in multithreaded workloads, its per-thread threshold and counters do support heterogeneous workloads and can potentially be extended and used in a multiprogramming environment.

13.2.3 CMP Cache Partitioning Schemes

Below we survey CMP cache partitioning proposals that prevent destructive inter-thread interference via resource isolation, according to their different optimization purposes.

Miss reduction. Stone et al. [121] studied the problem of partitioning cache capacity between different reference streams, and identified LRU as the near-optimal policy for their workloads. They also showed empirically that LRU can swiftly adapt to working set changes, without explicit repartitioning support. Liu et al. [84] proposed Shared Processor-based Split L2 cache that partitions the shared L2 space in units of 'split.' Their scheme can be configured to suit for both inter-application and intra-application non-uniform caching requirements, but involves both profiling support and operating system modification to determine and enforce cache partitioning. Suh et al. [123, 124] applied way partitioning to shared CMP caches. Using in-cache monitoring mechanism, their partitioning algorithm assumes convex miss rate curves and allocates extra capacity to threads having the best marginal miss rate reduction. Qureshi and Patt [106] proposed UMON sampling mechanism to provide more precise measurement, and lookahead partitioning algorithm to handle workloads with non-convex miss rate curves. Dybdahl et al. [32] extended way partitioning [124] by overbooking cache capacity to account for non-uniform per-set requirements, and evaluated its effectiveness using private L1/L2 caches with a shared L3 cache. Dybdahl and Stenstrom [31] further extended CC with an adaptive shared/private partitioning scheme to avoid inter-thread interference. Their partitioning algorithm is essentially the same as Suh's proposal, but instead of in-cache monitors, 'shadow tags' are used to measure the benefit of having one extra cache way.

Fairness improvement. Kim et al. [73] emphasized the importance of fair CMP caching, discussed the implication of unfairness (such as priority inversion) and proposed a set of fairness metrics as their goal of optimization. They evaluated both static and dynamic partitioning (both requiring profiling information), and discovered that, under heavy caching pressure, fair caching often improves overall throughput. Yeh and Reinman [137] proposed fast and fair partitioning, based on a NUCA cache consisted of ring-connected distributed banks. Their scheme ensures the 'baseline fairness' by guaranteeing QoS for all co-scheduled threads. To enforce partitioning decision, each NUCA bank is divided between portions used by the local thread and remote threads. Such partitioning is dynamically adjustable based on program requirements and phase changes. Instead of cache partitioning, Kondo et al. [74] applied dynamic voltage and frequency scaling (DVFS) to maintain CMP fair caching. Using the same metrics and policies as Kim et al., they also observed throughput improvement over shared cache for many cases and energy saving due to decreased voltage and frequency.

QoS provisioning. Iyer [63] motivated the importance of QoS guarantee and prioritization, not only between different users but also different types of access streams (for example, demand vs. prefetch requests) generated by the same thread. He also proposed the CQoS framework that defines and implements QoS via priority classification, assignment and enforcement. Yeh and Reinman [137] focused on throughput improvement on top of data QoS guarantee. Kannan et al. [70] studied CMP resource management to support flexible QoS. Their work demonstrated the feasibility of QoS-aware hardware and software using prototypes and showed predictable performance in multiprogramming and virtualization environments. Vardarajan et al. [129] proposed molecular caches to accommodate the diverse requirements of multiprogrammed workloads. By varying the number of allocated tiles (or molecules) and the per-tile management policies, molecular caches can provide varied cache line sizes, associativities and cache sizes for different co-scheduled threads. Molecular caches achieve power efficiency via private tile based organization, and provides software-defined QoS (specified as target miss rates) by dynamically adjusting per-application capacity.

Generic support. Rafique et al. [108] and Petoumenos et al. [104] proposed spatially fine-grained partitioning support, which can be used by various partitioning policies (such as miss rate reduction, fair caching and QoS provision). Hsu et al. [55] studied various partitioning metrics and policies. Their study focused on three caching paradigms (communist caching for fairness, utilitarian caching for overall throughput and uncontrolled capitalist caching), and recognized the difficulties in improving both overall throughput and fairness using a single partitioning scheme.

13.2.4 A Taxonomy of CMP Caching Techniques

Table 13.2 provides our taxonomy of related hardware CMP caching proposals. These schemes are classified along three dimensions: (1) goals (latency reduction, interference isolation, or their combinations), (2) target workloads (multithreaded or multiprogrammed) and (3) baseline cache organizations (shared or private). A desirable CMP caching scheme should be able to simultaneously achieve multiple optimization goals, perform robustly for a wide range of workloads, while being amenable to simple and modular hardware implementations.

The first group of proposals in Table 13.2 aims at memory latency reduction, but 5 out of 9 of them are limited in only supporting multithreaded workloads or being ineffective for workloads with significant data sharing. The schemes that do support both multithreaded and multiprogrammed workloads are all based on a shared cache design, and they either rely on specific hardware implementations (for example, victim migration [139] requires a directory protocol and CMP-NuRapid [20] depends on a snooping bus) or have to cooperate with software for cache management (for example, NUCA substrate [57] and OS-level page mapping [21]).

The schemes in the second group use cache partitioning to achieve some of the interference isolation goals. Among them, the first five schemes only minimize off-chip miss rate to improve overall throughput. Except for profile-based partitioning [84], these schemes are only applicable to multiprogrammed workloads. Share/private partitioning [31] is the only proposal that combines cache partitioning with latency reduction techniques (it uses CC as the baseline design). Other cache partitioning schemes either focus only on fairness or QoS improvement, or only provide generic partitioning mechanisms. Except for fast and fair [137] which improves throughput while maintaining QoS, none of these proposals support multiple optimization goals. Finally, CC is the only hardware CMP caching scheme that addresses all aspects of latency reduction and interference isolation optimizations, supports both multithreaded and multiprogrammed workloads and exploits the latency, power and design modularity benefits of private caches.

13.3 CMP Cooperative Caching Framework

Consider the task of building a CMP cache hierarchy that not only has high capacity and low latency, but also is fair, reliable, power-efficient and easy to design and verify. Each of these requirements may prefer a different way to organize the available resources and a different policy to strike the balance between resource sharing and isolation, while all optimizations have to fit in the same design. As shown previously, existing proposals do address some of the key challenges of CMP caching. However, without a unified solution, these schemes cannot satisfy all of these important, yet potentially conflicting, requirements for workloads with diverse caching characteristics. We need a holistic approach to address the challenges of CMP caching, which can

Table 13.2 A Taxonomy of Hardware CMP Caching Schemes. Reproduced by permission from Doctoral Dissertation 'Cooperative Caching for Chip Multiprocessors'. University of Wisconsin at Madison, Madison, WI, USA © 2007

Goals			Target Workloads	Baseline	Schemes
Latency Reduction			Multithreaded	Private	Adaptive L2 snarfing [119], ASR [10], Synergistic cache [49]
			Ineffective for shared data and multiprogrammed	Shared	CMP-DNUCA [11], Victim replication [140]
			Multithreaded and multiprogrammed	Shared	Victim migration [139], CMP-NuRapid [20], NUCA substrate [57], OS page mapping [21]
Interference isolation	**Miss reduction**		Multithreaded and multiprogrammed	Shared	Profile-based partitioning [84]
			Multiprogrammed	Shared	Dynamic partitioning [124], Utility-based partitioning [106], Overbooked partitioning [32]
				Private	Shared/private partitioning [31]
	Fairness and QoS		Multiprogrammed	Private	Fast and fair (QoS) [137], Molecular caches [129]
				Shared	Fair caching [73], Fair caching via DVFS [74], CQoS [63], QoS prototypes [70]
	Mechanism		Multiprogrammed	Shared	STATSHARE [104], OS-driven partitioning [108], Hsu et al. [55]
All			Multithreaded and multiprogrammed	Private	Cooperative Caching [17]

integrate and trade off among available optimizations, preferably based on a unified supporting framework. The questions are whether such a framework exists and, if so, how to use it to accommodate conflicting requirements. This section answers the first question with the cooperative caching framework for efficient organization and management of CMP cache resources, while we will demonstrate in later sections its uses in latency reduction and interference isolation, respectively. Below, we introduce the three-layer model of CC framework, the mechanism and implementation layers and the extension and implementation of CC for large-scale CMPs.

13.3.1 CMP Cooperative Caching Framework

The basic idea of CC is to form a globally-managed, aggregate on-chip cache via cooperative resource sharing among private caches. CC is inspired by software cooperative caching algorithms [25], which have been proposed and shown to be effective in the context of file and web caching [25, 37]. The key principle of CC is to support a wide spectrum of sharing behaviours via the combination of private cache's resource isolation with various forms of cooperative capacity sharing/throttling. Such capabilities will be exploited by high-level cache cooperation policies to satisfy different sets of optimization goals. Figure 13.1 shows a picture of the CC concept

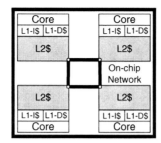

Figure 13.1 Cooperative caching (the shaded area represents the aggregate cache formed via cooperative private caches.). Reproduced by permission from Doctoral Dissertation 'Cooperative Caching for Chip Multiprocessors'. University of Wisconsin at Madison, Madison, WI, USA © 2007.

for a CMP with four cores. To simplify discussion, we assume a CMP memory hierarchy with private L1 instruction and data caches for each processing core. We focus on using L2 cache as the last level on-chip cache, although the ideas of CC are equally applicable if there are more levels of caches on the chip (for example, Dybdahl and Stenstrom [31] evaluated CC with L3 caches). Each processor core's local L2 cache banks are physically close to it, and privately owned by it such that only the processor itself can directly access them. Local L2 cache misses can possibly be served by remote on-chip caches via cache-to-cache transfers as in a traditional cache coherence protocol. The key difference between CC and conventional private caches is that here the private caches are not isolated from each other; instead they act as parts of a logically shared cache by sharing information and resources. Such sharing activities are enabled by CC's cooperation and throttling mechanisms, and orchestrated by cooperation policies to achieve specific optimization goals.

Before elaborating on why we choose a private cache based organization and how to support inter-cache sharing in the CC framework, we now introduce the layered structure of CC framework. As depicted in Figure 13.2, the idea of cooperative caching is supported by three layers focusing on implementation, mechanism and policy, respectively. This separation of layers not

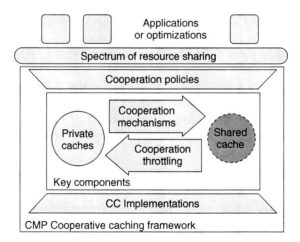

Figure 13.2 The CMP Cooperative caching framework. Reproduced by permission from Doctoral Dissertation 'Cooperative Caching for Chip Multiprocessors'. University of Wisconsin at Madison, Madison, WI, USA © 2007.

only allows their independent extension, but also simplifies the design and analysis of CMP cache hierarchy by isolating and addressing system properties at appropriate levels of abstraction.

Implementation layer. At the lowest level, the layer contains various implementations of cooperative cache structures and resource sharing/isolation mechanisms. This layer encapsulates the implementation details such as cache sub-banking, chip layout, on-chip networks and coherence protocols, so that the key mechanisms can be supported by different implementations and protocols. Correctness and scalability are the main system level properties maintained at this layer.

Mechanism layer. In the centre of the framework, the mechanisms layer provides abstractions of cache organization and resource management to support cooperative caching policies. This layer consists of three key components: (1) **private cache based organization** that provides on-chip locality and resource isolation; (2) **cooperation mechanisms** used by co-scheduled threads to explicitly share the aggregate cache resources; and (3) **cooperation throttling mechanisms** used to control and orchestrate cooperative resource sharing. These components can each be populated with concrete mechanisms, and the key property of this layer is composability. When these mechanisms are combined by caching policies, a wide spectrum of sharing behaviours should be available to suit the needs of specific workloads and cache optimizations.

Policy layer. This layer optimizes the cache hierarchy's high-level properties such as performance, fairness, QoS and reliability, to fit for its intended use. A specific policy can optimize for one property such as fairness, and potentially impact the other properties in positive or negative ways. CMP caching optimizations (for example, power or reliability optimization) or applications (for example, QoS provision) can further select from available policies according to their strengths and shortcomings.

13.3.2   CC Mechanisms

Below we describe the key components in CC's mechanism layer. This layer plays a similar role as the IP layer in the Internet protocol suite [22, 23] because they both aim to support diverse higher level applications with a small set of abstractions that are open to many possible implementations.

13.3.2.1   Private Cache Organization

Compared with many shared cache based CMP caching schemes, CC uses a private cache based organization because it has the following advantages that are likely to be of increasing importance for future CMPs.

1. **Latency.** Private caches reduce on-chip access latency by keeping frequently referenced data locally for fast later reuse.
2. **Bandwidth.** Locally cached data can filter out accesses to remote caches, significantly lowering the bandwidth requirement on the cross-chip interconnection network. This can translate into a simpler and potentially faster network, while the saved area and power budget can be used for other purposes.
3. **Associativity.** Instead of building a highly-associative shared cache to avoid inter-thread conflict misses, the same set-associativity is available for an aggregate cache formed by private caches each with much lower associativity, thus reducing power, complexity and latency overhead.

4. **Modularity.** Compared with a shared cache whose directory information is stored with cache tags and distributed across all banks, a private cache is more self-contained and thus serves as a natural unit for resource management (for example, power off to save energy).

5. **Encapsulation.** Because private caches interact only through exchanges of cache coherence messages, the internal organization and operation of individual caches are encapsulated and thus hidden from other caches and processors. Encapsulation enables CMP caches to have different sizes, associativities, caching policies, voltage/frequency configurations and reliability characteristics to either support heterogeneous processors, or accommodate workloads with diverse performance, power and reliability requirements.

6. **Explicit sharing.** Because encapsulation forces inter-cache resource sharing to be explicit, cooperation throttling is thus easy to implement. Because a cluster of cooperative private caches can be viewed as one larger private cache, explicit sharing based cooperation mechanisms and policies are reusable at the granularity of cache-clusters, thus simplifying the task of composing a large-scale CMP from small-scale clusters of cores and caches.

7. **Adoption of existing hardware optimizations.** Because many microarchitectural optimizations assume a uniprocessor with an exclusively owned cache hierarchy, private cache based designs allow a smooth adoption of such techniques into CMP systems.

8. **Software portability.** Existing parallel software (in particular the operating system) is written for processors with private caches, and thus directly portable for private cache based CMP systems. On the other hand, breaking the assumption of an exclusively owned cache could cause unexpected problems (for example, fairness and security issues [103]).

But partaking of these potential advantages first requires a solution to the major, and predominant, drawback of private cache designs: the larger number of off-chip cache misses compared to the shared cache designs. CC attempts to make the ensemble of private L2 caches (the shaded area in Figure 13.1) appear like a collective shared cache via cooperative resource sharing. For example, cooperative caching will use remote L2 caches to hold (and thus serve) data that would generally not fit in the local L2 cache, if there is spare space available in a remote L2 cache. To support such capacity sharing, CC extends conventional private caches in two ways. It relaxes multi-level inclusion between L1 and L2 caches to enable flexible data placement, and it supports sharing of on-chip clean data to save unnecessary off-chip misses.

13.3.2.2 Non-Inclusive Caches

Capacity sharing among private caches requires decoupled data placement in the L1 caches and their companion L2 caches. Conventionally, multi-level cache hierarchies often employ inclusion, where the lower level caches (which are closer to the processors and often smaller) can only maintain copies of data in their companion higher level caches. Cache inclusion implies that a block will have to be invalidated in an L1 cache when it is evicted from or invalidated in the companion L2 cache. This can greatly simplify the implementation of the coherence protocol because invalidation requests can be filtered by the higher level on-chip caches. However with CC, since the objective is to create a global L2 cache from the individual private L2 caches, maintaining inclusion with only a single L2 cache bank unnecessarily restricts the ability of an L1 cache to buffer a variety of data that reside in on-chip L2 caches. An arbitrary L1 cache needs to be able to cache any block from the aggregate L2 cache, and not only a block from the companion L2 cache bank. Thus, for CC to be effective, the L1 caches have to be either non-inclusive (where a block may be present in either the L1 or companion L2 cache, or both) or exclusive (where a block can be present in either the L1 or companion L2 cache, but not both), that is, be able to cache a block that may not reside in the associated L2 cache bank.

13.3.2.3 Sharing Clean Data among Private Caches

With private L2 caches, memory access can be avoided on an L2 cache miss if the data can be obtained from another on-chip cache. Such inter-cache data sharing (via cache-to-cache transfers) can be viewed as a simple form of cooperation, usually implemented by the cache coherence protocol to ensure correct operation. Except for a few protocols that have the notion of a 'clean owner',[4] modern multiprocessors employ variants of invalidate-based coherence protocols that only allow cache-to-cache transfers of 'dirty' data (meaning that the data was written by a processor and has not been written back to the higher level storage). If there is a miss on a block that only has clean copies in other caches, the higher-level storage (usually the main memory) has to respond with the block, even though it is unnecessary for correct operation and can be more expensive than a cache-to-cache transfer within a CMP. There are two main reasons why coherence protocols employed by traditional multiprocessors do not support such sharing of clean copies. First, cache-to-cache transfer requires one and only one cache to respond to the miss request to ensure correctness, and maintaining a unique clean-owner is not as straightforward as a dirty-owner (that is, the last writer). Second, in traditional multiprocessors, off-chip communication is required whether to source the clean copy from another cache or from the main memory, and the latency savings of the first option is often not big enough to justify the complexity it adds to the coherence protocol. In fact, in many cases obtaining the data from memory can be faster than obtaining it from another cache. However, CMPs have made off-chip accesses significantly more costly than on-chip cache-to-cache transfers; currently there is an order of magnitude difference in the latencies of the two operations. Moreover, unlike traditional multiprocessors, (on-chip) cache-to-cache transfers do not need off-chip communication. Furthermore, a high percentage of misses in commercial workloads can be satisfied by sharing clean data, due to frequent misses to (mostly) read-only data (especially instructions) [7, 9, 85]. These factors make it more appealing to let caches share clean data.[5]

13.3.2.4 Cooperation Mechanisms

By sharing on-chip clean/dirty data and relaxing multi-level cache inclusion, an aggregate cache is formed by the collection of on-chip private caches. This aggregate cache differs from the baseline private cache organization in that, through cache cooperation, its content can be controlled to offer different capacities and latencies for different processor cores.

Cache cooperation, for example to share the aggregate cache capacity, is a new hardware caching opportunity afforded by CMPs. As we shall see, cooperation between caches will require the exchange of information and data between different caches, under the control of cooperation and throttling policies. Such an exchange of information and data, over and above the support of a basic cache coherence protocol, was not considered practical, or fruitful, in multiprocessors built with multiple chips. For example, for a variety of reasons, when a data block is evicted from the L2 cache of one processor, it would not be placed in the L2 cache of another processor. The complexity of determining which L2 cache to place the evicted block into, and transferring the block to its new L2 cache home, would be significant. Moreover, this additional complexity would provide little benefit, since the latency of getting a block from memory would typically be lower than getting it from another L2 cache. But, the situation for CMP on-chip caches

---

[4] For example, the Illinois protocol [99], EDWP protocol [4] and Token Coherence [88].

[5] IBMPower4 [128] and Power5 systems add two states (SL and T) to their MESI-based coherence protocol to select and transfer clean ownership. The first module that fetches a clean data naturally becomes its owner and the ownership transfers to the next requester when the data is forwarded to a different cache. If the data is replaced before requested, the ownership is lost. But with the help of highly associative (8-way) L2 caches, this should happen only infrequently.

is very different: the transfer of information and data between on-chip caches can be done relatively easily and efficiently, while the benefit of cooperation (for example, avoiding costly off-chip misses) can be significant. The CC framework provides cache placement and replacement based cooperation mechanisms to exploit the aforementioned opportunities. These mechanisms add local extensions to each cache in two aspects: (1) what blocks are placed into it and (2) what blocks are replaced (or displaced) from it. CC policies can combine these mechanisms to determine what data are kept in the aggregate on-chip cache, and specifically, in which individual caches.

Cache Placement Based Cooperation
Cache placement based cooperation treats each L2 cache as a black-box: without changing the internal caching operation, it affects cache content by modifying what data can be placed into the cache. This is achieved by modifying an L2 cache's local and remote request streams, as illustrated in Figure 13.3. An L2 cache's local request stream consists of demand misses and prefetches generated by its associated L1 caches. Placement based cooperation can filter such requests to avoid L2 cache pollution. One such example is cache bypassing [69]: by using compiler-generated hints or runtime statistics to predict data locality in large chunks, it only inserts data with good temporal locality in the cache. Similarly, CMPNuRapid [20] reduces unnecessary data replication by only placing data in a local L2 when it has been recently reused (thus is likely to be reused in the future). Both examples selectively bypass L2 placement to improve the effective cache capacity. CC also introduces a remote request stream by placing (or 'spilling') locally evicted blocks into remote on-chip caches (labeled as spill streams in Figure 13.3). Spill allows capacity sharing among peer caches, making all on-chip storage resources available for one processor core to use. Through spill, each cache can observe the reference and reuse streams from both local and remote computation threads. Such information can be used to devise approximate global cache management policies (such as global LRU replacement) without paying the overhead of global coordination and synchronization.

The spill mechanism can be tailored in different ways. For example, cooperation policies can vary in deciding (1) what data can be spilled, (2) which cache to host the spilled data, (3) what data should be replaced to make room for the spilled data, (4) whether spilled data can trigger further spills, and (5) how to balance the competition between local and remote data, and so on. We leave policy decisions for later discussion, but make two mechanism-level decisions here.

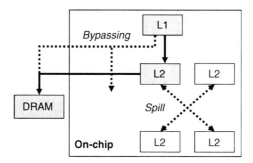

Figure 13.3 Examples of cache placement based cooperation (dotted lines represent data paths introduced by placement based cooperation, while solid lines indicate the original data paths). Reproduced by permission from Doctoral Dissertation 'Cooperative Caching for Chip Multiprocessors'. University of Wisconsin at Madison, Madison, WI, USA © 2007.

First, by default, we choose to randomly pick a host cache while giving higher probabilities to close neighbours. The random algorithm requires no global coordination, and allows one cache's victims to reach all other on-chip caches without requiring 'rippled spilling'. Keeping spilled blocks close to their previous caches can reduce both spilling time and access latency for later reuse. Second, because one of the purposes of spill is to extend the on-chip life cycle of local victims, the host cache should handle a newly-arrived spilled block in the same way as a demand miss. This implies that for LRU-based replacement policies, the spilled block is initially set as the most recently used (MRU) entry in the host cache.

Cache Replacement Based Cooperation

Through placement based cooperation, each on-chip cache can be potentially shared by many computation threads generating heterogeneous memory access streams and having different locality, communication and sharing properties. Consequently, conventional cache replacement policies (for example, LRU, random or pseudo- LRU) that treat all references equally can cause poor QoS and sub-optimal performance. To recognize and exploit data heterogeneity, cache replacement based cooperation combines the default cache replacement with data priority. This mechanism augments each cache block with a few bits for data classification, which can represent compiler generated locality hints [135] or dynamic sharing and communication properties [16, 140]. The cache replacement logic is extended to use such information and always evict blocks with the lowest priority. For example, giving lower priority to data with on-chip replicas can quickly yield space for unique data copies, thus increasing the aggregate cache's effective capacity. If multiple candidates exist in the lowest priority, the default cache replacement policy is used to break even. Because replacement based cooperation only changes how victim blocks are selected, it has no correctness implication and allows more flexible implementations (e.g., trading accuracy or even correctness for simplicity). Replacement based cooperation basically allows the customization of individual caches to prefer different types of data. This mechanism can be combined with spill in various ways to manage the content of the aggregate on-chip cache as well as individual on-chip caches.

Below we discuss three examples. First, caches with the same priority setting can be composed via spilling to collectively replace undesirable data. Consider a 2-core CMP where one core's L2 cache contains only high priority data while the other L2 cache only has low priority data. When isolated, each individual cache has only one class of data, so the prioritized replacement policy falls back to the default replacement policy. However, with the help of spill, the two caches can be glued together as an aggregate cache with two data classes. High priority data spilled from its original cache will cause global eviction of low priority data, leading to better utilization of the aggregate cache. Second, caches with opposite priority settings can be connected via spilling as producer-consumer pairs, where the cache blocks updated by a producer cache are evicted and spilled directly into its consumer cache. Such capability can be exploited by computation with task-level pipeline parallelism for streamlined data communication [44]. Lastly, caches with different performance, reliability and power characteristics can potentially be configured with different priority settings, so that heterogeneous cache resources can be matched to data classes with different properties [58]. On the other hand, priority-based replacement should be used wisely to avoid cache resources being occupied entirely by high priority data, potentially leading to a Denial-of-Service (DoS) attack for low priority data. DoS is not an issue here because we use priority-based replacement only in L2 caches, and only to reduce the amount of data replication (Section 4.2.1). Other uses of priority-based replacement can avoid the DoS problem with a probabilistic implementation, where high priority data have a small chance of being evicted while low priority blocks exist.

13.3.2.5   Cooperation Throttling Mechanisms

Because CC's cooperation is based on modifications of cache replacement and placement logic, it simplifies cooperation throttling to a small set of control points. Below we consider two classes of throttling mechanisms.

Probability Based Throttling
To adjust the amount of cooperation, cooperation probabilities can be specified at the following three control points: (1) local L1 to L2 data path, (2) L2 replacement logic, and (3) L2 spill logic. These probabilities are used to decide how often to apply cooperation instead of taking the default action. Probability based throttling allows CC to provide a wide spectrum of sharing behaviours. If all three probabilities are set to 0, CC defaults to private caches (albeit with support for cache-to-cache transfer of clean data). CC's behaviour moves towards more aggressive resource sharing as these probabilities increase. Probability based throttling can be used for different workloads. For homogeneous workloads where all threads have similar caching behaviour, only one set of system-wide probabilities is needed. However, heterogeneous workloads demand thread-specific probabilities to suit the caching requirements of individual threads. For example, programs with larger working sets can have higher probability for spill and cooperative replacement, while smaller programs can have higher probabilities on L2 bypassing. This way the smaller programs can save space in their local caches for larger programs, and the aggregate cache resource are more efficiently shared.

Quota Based Throttling
CC also supports quota based throttling: each thread's maximum resource consumption can be specified and CC will make sure these quotas are honoured. Resource quota can be either coarse-grained or fine-grained. Many cache partitioning schemes allocate capacity in large chunks, based on way partitioning [124]. Fine grained quota, on the other hand, can be an arbitrary number of cache blocks. CC maintains quota by tracking each thread's resource consumption and replaces data from over-quota threads with data from other threads. CC throttles a thread's data placement decisions based on whether the thread has used up its capacity quota. Specifically, CC disallows a thread to spill locally evicted data to other caches if the thread's current capacity exceeds its quota. As data spilled by another thread replaces data stored in its local cache, an over-quota thread's capacity usage will be decreased. On the other hand, CC avoids selecting an under-quota thread's private cache as a recipient of spilled data, so that this thread's capacity usage will only gradually increase.

13.3.3   CC Implementations

In this section, we present the implementation of the CC framework. Section 3.3.1 enumerates CC's functional requirements, and Section 3.3.2 proposes a possible implementation that exploits a CMP's high-bandwidth, low-latency, on-chip communication network and flexible topology to reduce space, latency and complexity overhead. More scalable CC implementations are possible by extending various existing implementations, which are discussed in Section 3.3.3. Sections 3.3.4 and 3.3.5 extend CC implementation and policy to support large-scale CMPs.

13.3.3.1   General Requirements

The functional requirements for CC are described before and summarized below.

- **Cache coherence extensions.**   Beyond a conventional cache coherence protocol for non-inclusive caches (for example, implemented by Piranha [8]), CC also requires support for cache-to-cache transfer of clean blocks and block spill. As part of the coherence protocol, this support has to be implemented correctly.

- **Cache replacement extensions.** Data classification information needs to be created, exchanged and maintained, while L2 cache replacement logic uses such information to support priority-based replacement. Because these modifications only affect the selection of eviction candidates, an incorrect or slow implementation should only cause performance degradation rather than correctness problems.

- **Extensions to support throttling.** To allow L2 cache bypassing, L1 caches need to directly write to the L2 write-back buffer (assuming write-back caches), which is straightforward to implement. CC also adds (1) extra states to track the amount of capacity used by each core (which can be imprecise) and (2) extra logic to decide whether to use cooperation and whether/where to spill. Such extensions do not affect correctness, thus can be imprecise or slow.

13.3.3.2 Cluster-Based CMP Organization

In this section we detail our proposed implementation of CC using a specialized, on-chip, centralized directory, which will be used to evaluate the performance of both private cache organization and CC in Section 4. We focus on the implementation of cache coherence extensions because they are critical for correctness. It should be noted that this design can be used for other CMP systems while CC can also be implemented in various other ways.

Centralized On-Chip Directory
Our design is based on a directory protocol, which has two advantages over a snooping protocol. (1) Latency. In a snooping protocol, every L2 miss incurs long-latency arbitration overhead to gather responses from all on-chip caches. A directory protocol can reduce such overhead into a request transfer and a directory lookup. The reduced network switching activities can also save active power. (2) Bandwidth. Compared with snooping, a directory protocol can significantly reduce the number of broadcast requests and network bandwidth requirement. However, implementing a directory protocol has to solve two challenges: directory storage overhead and protocol design complexity.

The proposed implementation is based on a MOSI directory protocol to maintain cache coherence, but improves over a traditional directory-based system in several ways: (1) To reduce storage requirements, the directory memory for private caches is implemented by duplicating the tag structures of all private caches, requiring only 3% extra cache space (Table 13.3); (2) The directory is centralized to serve as the only serializing point for cache coherence, which can greatly simplify the implementation of the directory protocol; (3) Located at the centre of the chip, the directory can provide fast access to all caches; (4) The directory is connected to individual cores using a special point-to-point ordered request network, separate from the network connecting peer caches for data transfers.

Figure 13.4 illustrates the major on-chip structures for an 8-core CMP. The Central Coherence Engine (CCE) embodies the directory memory and coherence engine, whose internal structure and directory memory organization is shown in Figure 13.5. The CCE consists of spilling buffers and the directory memory connected with router queues for incoming and outgoing messages. The spilling buffer is organized as a circular buffer, where each valid entry stores an in-flight spilling block (data and state) and its host cache ID. The lookup, insertion and deletion of spilling buffer entries will be discussed later (in Section 3.3.2). The CCE's directory memory is organized as a duplication of all private caches' tag arrays, similar to [95]'s remote cache

Table 13.3 CCE Storage Overhead for an 8-Core CMP with 1 MB 4-Way Per-Core L2 Cache. Reproduced by permission from Doctoral Dissertation 'Cooperative Caching for Chip Multiprocessors'. University of Wisconsin at Madison, Madison, WI, USA © 2007

Component	Location	Size (KB)
Tag extension (2-bit)	Caches	17.0
Processor ID (3-bit)	Caches	25.5
Tag duplication	Directory	221.0
Spilling buffers	CCE	8.0
Total (3.12%)		271.5

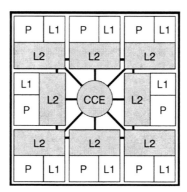

Figure 13.4 Private caches with a centralized directory. Reproduced by permission from Doctoral Dissertation 'Cooperative Caching for Chip Multiprocessors'. University of Wisconsin at Madison, Madison, WI, USA © 2007.

shadow directory. Because CC requires the private caches to be non-inclusive, the CCE has to duplicate the tags for all cache levels. The tags are indexed in exactly the same way as in an individual core. A tag entry consists of both the tag bits and state bits.

In our implementation, the directory memory is multi-banked using low-order tag bits to provide high throughput. Incoming coherence requests trigger lookups in both the spilling buffer and the directory memory. A directory lookup will be directed to the corresponding bank, and to search the related cache sets in all cores' tag arrays in parallel. The results from all tag arrays are gathered to form a state/presence vector as in a conventional directory implementation. An individual block's state may be updated according to the request type and current coherence state. The coherence engine will finish processing the request by generating requests for invalidation, data forwarding, or replies with data or acknowledgement messages. The latency of a directory lookup is expected to be almost the same as a private L2 cache tag lookup, since the extra gather step should only marginally increase the latency.

The coherence engine maintains each processor's duplicate tag arrays to reflect what blocks are stored in its corresponding local caches and what their states are. For non-inclusive caches, a cache block can remain in a processor core's private cache hierarchy (L1 and L2 caches) while moving frequently between local caches. CCE only keeps track of block installation and eviction from one processor's entire private cache hierarchy, because its correct operation only

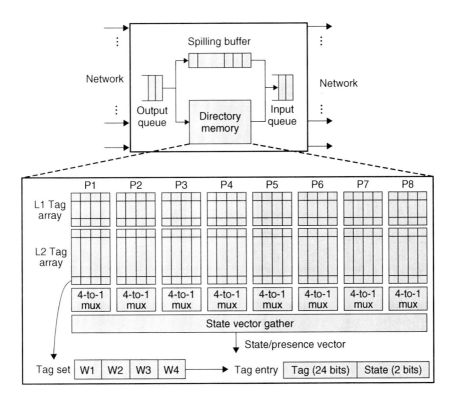

Figure 13.5 CCE and directory memory structure (8-core CMP with 4-way associative L2 caches. Reproduced by permission from Doctoral Dissertation 'Cooperative Caching for Chip Multiprocessors'. University of Wisconsin at Madison, Madison, WI, USA © 2007.

requires knowledge on whether a processor's private cache hierarchy has a block, but not its precise location within the hierarchy. However, lack of such information requires the CCE to carefully manage its tag arrays to avoid conflicts. Specifically, when allocating a new block in the directory, the CCE first attempts to allocate in the L2 tag array before filling the L1 tags. This is because for large L2 caches, blocks from multiple L2 sets can be mapped to the same L1 cache set and potentially cause an overflow in the L1 cache set. Filling the L2 tags first will guarantee the directory can find a free L1 tag when it is needed. Conversely, when CCE evicts a block from an L2 tag array, it also checks whether any L1 tags can be mapped and immediately moved to the corresponding L2 set. By keeping the L2 tag arrays as full as possible, the CCE ensures that no overflow occurs in its L1 tag arrays.

Table 13.3 lists the storage overhead for an 8-core CMP with a 4-way associative 1 MB per-core L2 cache, 2-way 32K split L1 instruction/data caches, and 8-entry per-core spilling buffers. The tag bits storage overhead is estimated assuming a system having 4 Terabytes of physical memory, and a 128-byte block size. The total storage needed for extra tag bits (recording information used for cache cooperation), processor ID (used for quota-based cooperation throttling), duplicate tag arrays and spilling buffers is 271.5 KB, increasing the on-chip cache capacity by 3.12% (or 6.10% for a 64-byte block size). This ratio is similar to Piranha [8] and lower than CMP-NuRapid [20]. Table 13.4 shows CCE's relative space overhead for several different CMP configurations. Although the absolute storage size increases with the number of cores and per-core cache size, the relative overhead remains stable and actually slightly decreases. We do not model the area of

Table 13.4   CCE Storage Overhead under Different CMP Configurations. Reproduced by permission from Doctoral Dissertation 'Cooperative Caching for Chip Multiprocessors'. University of Wisconsin at Madison, Madison, WI, USA © 2007

Configuration	Variable Parameter Value (Overhead)		
8-core, 4-way (varied L2 size)	512 KB (3.30%)	1 MB (3.12%)	2 MB (2.98%)
1 MB, 4-way (varied CMP size)	4-core (3.20%)	8-core (3.12%)	16-core (3.07%)

the separate point-to-point network as it requires the consideration of many physical constraints, which is not the focus in this book. However, we believe it should be comparable to that of existing CMP's on-chip networks.

Cache Coherence Extensions
Besides maintaining cache coherence, the CCE also needs to support cooperation-related on-chip data transfers – (1) cache-to-cache transfers of clean data and (2) spills. The implementation of these functions is discussed below.

Sharing of clean data. To support cache-to-cache transfers of clean data, the CCE needs to select a clean owner for a miss request. By searching the CCE directory memory, the CCE can simply choose any cache with a clean copy as the owner and forward the request to it, which will in turn forward its clean block to the requester. This implementation requires no extra coherence state or arbitration among the private caches. On the other hand, the CCE has to be notified when private caches replace clean blocks, in order 41 to keep the directory state updated. This requirement is met by extending the baseline cache coherence protocol with a 'PUTS' (or PUT-Shared) transaction, which notifies the CCE about the eviction of a clean block. On receiving such a request, the CCE will invalidate the block in the corresponding core's tag arrays.

Spill. Figure 13.6 illustrates two implementations of spill using coherence messages communicated among the spilling cache, CCE and host cache. In a pull-based implementation

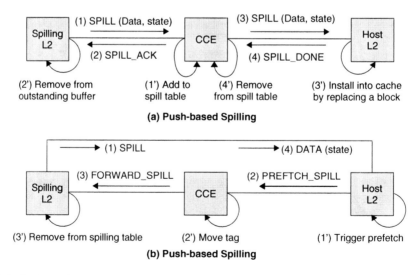

Figure 13.6   Push- and pull-based spill. Reproduced by permission from Doctoral Dissertation 'Cooperative Caching for Chip Multiprocessors'. University of Wisconsin at Madison, Madison, WI, USA © 2007.

(Figure 13.6(a)), a local victim is locally buffered while the evicting cache notifies a randomly chosen host cache to fetch the block. The host cache then issues a special prefetch request to pull the block. In addition to serving the prefetch request, the spilling cache also transfers the state bits and removes its local copy (thus migrating the block to the host cache). The implementation of pull-based spill is straightforward as most modern processors support prefetching. As shown in Figure 13.6(b), push-based spill consists of two pairs of data transfer and acknowledge messages. The first transfer is the same as a normal write-back initiated by the private cache. Upon receiving a spilled block, the CCE temporarily buffers the data in its spilling buffer and acknowledges the sender. The second transfer ships the block from the CCE to the chosen host cache. The host cache treats the incoming data similarly as a demand request, allocates space for it by possibly replacing another block, then acknowledges the directory to release the spill buffer. Race conditions can occur when the host cache issues a request for the spilled block during the second data transfer, in which case the CCE will receive the request message instead of the acknowledgement. The CCE handles it by searching the spilling buffer and releasing the entry with both matching block address and host cache ID. Similar as update based coherence protocols, push-based spill can have deadlock issues, which are often solved by using different virtual channels for different types of communications. We have implemented push-based spill in our simulator to prove its feasibility, and avoided deadlock by using a dedicated virtual channel for block spilling.

Data Classification. Because the coherence engine can observe all on-chip transactions, it has been used extensively to detect sharing, communication and synchronization patterns [24, 59, 71, 77, 78, 80, 81, 93, 109, 120]. The CCE can classify data according to their coherence states or behaviours, and use such information for CC's prioritized cache replacement. For example, Section 4 presents replication-aware cache replacement that tries to keep unique data on chip. CCE detects unique on-chip copies when a write-back leaves only one cache holding the data, and communicates this information to that cache with a notification message.

13.3.3.3 Other Implementation Options

This section discusses other possible implementations of the three key components in CC's mechanism layer: the private cache organization, cooperation mechanisms and throttling mechanisms.

Cache Coherent On-chip Private Caches

Cache coherence among on-chip private caches can be maintained by snooping or directory based protocols, as well as token coherence [88]. Multiple private caches can be connected via snooping buses as in traditional SMPs, and CMPs such as IBM Power4 and Power5 systems [128]. Non-inclusive L1/L2 caches are also supported by previous CMP designs. For example, Piranha [8] uses shadow tags to encode L1 cache states on the shared L2 cache side. Similarly, each private L2 cache can include duplicated L1 tags (with cache states) to simplify the implementation of cache coherence. Snooping requests can be filtered by simultaneously looking up both L2 and duplicate L1 tags, and forwarded to an L1 cache only if the data is actually stored in it. A similar implementation can be used for token coherence, as demonstrated in [9], while further optimizations can use soft-state directory information to reduce the number of broadcast requests.

Implementing private caches with a directory protocol is less straightforward, because the naive implementation of on-chip directory memory can incur prohibitive storage overhead. A concise on-chip directory can be implemented by duplicating private L1 and L2 cache tags (as shown in Section 3.3.2), or using a sparse directory cache [46]. The latter approach works well for workloads with good directory reference locality, but misses in the directory cache either incur expensive off-chip misses or require eviction of valid on-chip blocks.

Many options exist to support on-chip sharing of clean data. The clean owner can either be encoded in the protocol as a new state, or selected via arbitration of peer caches in a snooping protocol [128], or chosen by the directory if it keeps track of clean data write-backs (Section 3.3.2).

Supporting Cooperation

Spill-based cooperation can be implemented in either a 'push' or 'pull' strategy, which have been discussed in Section 3.3.2. Both placement and replacement based cooperation need to associate policy-specific information with individual cache blocks, which can be recorded in the cache tag with a few bits. These bits are initialized upon cache allocation, possibly using information associated with the newly arrived data block. Updates of these bits are triggered by external events, which can either be actively observed by caches in a snooping protocol or generated by the directory when it detects state changes. Because these bits are only used to make caching decisions, and not involved in cache coherence and computation efforts, they are allowed to be imprecise or out-dated without causing correctness problems. Prioritized replacement can be implemented with extra circuitry in the cache replacement logic. Figure 13.7 illustrates its integration with two representative cache replacement policies, assuming N classes of data are prioritized in an M-way associative cache.

- **LRU replacement.** As shown in Figure 13.7(a), each cache block is associated with N-1 priority bits, each bit indicating whether it belongs to a certain data class between priority 1 and N-1. A block belongs to class N (the highest priority level) if all priority bits are 0. We can also view the array of priority bits in a cache set as N-1 class membership vectors (each vector has M bits corresponding to the M blocks). In a stack-based implementation [90], a block's priority bits move along with it to reflect changes of the block's position in the LRU

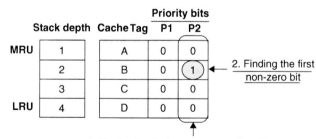

(a) LRU replacement

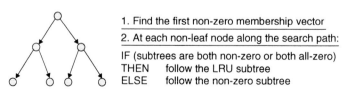

(b) Pseudo-LRU replacement

Figure 13.7 Implementing priority-based replacement (N = 3, M = 4). Reproduced by permission from Doctoral Dissertation 'Cooperative Caching for Chip Multiprocessors'. University of Wisconsin at Madison, Madison, WI, USA © 2007.

stack. The replacement candidate is selected from the lowest-priority non-zero membership vector, and the victim should have the lowest stack position among the blocks belonging to the selected priority class. This implementation can also be used for other stack-based replacement algorithm (such as random replacement [90]).

- **Pseudo-LRU replacement.** Figure 13.7(b) shows a tree-based pseudo-LRU implementation [117]. To integrate with priority-based replacement, we first select the lowest-priority non-zero membership vector. At each non-leaf node along the binary search path, the LRU-based search logic is augmented to also consider priority information. Specifically, the augmented logic selects the LRU subtree if the membership vectors for both subtrees are simultaneously non-zero or all-zero; otherwise, it selects the subtree with a non-zero membership vector. The size and complexity of these extra circuits grow with both levels of priority N and cache associativity M. We set (N, M) to be (2, 4) here, while expecting N to be less than 4 and M no more than 8, so these changes can only add minimal storage and latency overhead. More aggressive assumptions are made by other CMP caching proposals because write-backs are not on the critical path and they have negligible performance impact. For example, victim replication [140] uses a 4-level prioritized replacement in a 16-way set-associative shared cache.

Supporting Throttling

To support cooperation throttling, each private cache should include two sets of registers: the 'knob' registers to store specified probabilities or quota limits; while the 'measurement' registers to save performance measurements which are fed back to adjust the degree of cooperation throttling. For quota-based throttling, every cache block includes a processor-ID field to indicate for which core's data it stores. Each private cache maintains a set of counters to reflect the number of cache blocks used by each thread, as well as the number of invalidated/unused blocks. By periodically sharing such information among different cores, CC can monitor the capacity usage of different cores.

Cooperation throttling can be either static or dynamic. Setting the throttling knobs statically is straightforward and requires no special support. Dynamic throttling consists of a feedback loop where current throttling performance and program behavior changes are used to adjust the degree of future throttling. We assume that throttling decisions are made by the hardware, but CC itself has the flexibility to support software controlled adaptation. The software can periodically read the 'measurement' registers, make adaptation decisions and update the 'knob' registers, while CC is responsible for enforcing the specified throttling decisions.

13.3.4 CC for Large Scale CMPs[6]

The advent of CMPs has changed the scaling trend from boosting frequency into increasing the number of on-chip cores [102]. With Intel announcing its 80-core CMP [62] and Tilera's 100-core general purpose processor [110], computer architects are now starting to consider how to build and use CMPs with 1000 cores in a few technology generations [5]. In this section, we discuss several directions in improving the scalability of CC and outline the distributed implementation of CC (DCC) for large scale CMPs.

We believe that CC's private cache organization is essential for highly scalable CMPs due to its modularity and locality benefits. It will be difficult for a shared cache to support hundreds of threads, because the L1 miss traffic can saturate the on-chip network, and the needed cache associativity to avoid inter-thread conflict misses may incur prohibitive overhead. CC

---

[6] Reproduced by permission from Doctoral Dissertation 'Cooperative Caching for Chip Multiprocessors'. University of Wisconsin at Madison, Madison, WI, USA © 2007.

addresses the off-chip bandwidth bottleneck of private caches through inter-cache cooperation, which may limit the scalability in two aspects. The first bottleneck for current CC design is its central-directory based implementation. As the number of cores managed by the CCE increases, contention within CCE will delay cache coherence operations and cooperation activities.

Scalability barriers also exist at the policy level even if we have a scalable cache coherence protocol. As the number of cores increases, the average latency for cooperative capacity sharing among all cores (for example, via spilling/reusing) will also grow, eventually hitting a point where the capacity benefit provided by global sharing is offset by the growing on-chip communication overhead. Cooperative capacity sharing policies should consider such tradeoffs and limit cooperation within a group of closely located caches. Other cache optimizations may also prefer such a scoping policy if their algorithms cannot scale to hundreds of cores. A natural way to accommodate these requirements is to build large-scale CMPs via composition of small scale clusters (for example, 4–8 cores) and reuse the current CC design within each cluster. Comparing to a flat directory protocol possibly embedded in a mesh-base on-chip network, the hierarchical design can significantly improve latency and reduce bandwidth by exploiting intra-cluster data/communication locality. This approach provides a smooth transition path for small-scale workloads because it requires no extra modifications of cooperation policies and incurs little extra performance overhead. Figure 13.8 illustrates such a hierarchical design for a 128-core CMP with 16 8-core clusters. Within each cluster, the CCE is augmented to maintain cache coherence at two levels. Intra-cluster coherence is provided by the central directory (CCE) as discussed in Section 3.3.2, and inter-cluster coherence is achieved with a directory organization where, through address space partitioning, each CCE serves as the home node for a fraction of physical addresses. Multiple memory controllers (MC) are used, each responsible for servicing DRAM accesses generated by one or more neighbouring inter-cluster directories.

This design mitigates the implementation bottleneck by limiting the number of cores within each cluster, while encapsulating each cluster as a single core with private caches to build

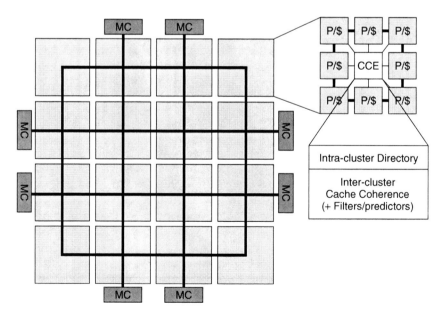

Figure 13.8   128-core CMP with 16X 8-core clusters. Reproduced by permission from Doctoral Dissertation 'Cooperative Caching for Chip Multiprocessors'. University of Wisconsin at Madison, Madison, WI, USA © 2007.

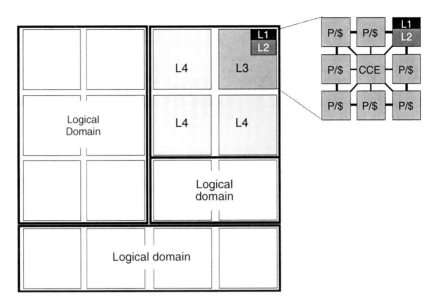

Figure 13.9 Logical cooperation domains (L3 = the aggregate cache within a cluster; L4 = the aggregate cache within a logical domain). Distributed Cooperative Caching, Enric Herrero et al., Proceedings of the 17th International Conference on Parallel Architectures and Compilation Techniques, © 2008 Association for Computing Machinery, Inc. Reprinted by permission.

larger systems. Such encapsulation not only reduces inter-cluster traffic for data forwarding, block invalidation and write-back, but also decouples intra- and inter-cluster coherence, allowing flexible combination of coherence protocols at different levels.

Treating a cluster as a single core (by aggregating reference streams and cache resources) also allows the reuse of previously proposed cooperation mechanisms and policies across multiple clusters. Instead of using a fixed scope of cooperation defined by the cluster boundary, inter-cluster cooperation will take place in 'logical domains' (as shown in Figure 13.9), which can be statically or dynamically formed according to management domains, communication patterns, or data locality. With cooperative caching at different levels, a single core can have a deep on-chip cache hierarchy consisting of its private L1 and L2 caches, as well as the L3 and L4 caches formed through intra- and inter-cluster cooperation.

13.3.5 Distributed Cooperative Caching[7]

Herrero et al. proposed the Distributed Cooperative Caching (DCC) [51] to addresses the CC's scaling issues in many-core CMPs. In such a scenario, the coherence engine must be able to distribute requests across all nodes to avoid the bottleneck of accessing a centralized architecture and avoid the limit on the number of tag checks that must be done on each request.

The Central Coherence Engine is partitioned into multiple Distributed Coherence Engines (DCE) each is responsible for a portion of the memory address space. The number of DCEs and the number of entries per DCE are independent on the cache sizes. Figure 13.10 shows the memory structure for a dual-core CMP (for simplicity). Addresses in the Coherence Engines are mapped in an interleaved manner. On a local L2 cache miss, the corresponding DCE for

_____

[7] Distributed Cooperative Caching, Enric Herrero et al., Proceedings of the 17th International Conference on Parallel Architectures and Compilation Techniques, © 2008 Association for Computing Machinery, Inc. Reprinted by permission.

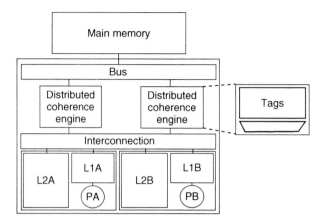

Figure 13.10 DCC memory structure. Distributed Cooperative Caching, Enric Herrero et al., Proceedings of the 17th International Conference on Parallel Architectures and Compilation Techniques, © 2008 Association for Computing Machinery, Inc. Reprinted by permission.

that address is accessed and, if the cache entry is found in the DCE, the request is redirected to the owner.

The organization of the directory in a DCE is completely different to the one in the CCE. In the DCC framework, tags are distributed in an interleaved way across the DCEs in order to distribute DCC requests across the network and hence avoid bottlenecks. This distribution implies that, unlike the centralized configuration that has a tag for each cache entry, tag entries are allocated just in one DCE depending on its address. As a result, if the entries are not perfectly distributed in the address space, we can have more entries in the caches than in the DCE and a DCE replacement is going to be triggered. For this reason, it is necessary to extend the cache coherence protocol to handle the invalidation of cache blocks due to DCE replacements.

Figure 13.11 shows the organization of the Coherence Engine for both Centralized and Distributed versions of Cooperative Caching. We can see that the organization of the centralized version is formed by a unique structure that has the replicated tags distributed in banks, each one representing a cache. In the DCC, we have an arbitrary number of Coherence Engines that store tags from all caches. The number of tags compared for every request is also significantly smaller in the DCC scheme. In this distributed implementation, the number of checked tags depends on the associativity that the designer assigns to the DCEs as compared to the aggregate associativity of all private L2 caches in the original CCE design. The example of Figure 13.11 shows the Coherence Engines of an 8-core CMP with 4-way associative caches. We can observe that for the centralized version 64 tags are compared while for the distributed version only 4 are compared.

In addition, the Distributed Cooperative Caching also allows hardware design reuse since its modular and scalable structure can be replicated as we add more processors on a chip. As an example of possible configurations in a 16-core CMP, Figure 13.12(a) shows a configuration with one DCE per core and Figure 13.12(b) shows a configuration with one DCE shared by all 4 processor cores.

Although the performance of each configuration will greatly depend on the number of entries and associativity of the DCEs, for a given overall equal size and associativity, the configuration with less DCEs than nodes increases the performance/Watt efficiency metric by 5%. This improvement is explained by the reduction in the average distance to the DCE. Other solutions have also appeared [53] to increase further the energy efficiency by a Distance-Aware spilling or by a Selective Spilling.

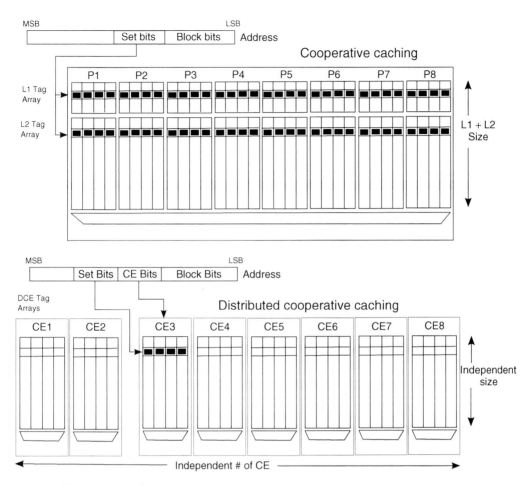

Figure 13.11 Coherence engine structures. Distributed Cooperative Caching, Enric Herrero et al., Proceedings of the 17th International Conference on Parallel Architectures and Compilation Techniques, © 2008 Association for Computing Machinery, Inc. Reprinted by permission.

13.3.5.1 DCE Replacement Policy

To show the benefits of the DCE tag replacement policy, Figure 13.13 illustrates the working principle of the Centralized and the Distributed versions of Cooperative Caching. The situation depicted shows the L2 caches and Coherence Engines of a system with two nodes (A and B) for simplicity. It considers the situation of two threads, one per node, that make an extensive use of their caches. It is also considered that node A always makes requests slightly before than node B. Blocks in the cache are represented by the letter of the requesting node and a number that indicates the time when that block was requested. We start in a warmed-up situation where both caches are full to see how replacements are handled.

In the upper part of the figure the behaviour of the Centralized Cooperative Caching is shown. Suppose that node A makes a request for a new block (Action 1), since the block is not in the local L2, the CCE is checked. Because the block is neither in any other cache, memory is then accessed (Block A5 in Action 2). Since there is not enough space, block A1 is spilled to node B.

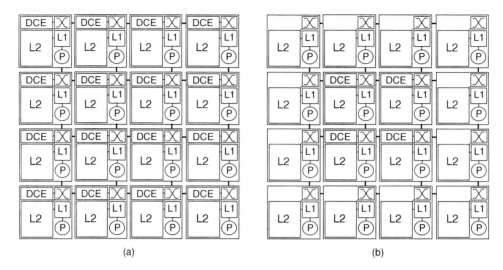

Figure 13.12 (a) DCC16CE and (b) DCC4CE organizations for an example 16-core CMP. Distributed Cooperative Caching, Enric Herrero et al., Proceedings of the 17th International Conference on Parallel Architectures and Compilation Techniques, © 2008 Association for Computing Machinery, Inc. Reprinted by permission.

Block B1 is evicted from the chip since rippled spillings are not allowed. In the second request, node B asks also for a block to the CCE (Action 3). Request is forwarded to memory that sends the block to the requester (Action 4). Since there is not enough space, a replacement is done. The locally oldest block, B2, is spilled to node A; evicting from the chip A2.

In the bottom part of Figure 13.13 the behaviour of the DCE is depicted. As in the previous case block A5 is requested (Action 1), but now to the corresponding DCE. Since the block is not in any other cache, memory is accessed. In this configuration when the block is sent to the DCE it generates an eviction. In order to make the example more interesting, although the result is the same, block B1 is replaced, invalidating the entry in the L2 (Action 2). Then the block is allocated in the corresponding DCE and sent to the requesting node (Action 3). Since the cache is full, block A1 is spilled to node B and is placed in the invalidated entry.

In the second request node B accesses also the DCE and memory asking for the block (Action 4). When block B5 is allocated in the DCE, it triggers also another replacement. In this case the oldest block of the set in the DCE is evicted (Action 5), this is A1. Finally B5 is sent to cache B and allocated where the invalidated block was.

The right parts of these sub-figures show the final state of caches after requesting blocks A6 and B6. It is clear from the result that in the distributed version cache blocks are closer to the requesting node, improving access latency. We can also see that the distributed version also keeps all the newer blocks in the cache, reducing the number of off-chip accesses. This is because the DCE may have inherently data from all cores so that the oldest blocks in the CMP from that set are replaced first.

13.3.5.2 Summary of DCE vs. CCE

The main differences between these two implementations are:

• In the centralized version tags are replicates of their corresponding caches while in the distributed version tags are ordered like a big shared cache in the DCEs. Since tag entries are not

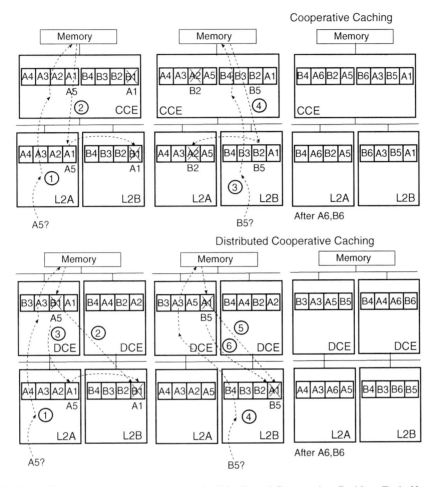

Figure 13.13 DCE operations: a working example. Distributed Cooperative Caching, Enric Herrero et al., Proceedings of the 17th International Conference on Parallel Architectures and Compilation Techniques, © 2008 Association for Computing Machinery, Inc. Reprinted by permission.

restricted to represent only one cache entry, this organization makes a more efficient use of them. Furthermore, the distributed organization does not require the reallocation a tag when a block is spilled or allocated in another cache. It is only necessary to update the tag, which has associated the energy saving benefits.

- The number of tags checked per request in the DCE is equivalent to its own associativity and independent of the associativity of the L1 and L2 caches. In the CCE the number of tags checked is (#L1*L1 Associativity + #L2*L2 Associativity). This directly translates into reduced energy consumption.
- DCE implements a LRU replacement policy that favours a broad view of evicted blocks instead of the individual replacement of private caches in the centralized version. This allows a more efficient use of cache capacity.
- The size of the DCE is independent of the sizes of the L1s and L2s while in the CCE the number of tags must be the same as the number of L1 and L2 cache entries. On the other hand, the coherence protocol of the DCE framework needs to be able to handle DCE replacements.

- DCE makes use of several coherence engines that can be distributed across the chip. Such an organization avoids bottlenecks in the on-chip network and allows parallel request handling. This will become more important as the number of processors on a chip increases.

13.3.5.3 DCE Performance and Power Evaluation

The benefits of DCE have been evaluated using the Simics [87] full-system execution-driven simulator extended with the GEMS toolset [89] and a power model based on Orion [134] and Cacti [113]. Table 13.5 shows the values for the important configuration parameters. SPECOMP2001 benchmarks [6] are used with the reference input sets. The Distributed Cooperative Caching framework has been compared against traditional organizations such as shared or private last level cache as well as cooperative caching with CCE. Two-level caches are connected with a MOESI cache coherence protocol. All simulations use a local and private L1 cache and a shared/private L2 cache for every processor.

Figure 13.14 shows the speedups of the various schemes over the shared cache organization for CMPs with 8, 16 and 32 cores. CCE outperforms private and shared caches for smaller number of cores, but becomes less competitive than shared cache for 32-core CMPs. However, DCE outperforms all other schemes for all core counts and its performance advantage increases with core count.

The power contribution of the cache and the on-chip network subsystems vary by the core count. For a small number of complex cores, the power consumption of cores dominate, but for many simple cores the percentage of interconnect power grows. Figure 13.15 shows the power/performance tradeoff for the caches and interconnect subsystems only. Private caches have power/performance benefits for a limited number of cores. As the core count increases, however, the performance of private caches degrades dramatically due to off-chip misses. On the other hand, the power consumption of shared caches and CCE increases significantly for large number of cores, due to network traffic growth for shared cache and CCE energy consumption for CC. Across various configurations, DCE is the most energy efficient.

Figure 13.16 shows the power/performance results when including the CPU power component. Although the memory system power differences become less important when considering the entire chip power, DCE is still the most energy efficient for a 32-core CMP. It shows an average improvement of 2.61X over a traditional shared memory configuration and a 3.33X over the Cooperative Caching framework. The performance of DCE for 32-core CMPs is 21% better than a shared cache and 57% better over CC with centralized coherence engine.

Table 13.5 Configuration Parameters. Distributed Cooperative Caching, Enric Herrero et al., Proceedings of the 17th International Conference on Parallel Architectures and Compilation Techniques, © 2008 Association for Computing Machinery, Inc. Reprinted by permission

Parameter	Value	Parameter	Value
Number Processors	8-16-32	**Block size**	64 bytes
Instr Window/ROB	16/48 entries	**L1 I/D Cache**	16 KB, 4-way
Branch predictor	YAGS	**L2 Cache**	512-256-256 KB, 4-way
Technology	70 nm	**Network type**	Mesh with 2 VNC (3-cycle per hop)
Frequency	4 GHz	**Link BW**	16 bytes/cycle
Voltage	1.1 V	**Memory bus latency**	250 cycles

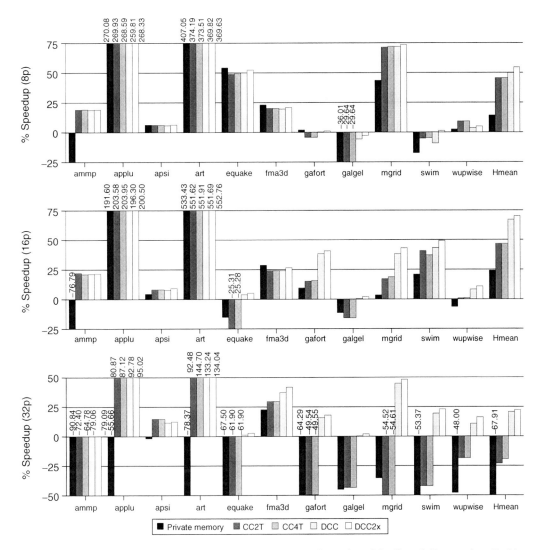

Figure 13.14 Normalized speedup over shared memory configuration. Distributed Cooperative Caching, Enric Herrero et al., Proceedings of the 17th International Conference on Parallel Architectures and Compilation Techniques, © 2008 Association for Computing Machinery, Inc. Reprinted by permission.

13.3.6 Summary

To meet both performance and non-performance related, yet potentially conflicting cache requirements, a wide spectrum of application/optimization specific cache resource sharing behaviours are needed. Because neither private nor shared cache organization can answer these challenges, we advocate a unified framework to manage the aggregate CMP on-chip cache resources. The proposed CC framework includes three key mechanisms. (1) Private cache based organization provides both latency/bandwidth benefits and resource isolation. CC also removes the inclusion restriction within a processor core's multi-level private cache hierarchy for flexibility and supports cache-to-cache transfers of clean data to avoid unnecessary off-chip misses. (2) Cache

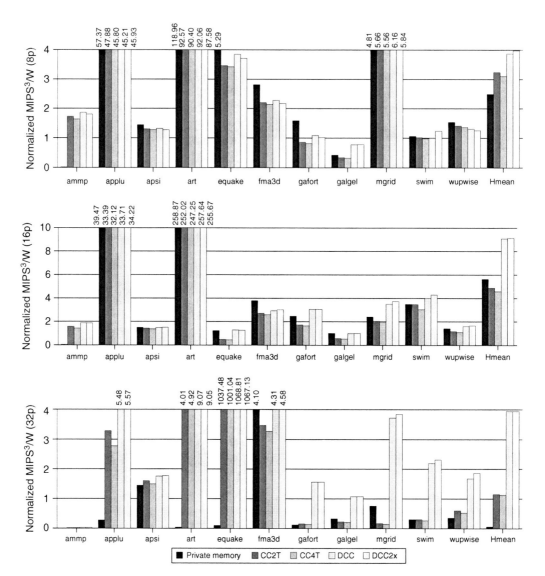

Figure 13.15 Normalized MIPS/W (cache) over shared cache. Distributed Cooperative Caching, Enric Herrero et al., Proceedings of the 17th International Conference on Parallel Architectures and Compilation Techniques, © 2008 Association for Computing Machinery, Inc. Reprinted by permission.

placement and replacement based cooperation mechanisms enable inter-cache resource sharing. (3) Probability and quota based throttling mechanisms can orchestrate and control cooperative resource sharing. Cooperative policies can thus combine these core mechanisms in various ways to suit the resource sharing needs for specific workloads and optimization goals. CC can also be implemented in different ways. Cluster-based CMP organization and Distributed Coherence Engine are two example implementations to support large-scale CMPs. Detailed performance and model evaluation shows that DCE indeed performs significantly better for 32-core CMPs.

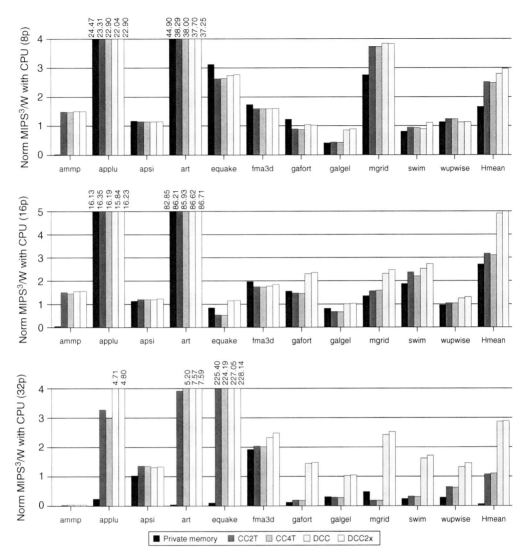

Figure 13.16 Normalized MIPS/W over shared cache (entire chip power). Distributed Cooperative Caching, Enric Herrero et al., Proceedings of the 17th International Conference on Parallel Architectures and Compilation Techniques, © 2008 Association for Computing Machinery, Inc. Reprinted by permission.

13.4 CMP Cooperative Caching Applications

In this section we showcase three applications of CMP cooperative caching: memory latency reduction, adaptive cache partitioning and cache performance isolation.

- CC reduces the average memory access latency by balancing between cache latency and capacity optimizations. Based upon private caches, CC naturally exploits their access latency benefits. To improve the effective cache capacity, CC forms a 'shared' cache using replication control and LRU-based global replacement policies. Via cooperation throttling, CC provides a spectrum of caching behaviours between the two extremes of private and shared caches, thus enabling dynamic adaptation to suit workload requirements. We show that CC can achieve a robust performance advantage over private and shared cache schemes across different processor, cache, and memory configurations, and a wide selection of multithreaded and multiprogrammed workloads.

- To isolate inter-thread caching interference a configuration with dynamically repartitioned shared/private caches is shown. This technique relies on the distributed organization of the Distributed CC and also adds distributed cache repartitioning units to adapt the resource assignation to cache requirements. Nodes with high cache requirements are given a big local private cache space and nodes with low data reuse contribute to increase the size of the global shared cache for spilled blocks.

- Another technique to isolate inter-thread caching interference relies on adding a time-sharing aspect on top of spatial cache partitioning. This approach uses Multiple Time-sharing Partitions (MTP) to simultaneously improve throughput and fairness while maintaining QoS over the longer term. Each MTP partition unfairly improves at least one thread's throughput, and different MTP partitions favouring different threads are scheduled in a cooperative, time-sharing manner to either maintain fairness and QoS, or implement priority. MTPs are also integrated with CC's LRU-based capacity sharing policy to combine their benefits. The integrated scheme – Cooperative Caching Partitioning (CCP) – divides the total execution epochs into those controlled by either MTP or the baseline CC policy, respectively, according to the fraction of threads that can benefit from each of them. Our simulation results show that for a wide range of multiprogrammed workloads, CCP can improve throughput, fairness and QoS for workloads suffering from destructive interference, while achieving the performance benefit of the baseline CC policy for other workloads.

13.4.1 CMP Cooperative Caching for Latency Reduction[8]

We try to optimize the average latency of memory requests with CC by combining the strengths of private and shared caches adaptively. This is achieved in three ways: (1) by using private caches as the baseline organization, CC attracts data locally to reduce remote on-chip accesses, thus lowering the average on-chip memory latency; (2) via cooperation among private caches, it can form an aggregate cache having similar effective capacity as a shared cache, to reduce costly off-chip misses; (3) by cooperation throttling, it can provide a spectrum of choices between the two extremes of private and shared caches, to better suit the dynamic workload behaviour. Our approach attempts to manage the aggregate on-chip caches with a set of unified heuristics. By mimicking a shared cache, CC does not distinguish between different sharing types (for example, private, read-only, read-write) or treat individual threads separately (according to their different working set sizes and locality characteristics). The proposed cooperation policies are conceptually simple, only requiring modifications to the default cache placement and replacement policies, and are easily supported by the CC framework.

13.4.1.1 Policies to Reduce Off-Chip Accesses

Because CC's baseline organization already uses private caches to reduce cache latency, we now consider cooperation policies to efficiently use the aggregate on-chip cache resources and thereby reduce the number of off-chip accesses. We choose to mimic the caching behaviour of a shared cache with a group of cooperative private caches. Compared with private caches, a shared cache makes more efficient use of available capacity in three ways, corresponding to the

---

[8] Reproduced by permission from Doctoral Dissertation 'Cooperative Caching for Chip Multiprocessors'. University of Wisconsin at Madison, Madison, WI, USA © 2007.

three cooperation polices discussed in this section. First, a shared cache uses all on-chip data (both dirty and clean) to satisfy processor requests. On the contrary, traditional private caches only support the sharing of on-chip dirty data. As discussed in Section 3.2.1, CC matches a shared cache by supporting cache-to-cache transfer of clean data. Because unnecessary off-chip accesses are removed when there are clean copies residing elsewhere on the chip, we show later on that this baseline CC design can significantly outperform conventional private caches. Second, a shared cache eliminates replication by storing only one copy of each unique data, while private caches may keep multiple copies of the same data on-chip. We introduce a cooperation policy that replaces replicated data blocks to make room for unique on-chip copies (called singlets), thereby making better use of the available cache resources. Third, a shared cache observes references from all processor cores and chooses replacement victims globally. Consequently, different cache capacities are allocated to different threads according to their requests. On the other hand, cache capacities are statically allocated to different threads in a private cache organization, and each private cache can only observe requests and select replacement victims locally. Using CC's spill mechanism, the last cooperation policy combines local replacement policies with global spill/reuse history to approximate a global replacement policy, thereby keeping potentially more useful data on the chip. Because in the previous section we have described the details of cache-to-cache transfer of clean data, below we will only discuss the two new cooperation policies and their throttling.

Replication-Aware Cache Replacement
The baseline private L2 caches employed by CC allow replication of the same data block in multiple on-chip caches. When cache capacity can sufficiently satisfy the program's working set requirement, replication reduces cache access latency because more requests can be satisfied by locally replicated copies. However, when cache size is dwarfed by the working set size, replicated blocks will compete for the limited capacity with unique copies. CC uses replication-aware data replacement to optimize capacity, which discriminates against replicated blocks in the replacement process. This policy aims to increase the number of unique on-chip blocks, thus improving the probability of finding a given block in the aggregate on-chip cache. We define a cache block in a valid coherence state as a **singlet** if it is the only on-chip copy, otherwise it is a **replica** because multiple on-chip copies exist. We employ a simple policy to trade off between access latency and capacity: evict singlets only when no replicas are available as victims. This can be implemented by CC using prioritized cache replacement. All on-chip data are classified as either singlets or replicas, and replicas are first selected when choosing victims. With CC, a singlet block evicted from a cache can be further spilled into another on-chip cache. Using the aforementioned replacement policy, both invalidated and replica blocks in the receiving cache are replaced first, again reducing the amount of replication. By giving priority to singlets, all private caches cooperate to replace replicas with unique data that may be used later by other caches, further reducing the number of off-chip accesses.

The model here is that a thread runs on a processor core, whose requests are filtered by L1 private caches and triggers capacity allocation in the shared cache. To indicate whether a block is a singlet, each cache tag is augmented with a singlet bit. This bit is advisory and not needed for correctness. Figure 13.17 describes the state diagram for the singlet bit, which can be initialized in two ways: (1) it is set to 0 if the block is first fetched from off-chip memory or as a result of write miss (assuming an invalidation based coherence protocol), or (2) it is set as 1 if the block is forwarded from other caches. In the second case, the forwarding cache also resets its singlet bit to 0, indicating the data now has replicas. The singlet information is also communicated from the directory to on-chip caches: when the directory receives a write back message, or a PUTS message which indicates the eviction of a clean block, it checks the presence vector to see if

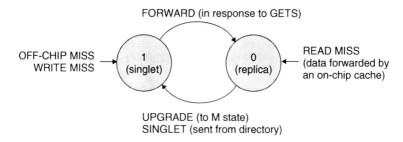

Figure 13.17 State diagram for the singlet bit. Reproduced by permission from Doctoral Dissertation 'Cooperative Caching for Chip Multiprocessors'. University of Wisconsin at Madison, Madison, WI, USA © 2007.

this action leaves only one copy of the data on-chip. If so, an advisory notification message (SINGLET) is sent to the cache holding the last copy of the block, which can set the block's singlet bit to 1.

Global Replacement of Inactive Data

Spilling a victim into a peer cache both allows and requires global management of cooperative private caches. The aggregate cache's effective associativity now equals the aggregate associativity of all caches. For example, 8 private L2 caches each with 4-way associativity effectively offers a 32-way set associativity for CC to exploit. Similar to replication-aware data replacement, we want to cooperatively identify singlet but inactive blocks, and keep globally active data on-chip. This is especially important for multiprogrammed workloads with heterogeneous access patterns. Because these applications do not share data and have little operating system activity, almost all cached blocks are singlets after the initial warm-up stage. However, one program with poor temporal locality may touch many blocks which soon become inactive (or dead), while another program with good temporal locality but large working set will have to make do with its fixed, private cache space, frequently evicting active data and incurring misses. Implementing a global-LRU policy for CC would be beneficial but is also difficult because all the private caches' local LRU information has to be synchronized and communicated globally. Practical global replacement policies have been proposed to approximate global age information by maintaining reuse counters [107] or via epoch-based software probabilistic algorithms [41]. We modify N-Chance Forwarding [25], a simple and fast algorithm from cooperative file caching research, to achieve global replacement.

N-Chance Forwarding was originally designed with two goals: it tries to avoid discarding singlets, and it tries to dynamically adjust each program's share of aggregate cache capacity depending on its activity level. Specifically, each block has a recirculation count. When a private cache selects a singlet victim, it sets the block's recirculation count to N, and forwards it to a random peer cache to host the block. The host cache receives the data, set it as the most recently used (MRU) entry in the chosen cache set and evicts the least active block in its local cache. The life cycle of the spilled block is thus extended, giving it new chances to compete with other cache blocks for on-chip space. If a recirculating block is later evicted, its count is decremented and it is forwarded again unless the count becomes zero. If the block is reused, its recirculation count is reset to 0. To avoid a ripple effect where a spilled block causes a second spill and so on, a cache that receives a spilled block is not allowed to trigger a subsequent spill. The parameter N was set to 2 in the original proposal [25]. A larger N gives singlet blocks more opportunities to

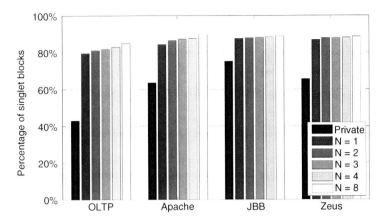

Figure 13.18 Percentages of unique cache blocks in different schemes. Reproduced by permission from Doctoral Dissertation 'Cooperative Caching for Chip Multiprocessors'. University of Wisconsin at Madison, Madison, WI, USA © 2007.

recirculate through different private caches, hence makes it more likely to reduce the amount of replication and improve the aggregate cache's effective capacity. We have studied CC schemes with different N values, and found that increasing N beyond 1 has little additional benefit for CMP caching.

To explain this, we compare the percentages of singlet blocks under various N values in Figure 13.18. It shows that, for 4 commercial workloads with significant data sharing, CC with N = 1 can achieve almost the same level of replication control as CC with N = 8, therefore providing the same level of performance.[9] Figure 13.18 also shows that, even with N = 8, N-Chance Forwarding cannot remove all replicas in the aggregate cache. This is because CC can only reduce replication in cache sets where singlets compete space with replicas, which may not cover all cache sets. We therefore sets N to 1 here, and call the modified policy **1-Fwd**. 1-Fwd dynamically balances each private cache's allocation between local data accessed by its own processor and global data stored to improve the aggregate cache usage. The active processors' local references will quickly force global data out of their caches, while inactive processors will accumulate global data in their caches for the benefit of other processors. This way, both capacity sharing and global age-based replacement are achieved. Implementing 1-Fwd in CC is also simple. Each cache tag needs to be extended with one bit to indicate whether the block was once spilled but has not been reused. Figure 13.19 illustrates the operation of the spilled bit with a state diagram. This bit is initialized to 0 for blocks installed due to local accesses. It is set to 1 when a cache receives the spilled block, and reset to 0 if the block is reused by either local or remote processors. Similar to the singlet bit, the spilled bit is advisory and not needed for correctness.

Cooperation Throttling
At this point, CC can operate in one of two extreme modes: (1) shared cache mode by always using its cooperative capacity improvement policies, and (2) private cache mode that never uses cooperation policies. Now we use probability based cooperation throttling to provide a wide

---
[9] For multiprogramming workloads with little replication, CC with larger N values perform essentially the same as CC with N = 1.

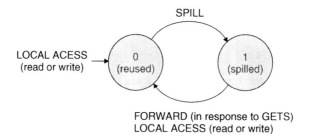

Figure 13.19 State diagram for the spilled bit. Reproduced by permission from Doctoral Dissertation 'Cooperative Caching for Chip Multiprocessors'. University of Wisconsin at Madison, Madison, WI, USA © 2007.

spectrum of caching behaviors between the two extreme modes, and discuss how to dynamically choose the best cooperation probabilities. As discussed in Section 3.2.3, probability based throttling controls how often to apply the cooperation policies. In the context of memory latency reduction, CC behaves more like as a shared cache with higher cooperation probabilities and more like a group of isolated private caches with lower probabilities. Although cooperation throttling is needed for both multithreaded and multiprogrammed workloads, we focus on multithreaded workloads here, and apply quota based throttling later in this section to enforce isolation among multiprogrammed applications.

Several options exist in choosing the best cooperation probabilities for a given workload. Static tuning sets the optimal probability based on profile information, and dynamic tuning adapts by predicting the costs/benefits of various throttling degrees. Beckmann et al. [9, 10] examined the tradeoffs in balancing latency and capacity, and proposed adaptive selective replication (ASR) mechanisms and policies to reach the best replication level. ASR can be integrated with CC to optimize both homogeneous and heterogeneous workloads. Alternatively, we can use dynamic set sampling (DSS) [105] to predict the memory latencies experienced under different cooperation probabilities simultaneously. The key intuition behind DSS is that a caching scheme's impact on the whole cache can be accurately predicted by sampling its impact on a small fraction of cache sets. As shown in Figure 13.20, we divide each L2 cache into 5 disjoint cache set groups: 4 small sampling groups (each having 3% of the total cache sets) and one large group consisting of all the remaining cache sets. Each sampling group uses a different cooperation probability (0%, 30%, 70% and 100%, respectively), and periodically a global selector will

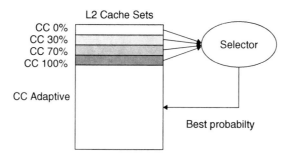

Figure 13.20 DSS-based adaptive throttling. Reproduced by permission from Doctoral Dissertation 'Cooperative Caching for Chip Multiprocessors'. University of Wisconsin at Madison, Madison, WI, USA © 2007.

choose the best performing sampling group (performance measured in average memory latency) and use its cooperation probability in the remaining cache sets. Here we compare the average memory latencies of various throttling options by assuming strong correlation between memory latency and performance, similar to previous proposals [9, 105]. Techniques such as out-of-order processors, prefetching and memory-level parallelism optimizations can break this assumption by partially or totally overlapping cache misses with useful computation or concurrent memory accesses. More accurate prediction can be made by sampling the direct performance measurement such as IPC or user-specified throughput metrics (as suggested in [2]), which is left as future work.

13.4.1.2 Performance Evaluation

Similar to Section 3.5.3 we use a GEMS-based full-system simulator to evaluate the benefits of CC's latency reduction policies. We use a mixture of multithreaded commercial workloads and multiprogrammed SPEC workloads.

Table 13.6 provides more information on the workload selection and configuration. The commercial multithreaded benchmarks include OLTP (TPC-C), Apache (static web content serving using the open source Apache server), JBB (a Java server benchmark) and Zeus (another static web benchmark running the commercial Zeus server) [1]. To compensate for workload variability, we measure the performance of multithreaded workloads using a work-related throughput metric [1, 2] and run multiple simulations with random perturbation to achieve statistically valid conclusions. The number of transactions simulated for each benchmark is listed in Table 13.6. Multiprogrammed workloads are combinations of heterogeneous and homogeneous SPEC CPU2000 benchmarks. We use the same set of heterogeneous workloads as [20] for their representative behaviours, and include two homogeneous workloads with different working set sizes to explore extreme cases. The commercial workloads are simulated with an 8-core CMP,

Table 13.6 Workload Descriptions. Reproduced by permission from Doctoral Dissertation 'Cooperative Caching for Chip Multiprocessors'. University of Wisconsin at Madison, Madison, WI, USA © 2007

Multiprogrammed (4-core)	
Name	**Benchmarks**
Mix1	Apsi,art,equake,mesa
Mix2	Ammp,mesa,swim,vortex
Mix3	Apsi,gzip,mcf,mesa
Mix4	Ammp,gzip,vortex,wupwise
Rate1	4 copies of twolf, small working set (<1 MB)
Rate2	4 copies of art, large working set (>1 MB)

Multithreaded (8-core)		
Name	**Transactions**	**Setup**
OLTP	400	IBM DB2 v7.2 EEE, 25000 warehouses, 128 users
Apache	2500	20000 files (500 MB data), 3200 clients, 25 ms think time
JBB	12000	Sun HotSpot 1.4.0, 1.5 warehouses per core, 44 MB data
Zeus	2500	Event-driven, other configurations similar to Apache

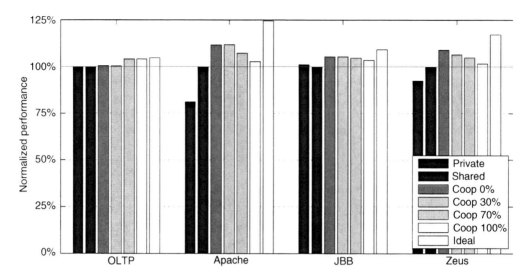

Figure 13.21 Multithreaded workload performance (the 'ideal' scheme models a shared cache with only the latency of a local bank). Reproduced by permission from Doctoral Dissertation 'Cooperative Caching for Chip Multiprocessors'. University of Wisconsin at Madison, Madison, WI, USA © 2007.

while the multiprogrammed workloads use a 4-core CMP, as we believe the scale of CMP systems may be different for servers vs. desktops.

Figure 13.21 compares the performance of private, shared, CC and the 'ideal' caching schemes, as transaction throughput normalized to the shared cache. Four CC schemes are included for each benchmark: from left to right they use system-wide cooperation probabilities of 0%, 30%, 70% and 100% respectively. As discussed in Section 3.2.3, the cooperation probability is used by L2 caches to decide how often to apply replication-aware data replacement and spilling of singlet victims. The baseline CC design (without capacity improvement policies) uses 0% cooperation probabilities, while the default CC scheme uses 100% cooperation probabilities to optimize capacity. We choose only four different probabilities as representative points along the sharing spectrum, although CC can support finer-grained throttling by simply varying the cooperation probabilities. For our commercial workloads, the default CC scheme ('CC 100%') always performs better than the private and shared caches. The best performing cooperation probability varies with different benchmarks, which boosts the throughput to be 5-11% better than a shared cache and 4-38% better than with private caches. CC achieves over half of the performance benefit of the ideal scheme for all benchmarks.

For multiprogrammed workloads, the aggregate IPCs of various schemes, normalized to the shared cache scheme, are shown in Figure 13.22. CC outperforms the private scheme by an average of 6% and the shared caches by 10%. For Rate2, CC performs better than the ideal shared scheme because it has lower miss rate than a shared cache. Private caches run faster than a shared cache for Mix1, Mix3 and Mix4 by reducing cross-chip references, while a shared cache improves Mix2's performance by having many fewer off-chip misses. CC combines their strengths and outperforms both for all heterogeneous workloads. For homogeneous workloads, Rate1 consists of four copies of twolf, whose working set fits in the 1 MB L2 cache, so private caches are the best choice while CC performs slightly worse. CC reduces the off-chip miss rate for Rate2, and consequently improves its performance by over 20%.

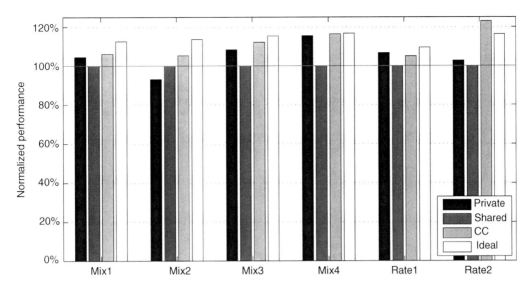

Figure 13.22 Multiprogrammed workload performance. Reproduced by permission from Doctoral Dissertation 'Cooperative Caching for Chip Multiprocessors'. University of Wisconsin at Madison, Madison, WI, USA © 2007.

13.4.2   CMP Cooperative Caching for Adaptive Repartitioning[10]

Cooperative Caching also can be used to manage coherence in an adaptive cache hierarchy able to repartition autonomously in private and shared regions. This allows the creation of big local private caches if all applications have similar cache requirements and a big shared cache if only a few take advantage of extra cache space. In addition, the usage of separate private and shared regions allows to prevent inter-thread interference if several applications with different cache requirements are executed simultaneously.

13.4.2.1   Elastic Cooperative Caching

The Elastic Cooperative Caching (ElasticCC) [52] is a distributed and dynamic memory hierarchy that adapts autonomously to application behaviour. The ElasticCC framework consists of several independent L2 cache memories that are logically divided into a shared and a private region that compete for the cache space. Private regions store all the evicted blocks from the local L1 and shared regions store spilled blocks from neighbouring caches. ElasticCC also adjusts the level of replication by repartitioning caches. Shared data is replicated when requested in the corresponding private regions but is never replicated in the shared region since it stores only unique blocks. Therefore, bigger private regions allow a higher replication and bigger shared regions limit it. To achieve the separation between private and shared cache space we have used the column caching technique that allows separation and dynamic repartitioning without having to invalidate any block.

[10] "Elastic Cooperative Caching: An Autonomous Dynamically Adaptive Memory Hierarchy for Chip Multiprocessors", Enric Herrero et al., Proceedings of the 37th Annual International Symposium on Computer Architecture, © 2010 Association for Computing Machinery, Inc. Reprinted by permission.

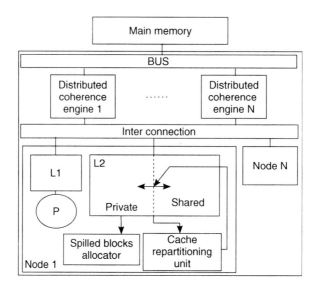

Figure 13.23 Elastic cooperative caching structure. "Elastic Cooperative Caching: An Autonomous Dynamically Adaptive Memory Hierarchy for Chip Multiprocessors", Enric Herrero et al., Proceedings of the 37th Annual International Symposium on Computer Architecture, © 2010 Association for Computing Machinery, Inc. Reprinted by permission.

Figure 13.23 shows the structure of this configuration. To be able to have scalable memory hierarchy organizations Distributed Coherence Engines (DCEs) are used to grant coherence. The extra hardware required for each node is a Repartitioning Unit and a Spilled Block Allocator.

The Cache Repartitioning Unit is responsible for dynamically adjusting the amount of cache that is going to be private or devoted to spilled blocks. The Cache Repartitioning Unit adjusts the proportion of private and shared space locally for every L2 cache, avoiding centralized structures that limit the scalability. Repartitioning is done every fixed number of cycles and the decision is based on the number of hits on the Least Recently Used (LRU) blocks of the shared and the private parts of the cache. Repartitioning is done at a given number of cycles to match program phases behaviour and thresholds are control registers set at boot time.

Since not all nodes have the same shared cache space, the Spilled Block Allocator is responsible for deciding to which node a locally evicted block is spilled. This part is also important because more blocks should be spilled to the nodes with more shared space and less to the ones highly used by the local node. The proposed Spilled Block Allocator uses a Round-Robin arbiter with a bit vector containing the cache partitioning information of each way of each cache. When a block is evicted from the local private partition the arbiter selects the next shared way and spills the block to the corresponding node. Therefore, all shared ways are used equally and in circular order and nodes with more shared space receive more spilled blocks.

This technique is also extended with the Adaptive Spilling mechanism, an extension to ElasticCC, to allow spilling only to applications that benefit from extra cache space (Private and Shared High Utility). Private High Utility applications are detected through block reuse. Since applications with high reuse will have a high number of private ways, spilling is allowed when 75% of the cache (6 ways) is private. For Shared High Utility applications, sharing is detected by monitoring cache-to-cache transfers. The spilling decision is done on a per-block basis, only allowing spilling for truly shared blocks.

13.4.2.2 Evaluation

The ElasticCC has been evaluated and compared against traditional configurations such as shared and private caches, Adaptive Selective Replication (ASR), Distributed Cooperative Caching (DCC) and an upper bound configuration with twice the amount of storage capacity.

The performance graph of Figure 13.24 shows that the Elastic Cooperative Caching outperforms the Distributed Cooperative Caching by an average of 27%, by 12% over the distributed version of Adaptive Selective Replication, by 52% over private caches, and by 53% over a distributed shared cache. The energy efficiency of Elastic Cooperative Caching is showed in Figure 13.24(b). Results in MIPS^3/W are normalized over the DCC configuration. ElasticCC+AS shows a 71% improvement over DCC and 24% over ASR. The more effective usage of shared regions with Adaptive Spilling can be seen in this graph. ElasticCC+AS improves the energy efficiency by 12% over ElasticCC without Adaptive Spilling. This improvement is produced by

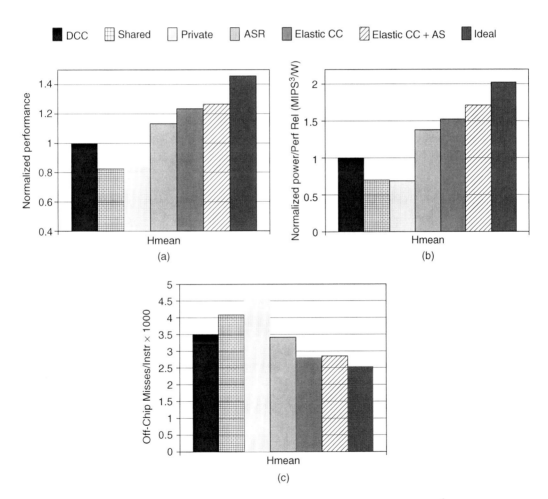

Figure 13.24   (a) Normalized performance, (b) Normalized Energy-Efficiency (MIPS^3/W), (c) Off-chip misses/Instrx1000. "Elastic Cooperative Caching: An Autonomous Dynamically Adaptive Memory Hierarchy for Chip Multiprocessors", Enric Herrero et al., Proceedings of the 37th Annual International Symposium on Computer Architecture, © 2010 Association for Computing Machinery, Inc. Reprinted by permission.

more effective spilling that reduces the network traffic and that avoids unnecessary realloca-
tions of blocks without reuse. Finally, Figure 13.24(c) shows that ElasticCC brings an average
reduction of off-chip misses per instruction of 18.6% over DCC and 16.4% over ASR.

13.4.3   CMP Cooperative Caching for Performance Isolation[11]

We view the sharing of CMP cache resources among threads as a resource management problem.
The properties of a good resource manager have been extensively examined by the operating
system researchers (for example, in [127]), which include (1) abilities to improve overall performance, (2) maintenance of fairness and QoS, and (3) suitability for a wide range of workloads
combinations. Fairness and QoS are especially important for CMP as it is used in consolidated
servers, shared computing clusters, embedded systems, and other platforms where meeting these
requirements is as important as improving overall throughput.

  Previous research in CMP cache management mainly focused on using cache partitioning to
achieve some of these requirements [57, 63, 73, 84, 104, 106, 124, 137]. However, none of these
proposals is sufficient to satisfy all CMP cache management requirements because of two limitations. (1) **Limited functionality**. Prior proposals cannot address all functional requirements,
including thrashing avoidance, fairness improvement, QoS guarantee and priority support, partially due to the difficulty of satisfying multiple, often conflicting, goals in a single cache partition.
(2) **Limited scope of application**. Cache partitioning can only outperform LRU-based latency-
reducing CMP caching schemes for some multiprogrammed workloads. An attempt to solely
use cache partitioning can cause sub-optimal performance for workloads that do not experience
destructive inter-thread interference.

13.4.3.1   Multiprogramming Metrics for CMP Caching

To compare the effectiveness of CMP caching schemes for multiprogramming, we first need to
find proper metrics to summarize the overall performance, fairness and QoS results for a thread
schedule. A multiprogrammed workload's throughput can be simply measured as the sum of
per-thread throughput (that is, IPC for our workloads), but quantifying QoS and fairness can be
hard, and requires an understanding of these notions in the context of CMP caching. Our notions
of performance, fairness and QoS are based on two principles: (1) proportional-share resource
allocation and (2) Pareto efficiency. The first principle states that QoS and fairness is achieved
when the shared resource is divided among sharers in proportion to their priorities or weights
[14, 116, 131, 133].[12] Using proportional-share allocation to maintain the baseline fairness, the
second principle further improves performance (efficiency) by allowing disproportional sharing if
it helps some sharers without hurting the others. These principles have been used to define min-
max fairness [13], which has wide applications in computer networks and scheduling policies
(for example, Generalized Processor Sharing [100]).

$$QoS(scheme) = \sum_{i=1}^{\#app} \min\left(0, \frac{IPC_i(scheme)}{IPC_i(base)} - 1\right)$$

[11] "Cooperative Cache Partitioning for Chip Multiprocessors", Jichuan Chang et al., Proceedings of the 21st annual international conference on Supercomputing, © 2007 Association for Computing Machinery, Inc. Reprinted by permission.
[12] Contention in other shared resources, especially the memory system, can also cause destructive interference. Here we
focus on the impact of destructive interference occurring in the last-level CMP caches, assuming a fair memory system as
proposed in [97].

QoS Metric. QoS is the ability to provide a thread with a guaranteed baseline performance (corresponding to a specific resource partition) regardless of the load placed on the shared resource from other co-scheduled threads [131]. We use **equal-share cache allocation** to define the performance bottom line for QoS, which corresponds to the special case of proportional-sharing when all threads have the same priority. Notice that equal-priority has been implicitly assumed by previous fair caching proposals [55, 73, 137]. The QoS metric is thus defined as the sum of per-thread slowdowns (as negative percentages) over this baseline. Same as [97, 137], we claim a caching scheme can guarantee QoS if this measurement is bounded within a user-defined threshold (for example, -5%). Other ways of measuring QoS exist (e.g., reporting the maximum slowdown or the number of threads that violate QoS), but we use the total slowdown because it captures the behavior of the entire workload and thus is a more stringent criteria.

Fair Speedup Metric. Summarizing the overall performance of multiple benchmarks (co-scheduled threads in our context) has been an extensively discussed topic [68, 114]. We adopt prior wisdom and define the **Fair Speedup (FS)** metric to quantify the overall performance of co-scheduled threads. FS is calculated as the harmonic mean of per-thread speedups over the equal-share allocation baseline. The notion of fair speedup is similar to the fair slowdown metrics proposed by Kim et al. [73], which is measured against a single-thread execution baseline where one thread has exclusive use of all cache resources. Such a baseline is borrowed from SMT processors [115], where it corresponds to the single thread execution mode that allocates all execution and cache resources to one thread. However, single thread execution in a CMP will waste the majority of execution resources. Instead, we choose to use the equal-share allocation baseline because it has better resource utilization by supporting multiple concurrently running threads and performs similarly as in traditional multiprocessors. For the same reason, two other SMT performance metrics using a single-thread execution baseline – weighted speedup (or WS, which is the sum of speedups) [115] and harmonic mean of speedups [86] – are not used.

$$FS(scheme) = \#app / \sum_{i=1}^{\#app} \frac{IPC_i(base)}{IPC_i(scheme)}$$

13.4.3.2 Multiple Time-Sharing Partitions (MTP)

Prior CMP cache partitioning policies use a single spatial partition to achieve their optimization goals. However, it is an intrinsically hard problem to satisfy multiple goals (for example, throughput, fairness and QoS) with a single partition when conflicts arise between competing threads. In this section we add a time-sharing aspect on top of multiple spatial partitions, which uses Multiple Time-sharing Partitions (MTP) to resolve such conflicts over the long term.

Thrashing Avoidance

We first discuss when cache partitioning is needed by examining when destructive interference occurs. Starting with the equal-share allocation baseline, if this configuration can satisfy the caching requirements of every co-scheduled thread, then cache partitioning is not needed because little inter-thread interference exists. Cache partitioning is needed only if some threads experience thrashing with their current capacity allocations. These threads will attempt to acquire extra cache resources from each other and from other threads, which leads to performance, fairness and QoS problems.

Consider partitioning a 4 MB 16-way L2 cache between 4 co-scheduled copies of art. With the equal share allocation of a 1 MB L2 cache, art has a low IPC of 0.066 due to thrashing (over 50 off-chip misses per thousand instructions) as previously shown in Figure 5.6. As more cache resources are allocated, its throughput increases quickly and reaches a saturating point of 0.215 IPC with 1.75 MB capacity. At this point, thrashing can be avoided for 2 threads by unfairly expand each of their capacity allocate to 1.75 MB, and shrink the capacities of the other threads to 256 KB each (0.05 IPC). This partition doubles the total throughput (0.215 * 2+0.05 * 2 = 0.52, which is two times of 0.066 * 4 = 0.264), but is unfair to the shrinking threads.

Fairness Improvement

Cache partitioning between four copies of art is an example of the throughput-fairness dilemma. When the available cache capacity cannot simultaneously satisfy the working set requirements of multiple large threads, compromise has to be made within a single spatial partition. In this example, fair partitions cause thrashing for all threads, while thrashing avoidance requires unfair partitioning. Existing cache partitioning schemes all face this dilemma, but differ in the way they trade off between throughput and fairness.

We resolve this dilemma by learning from a similar example in game theory [35]. Consider two officemates who commute to their workplace, performance is doubled when they carpool but it is unfair because the driver invests more effort and money. Not carpooling is a fair strategy, but is also inefficient. In real life, such games are played daily by the same players who often improve both performance and fairness by 'taking turns' to drive when they carpool. We adopt the same cooperative policy to simultaneously improve throughput and fairness with multiple time-sharing partitions (MTP). Instead of using a single partition that is either low-throughput or unfair, multiple unfair but high-throughput partitions are used in a time-sharing manner to also improve fairness.

Specifically, individual threads are coordinated to shrink and expand their cache allocations in different cache partitions. Within a partition, the spare capacity collected from shrinking threads is used by expanding threads, and different threads are expanded in different partitions. As a thrashing thread goes through shrinking and expanding partitions, its average throughput can be much better than its baseline throughput. This is because a thrashing thread's baseline performance is already low by definition, and shrinking its capacity usually only causes insignificant slowdown. However, it can achieve dramatic speedup in one expanding partition (when the allocated cache can hold its working set) and, on average, the speedup in one expanding partition is often more than what is needed to compensate the slowdowns in multiple shrinking partitions. Overall, the multiprogrammed workload's fair speedup measurement is improved because all expandable threads get a fair chance to speedup.

Figure 13.25 compares three cache partitioning schemes for four copies of art. Single spatial partition based schemes A and B provide the most fair and fast partitions, respectively. Based on MTP, scheme C can both maintain the same level of fairness as scheme A (by equalizing per-thread speedups) and achieve the same high throughput (IPC = 0.52) and weighted speedup (WS = 2.42) of scheme B. Such improvement is reflected by its high FS result (97% and 61% higher than scheme A and B, respectively), but can be overlooked by only comparing IPC or WS results.

Priority Support

MTP extends the option of cache partitioning from the single dimension of space-sharing into two-dimensional time-sharing between spatial partitions. The time-sharing optimization can be applied to any proportional sharing resource partition baseline, thus supporting priority if the priority levels of co-scheduled threads are reflected in the baseline. Priority can also be supported through time-sharing. Instead of giving different threads equal opportunity to speedup, different time-sharing priorities can be assigned to different unfair partitions to deliver differentiated

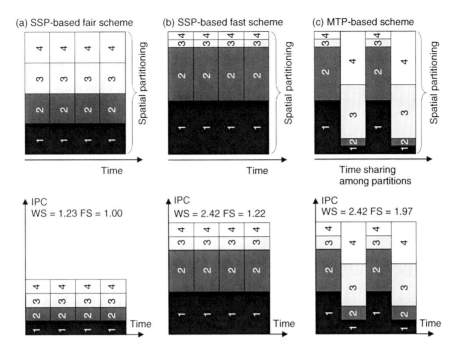

Figure 13.25 Cache partitioning options for a co-schedule of four Copies of art. "Cooperative Cache Partitioning for Chip Multiprocessors", Jichuan Chang et al., Proceedings of the 21st annual international conference on Supercomputing, © 2007 Association for Computing Machinery, Inc. Reprinted by permission.

levels of performance. Because time-sharing based priority support has been well understood and implemented by operating systems [50, 133], MTP can serve as the cache management primitive to the high-level software by focusing on the determination and enforcement of multiple unfair partitions.

QoS Guarantee

QoS can be guaranteed either in a real-time manner or over the long term to meet different application's timing requirements. Real-time QoS is needed only by certain applications (for example, real-time video playback or transaction processing systems), and is not needed for many other programs. For example, users of SPEC benchmarks, batching systems and scientific applications are mostly concerned about total execution time, and thus long-term QoS, often measured over hundreds of millions of cycles.

For single spatial partition (SSP) based cache partitioning schemes, the same cache partition is used repeatedly throughout a stable program phase, these schemes have to guarantee long-term QoS by guaranteeing QoS within every cache partition. In contrast, MTP's cooperative shrink/expand model can be used to guarantee long-term QoS with little loss of performance. To meet the QoS requirement, the MTP partitioning algorithm now uses multiple partitions to maximize FS, under the constraint that each thread's average throughput across multiple partitions is no worse than the equal-share baseline throughput.

To demonstrate MTP's advantage in guaranteeing long-term QoS, Figure 13.26 compares SSP vs. MTP based cache partitioning schemes that optimize FS under different types of QoS requirements: no QoS (SSPnoQoS and MTPnoQoS), real-time QoS (SSPQoS and MTPrtQoS),

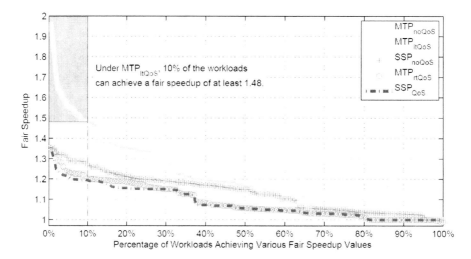

Figure 13.26 FS comparison of SSP and MTP under different QoS requirements. "Cooperative Cache Partitioning for Chip Multiprocessors", Jichuan Chang et al., Proceedings of the 21st annual international conference on Supercomputing, © 2007 Association for Computing Machinery, Inc. Reprinted by permission.

and long-term QoS (MTPltQoS). These results are obtained from offline analysis to demonstrate the performance potential of MTPltQoS implementation.

For each scheme, we plot the percentage of workloads that can achieve various metric values. These curves are essentially Cumulative Distribution Functions (CDF) being transposed, so that a higher curve indicates a better performing scheme. For example in Figure 13.26, each point (X%, Y) on the MTPltQoS curve indicates that, X% of workloads have FS measurements equal to or large than ($\geq$) value Y. Notice that for the same type of QoS guarantee, the ideal SSP scheme can never outperform the ideal MTP because single spatial partitioning is a special case in the MTP model, which is also empirically shown in the figure.

Figure 13.26 shows four distinct curves for five schemes because MTPltQoS and MTPno-QoS have almost the same FS results. With QoS results, we observe that (1) SSPnoQoS and MTPnoQoS cannot bound per-thread slowdown within the user-specified threshold (-5% here); (2) SSPQoS and MTPrtQoS are the worst performing policies (their curves overlap for workloads with smaller FS values), indicating that real-time QoS guarantee can restrict performance optimization; and (3) MTPltQoS can maintain long-term QoS while achieving almost the same performance as the best performing scheme MTPnoQoS. For its performance and QoS benefits, we now use MTPltQoS as the representative MTP policy, and denote it directly as MTP. MTP can be extended to support real-time QoS by reserving guaranteed partitions for real-time applications and optimizing the rest of programs with the remaining capacity.

13.4.3.3 Cooperative Cache Partitioning (CCP)

We now develop cooperative cache partitioning (CCP), a heuristic-based hybrid cache allocation scheme that integrates MTP with CC. CCP consists of three components: (1) a heuristic-based partitioning algorithm to determine MTP partitions; (2) a weight-based integration policy to decide when to use MTP or CC's LRU-based capacity sharing policy (1-Fwd, discussed in Section 4.1); and (3) modifications to the baseline CC design to enforce fine-grained cache partitioning decisions.

CCP Partitioning and Weighting Heuristics

Before MTP partitioning, CCP first gathers each thread's L2 cache miss rates under candidate cache allocations, and uses them to estimate the IPC curve. Miss rates are collected in our simulator in dedicated, online sampling epochs where each thread takes turns to use the maximum amount of cache. We use LRU stack hit counters to estimate miss rates under all possible cache associativities to reduce sampling overhead. Although such overhead can be avoided with the recently proposed UMON online sampling mechanism [106], we include it in our evaluation results.

Using IPC estimations, each thread's guaranteed partition (for real-time QoS guarantee) can be calculated. CCP also initializes each thread's Cexpand to the minimum capacity needed to achieve the highest speedup, and Cshrink to the minimum capacity that can ensure long-term QoS when cooperating with Cexpand. A thread is a supplier benchmark if its Cshrink is the same as its guaranteed partition.

The CCP partitioning algorithm (shown in Figure 13.27) then returns a set of MTP partitions that are likely to outperform CC, using the test of a thrashing benchmark as a simple heuristic. This algorithm has the following three steps: (1) filtering out supplier benchmarks which will not benefit from any partitioning schemes; (2) determine MTP partitions that each favors one thrashing benchmark by starving other thrashing benchmarks with their Cshrink capacity; (3) for MTP partitions where one expanding thread cannot use all the remaining space, expand other threads to further increase speedup. We will describe steps (2) and (3) in detail because step (1) is rather straightforward.

Step (2) determines the set of thrashing benchmarks by removing threads whose speedups are not large enough to guarantee long-term QoS. Each candidate thread is tested by the function

---

Inputs: capacity C, thread set TS, sample results (IPC[i][c], guaranteed partitions g[i]);
Outputs: expanded[i], MTP partitions MTP[p][i]: /* *Thread i 's capacity in partition p* */

/* **Step 1: Filter out supplier benchmarks** */
Identify supplier benchmarks SupplierTS, subtract their g[i] from C;

/* **Step 2: Determine the set of thrashing benchmarks ThrashTS** */
/* init stable = false; ThrashTS=TS-SupplierTS; */
while (ThrashTS is non-empty and ! stable)
    stable=true;
    foreach thread i∈ ThrashTS
        C_{expand}[i]=i's Capacity usage when other threads use their C_{shrink}[i];
        stable &= thrashing_test(i, size(ThrashTS), C_{expand}[i],C_{shrink}[i];

/* **Step 3: Merge multiple expanding thread** */
/* init p = 0; expanded[i] = false; MTP[p][i]=C_{shrink}[i]; */
foreach thread i∈ ThrashTS, p++
    foreach thread j, start from i, in circular order
    MTP[p][j] += minimal remaining capacity for j to achieve its best speedup;
    if (MTP)[p][j]≥C_{expand}[j]) expanded[j]=true;   /* *Expanded in MTP* */

---

    thrasing_test(i, nump, expand, shrink)   /***Key heuristic***/
    if (IPC[i][expand]-IPC[i][base])>(nump-1)*(IPC[i][base]-IPC[i][shrink])
        return true;   /* *large speedup* */
    C_{shrink}[i]=g[i]; C=C-g[i]; remove i from ThrashTS;
    return false;

Figure 13.27 CCP partitioning algorithm. "Cooperative Cache Partitioning for Chip Multiprocessors", Jichuan Chang et al., Proceedings of the 21st annual international conference on Supercomputing, © 2007 Association for Computing Machinery, Inc. Reprinted by permission.

thrashing_test to see if its speedup in one expanding partition can compensate for the total slowdown accumulated in other (shrinking) partitions. The threads that fail the thrashing_test are assigned with their guaranteed partitions and removed from the candidate set, which will reduce the number of candidate partitions, the amount of remaining capacity and possibly remaining candidates' Cexpand and speedups. Such tests are repeated until one of the two termination conditions is satisfied: (1) the candidate set is empty, or (2) all candidate threads pass the test. This step is guaranteed to terminate because each round of tests either reduces the candidate set size which leads to condition (1) in a finite number of steps, or satisfies condition (2).

After step (2), it is possible that in an MTP partition, the expanding thread does not need all the spare space provided by other shrinking threads. Step (3) merges multiple expandable threads in such a partition to further increase speedup. To be fair, the algorithm attempts to expand different sets of threads in different partitions. This algorithm also returns a vector expanded. A thread i benefits from MTP if it is allocated with Cexpand capacity in at least one partition (expanded[i] is true), otherwise it is likely to benefit from CC. This observation leads to the CCP integration heuristic: the execution time is broken into epochs managed by either MTP or CC's LRU-based capacity sharing policy (1-Fwd), weighted by how many threads can benefit from them respectively. For N concurrently running threads, if M of them can be expanded by MTP partitions, then CCP will use MTP for every M out of N epochs and use CC for other epochs. A special case is when no thread is expanded because step (2) cannot find any MTP partitions, in which case CC should be used throughout the execution.

Extending CC to Enforce Capacity Quota

CCP uses the quota-based throttling to enforce MTP partitions. Compared with way partitioning, CC's fine-grained cache-level quota enforcement can support threads with non-uniform capacity demands across different cache sets. For MTP partitions based on way partitioning miss rates, such threads only use part of its capacity quota in their expanding partitions while still achieving large speedups. CCP detects such cases and triggers the partitioning algorithm with newly collected capacity usage, which often leads to better results.

13.4.3.4 Evaluation and Results

CMP caching schemes should be compared using a wide range of multiprogrammed workloads to evaluate their performance robustness. For evaluation purpose, we model a CMP with 4 single-threaded cores and consider all 4-thread multiprogramming combinations (repetition allowed) from 7 representative SPEC2000 benchmarks (ammp, apsi, art, gcc, mcf, twolf, and vpr). There are 210 workloads because the number of K combinations (with repetition) selected from N objects is C(N+K-1,K), so selecting 4-thread combinations from 7 benchmarks can generate 210 workloads. We evaluate the effectiveness of different cache allocation schemes using the same Simics-based full system simulator as previously described. Here the same set of cache/memory/interconnect configuration parameters as in the default setup, but execution is driven by a single-issue, in-order processor model. The simpler processor model allows us to simulate all 210 multiprogrammed workloads in a manageable time frame.

We compare the online simulation results of realistic CCP implementation with offline analysis results of ideal cache partitioning policies (for example, MTP). Because the ideal MTP implementation results were shown to be the performance upper bound of existing SSP cache partitioning schemes in Figure 13.26, we do not compare CCP with realistic implementations of prior cache partitioning proposals.

In the previous section, MTP has already been shown to be better than CC for only a subset of workloads. Since the ideal MTP implementation represents the best cache partitioning results, we

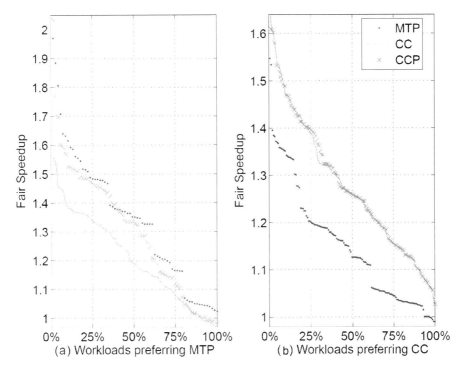

Figure 13.28   MTP, CC and CCP's using the fair speedup (FS) metric. "Cooperative Cache Partitioning for Chip Multiprocessors", Jichuan Chang et al., Proceedings of the 21st annual international conference on Supercomputing, © 2007 Association for Computing Machinery, Inc. Reprinted by permission.

now refer to workloads that prefer CC over MTP as workloads where cache partitioning could hurt performance, and the other workloads as workloads that need the help of cache partitioning. Figure 13.26 compares the performance of CCP (realistic) with MTP (ideal) and CC (realistic) on both classes of workloads. Only FS results are reported because both CCP and MTP can guarantee QoS. As in Figure 13.26, we use transposed CDF curves to show the percentage of workloads that can achieve various levels of performance. Here, a higher curve indicates a better scheme because it achieves better FS measurements across different fractions of the workloads, and the gaps between curves correspond to their performance differences.

Figure 13.28(a) shows that when cache partitioning is needed, CCP achieves comparable performance as MTP (the gap between CCP and MTP curves is small), and significantly better FS values than CC (the gap between CCP and CC is relatively large). The performance difference between CCP and MTP reflects the difference between our practical partitioning heuristic and a less realistic, offline exhaustive search of MTP partitions. For workloads where cache partitioning hurts, Figure 13.28(b) shows that CCP performs almost the same as CC and significantly better than MTP. Together they demonstrate that CCP effectively combines the strengths of both MTP and CC.

13.5   Summary

As CMPs become the mainstream processors, designs of on-chip cache hierarchy will play an important role to provide fast and fair data accesses for multiple, competing processor cores. To offset the negative impact of limited off-chip bandwidth, on-chip wire delay, and hardware/software design complexity, cooperative caching satisfies the two main goals of CM caching – latency reduction and shared resource management – with a unified supporting framework and two sets of cooperation policies. We have demonstrated the effectiveness of the CC framework with three important CMP caching applications, and we believe the cooperative sharing mechanisms and the philosophy of using cooperation for conflict resolution can be applied to many other resource management problems.

References

[1] A. R. Alameldeen, M. M. K. Martin, C. J. Mauer, K. E. Moore, M. Xu, D. J. Sorin, M. D. Hill, and D. A. Wood. Simulating a $2M commercial server on a $2K PC. *IEEE Computer*, 36(2): 50–57, Feb. 2003.

[2] A. R. Alameldeen and D. A. Wood. IPC Considered Harmful for Multiprocessor Workloads. *IEEE Micro*, 26(4): 8–17, 2006.

[3] AMD. AMD Multi-core. http://multicore.amd.com/en/, 2006.

[4] J. K. Archibald. A Cache Coherence Approach for Large Multiprocessor Systems. In Proceedings of the 2nd International Conference on Supercomputing, 1988.

[5] K. Asanovic, R. Bodik, B. C. Catanzaro, J. J. Gebis, P. Husbands, K. Keutzer, D. A. Patterson, W. L. Plishker, J. Shalf, S.W. Williams, and K. A. Yelick. The Landscape of Parallel Computing Research: A View from Berkeley. Technical Report No. UCB/EECS-2006-183, EECS Department, University of California, Berkeley, 2006.

[6] V. Aslot, M. J. Domeika, R. Eigenmann, G. Gaertner, W. B. Jones, and B. Parady. SPECOMP:A New Benchmark Suite for Measuring Parallel Computer Performance. In Proceedings of the International Workshop on OpenMP Applications and Tools, 2001.

[7] L. A. Barroso, K. Gharachorloo, and E. Bugnion. Memory System Characterization of Commercial Workloads. In Proceedings of the 25th Annual International Symposium on Computer Architecture, pages 3–14, 1998.

[8] L. A. Barroso, K. Gharachorloo, R. McNamara, A. Nowatzyk, S. Qadeer, B. Sano, S. Smith, R. Stets, and B. Verghese. Piranha: A Scalable Architecture Based on Single-Chip Multiprocessing. In Proceedings of the 27th Annual International Symposium on Computer Architecture, 2000.

[9] B. Beckmann. Managing Wire Delay in Chip Multiprocessor Caches. PhD thesis, University of Wisconsin-Madison, 2006.

[10] B. M. Beckmann, M. R. Marty, and D. A. Wood. ASR: Adaptive Selective Replication for CMP Caches. In Proceedings of the 39th Annual International Symposium on Microarchitecture, Dec. 2006.

[11] B. M. Beckmann and D. A. Wood. Managing Wire Delay in Large Chip-Multiprocessor Caches. In Proceedings of the 37th Annual International Symposium on Microarchitecture, Dec. 2004.

[12] L. A. Belady. A Study of Replacement Algorithms for a Virtual-Storage Computer. *IBM Systems Journal*, 5(2): 78–101, 1966.

[13] D. Bertsekas and R. Gallager. *Data networks* (2nd ed.). Prentice-Hall, Inc., 1992.

[14] J. Bruno, E. Gabber, B. Özden, and A. Silberschatz. The Eclipse Operating System: Providing Quality of Service via Reservation Domains. In USENIX 1998 Annual Technical Conference, 1998.

[15] R. W. Carr and J. L. Hennessy. WSClock – a Simple and Effective Algorithm for Virtual Memory Management. In 8th ACM Symposium on Operating Systems Principles, 1981.

[16] J. Chang and G. S. Sohi. Cooperative Caching for Chip Multiprocessors. In Proceedings of the 33th Annual International Symposium on Computer Architecture, June 2006.

[17] J. Chang and G. S. Sohi. Cooperative Cache Partitioning for Chip Multiprocessors. In Proceedings of the 21th ACM International Conference on Supercomputing (ICS'07), 2007.

[18] A. Charlesworth, N. Aneshansley, M. Haakmeester, D. Drogichen, G. Gilbert, R. Williams, and A. Phelps. The Starfire SMP interconnect. In Proceedings of the 1997 ACM/IEEE conference on Supercomputing, 1997.

[19] Z. Chishti, M. D. Powell, and T. N. Vijaykumar. Distance Associativity for High-Performance Energy-Efficient Non-Uniform Cache Architectures. In Proceedings of the 36th International Symposium on Microarchitecture, 2003.

[20] Z. Chishti, M. D. Powell, and T. N. Vijaykumar. Optimizing Replication, Communication and Capacity Allocation in CMPs. In Proceedings of the 32th Annual International Symposium on Computer Architecture, June 2005.

[21] S. Cho and L. Jin. Managing Distributed, Shared L2 Caches through OS-Level Page Allocation. In MICRO-39, 2006.

[22] D. D. Clark. The Design Philosophy of the DARPA Internet Protocols. ACM SIGCOMM Computer Communication Review, 18(4): 106–114, 1988.

[23] Corporate Computer Science and Telecommunications Board. Realizing the Information Future: the Internet and Beyond. National Academy Press, Washington, DC, USA, 1994.

[24] A. L. Cox and R. J. Fowler. Adaptive Cache Coherency for Detecting Migratory Shared Data. In Proceedings of the 20th Annual International Symposium on Computer Architecture, May 1993.

[25] M. Dahlin, R. Wang, T. E. Anderson, and D. A. Patterson. Cooperative Caching: Using Remote Client Memory to Improve File System Performance. In Proceedings of the First Symposium on Operating Systems Design and Implementation, 1994.

[26] P. J. Denning. The Working Set Model for Program Behavior. *Communication of ACM*, 11(5): 323–333, 1968.

[27] P. J. Denning. Thrashing: Its Causes and Prevention. In AFIPS 1968 Fall Joint Computer Conference, volume 33, pages 915–922, 1968.

[28] P. J. Denning. Virtual Memory. ACM Computing Survey, 2(3): 153–189, 1970.

[29] P. J. Denning. The Locality Principle. Communication of ACM, 48(7): 19–24, 2005.

[30] R. Drost and I. Sutherland. Proximity Communication and Time. In Proceedings of the 11th IEEE International Symposium on Asynchronous Circuits and Systems, 2005.

[31] H. Dybdahl and P. Stenstrom. An Adaptive Shared/Private NUCA Cache Partitioning Scheme for Chip Multiprocessors. In Proceedings of the 13th Annual International Symposium on High- Performance Computer Architecture, 2007.

[32] H. Dybdahl, P. Stenstrom, and L. Natvig. A Cache-Partition Aware Replacement Policy for Chip Multiprocessors. In Proceedings of 13th International Conference of High Performance Computing (HiPC), 2006.

[33] H. Dybdahl, P. Stenstrom, and L. Natvig. An LRU-based Replacement Algorithm Augmented with Frequency of Access in Shared Chip-Multiprocessor Caches. In Proceedings of the 6th Workshop on Memory Performance: Dealing with Applications, Systems and Architecture (MEDEA 2006), 2006.

[34] J. Edward G. Coffman and P. J. Denning. Operating Systems Theory. Prentice Hall, 1973.

[35] R. Fagin and J. H. Williams. A Fair Carpool Scheduling Algorithm. IBM Journal of Research and Development, 27(2): 133–139, 1983.

[36] B. Falsafi and D. A. Wood. Reactive NUMA: a design for unifying S-COMA and CC-NUMA. In Proceedings of the 24th Annual International Symposium on Computer Architecture, 1997.

[37] L. Fan, P. Cao, J. Almeida, and A. Z. Broder. Summary Cache: a Scalable Wide-area Web Cache Sharing Protocol. *IEEE/ACM Transactions on Networking*, 8(3): 281–293, 2000.

[38] A. Fedorova. Operating System Scheduling for Chip Multithreaded Processors. PhD thesis, Harvard University, 2006.

[39] A. Fedorova, M. Seltzer, C. Small, and D. Nussbaum. Performance of Multithreaded Chip Multiprocessors and Implications for Operating System Design. In USENIX 2005 Annual Technical Conference, April 2005.

[40] A. Fedorova, M. Seltzer, and M. D. Smith. A Non-Work-Conserving Operating System Scheduler For SMT Processors. In Workshop on the Interaction between Operating Systems and Computer Architecture (WIOSCA), in conjunction with ISCA-33, 2006.

[41] M. J. Feeley, W. E. Morgan, E. P. Pighin, A. R. Karlin, H. M. Levy, and C. A. Thekkath. Implementing Global Memory Management in a Workstation Cluster. In Proceedings of the 15th ACM Symposium on Operating Systems Principles, 1995.

[42] I. T. R. for Semiconductors. ITRS 2005 Update. Semiconductor Industry Association, 2005.

[43] A. Glew. MLP Yes! ILP No! Memory Level Parallelism, or, Why I No Longer Worry About IPC. Wild and Crazy Ideas Session of the 8th International Conference on Architectural Support for Programming Languages and Operating Systems, 1998.

[44] M. Gordon, W. Thies, and S. Amarasinghe. Exploiting Coarse-Grained Task, Data, and Pipeline Parallelism in Stream Programs. In Proceedings of the 12th International Conference on Architectural Support for Programming Languages and Operating Systems, Oct. 2006.

[45] M. Gschwind, P. Hofstee, B. Flachs, M. Hopkins, Y. Watanabe, and T. Yamazaki. A Novel SIMD Architecture for the Cell Heterogeneous Chip-multiprocessor. In Hot Chips 17, 2005.

[46] A. Gupta, W.-D. Weber, and T. C. Mowry. Reducing Memory and Traffic Requirements for Scalable Directory-Based Cache Coherence Schemes. In ICPP, 1990.

[47] E. Hagersten, A. Landin, and S. Haridi. DDM: A Cache-Only Memory Architecture. *IEEE Computer*, 25(9): 44–54, 1992.

[48] E. G. Hallnor and S. K. Reinhardt. A Fully Associative Software-Managed Cache Design. In Proceedings of the 27th International Symposium on Computer Architecture, 2000.

[49] S. Harris. Synergistic Caching in Single-Chip Multiprocessors. PhD thesis, Stanford University, 2005.

[50] J. L. Hellerstein. Achieving Service Rate Objectives with Decay Usage Scheduling. IEEE Transaction of Software Engineering, 19(8), 1993.

[51] E. Herrero, J. González and R. Canal. Distributed Cooperative Caching. In Proceedings of the 17th International Conference on Parallel Architectures and Compilation Techniques, pages 134–143, 2008.

[52] E. Herrero, J. González and R. Canal. Elastic Cooprative Caching. In Proceedings of the 37th International Symposium on Computer Architecture, 38(3), pages 419–428, 2010.

[53] E. Herrero, J. González and R. Canal. Power-Efficient Spilling Techniques for Chip Multiprocessors. In Proceedings of the International Conference on Parallel and Distributed Computing, 6271, pages 256–267, 2010.

[54] R. Ho, K. Mai, and M. A. Horowitz. The Future of Wires. In Proceedings of the IEEE, pages 490–504, 2001.

[55] L. R. Hsu, S. K. Reinhardt, R. Iyer, and S. Makineni. Communist, Utilitarian, and Capitalist Cache Policies on CMPs: Caches as a Shared Resource. In Proceedings of the 15th International Conference on Parallel Architecture and Compilation Techniques, 2006.

[56] J. Huh, D. Burger, and S. W. Keckler. Exploring the Design Space of Future CMPs. In Proceedings of the 2001 International Conference on Parallel Architectures and Compilation Techniques, 2001.

[57] J. Huh, C. Kim, H. Shafi, L. Zhang, D. Burger, and S. W. Keckler. A NUCA Substrate for Flexible CMP Cache Sharing. In Proceedings of the 19th ACM International Conference on Supercomputing, June, 2005.

[58] H. C. Hunter. Matching On-Chip Data Storage To Telecommunication and Media Application Properties. PhD thesis, University of Illinois, Urbana, 2004.

[59] IEEE. IEEE Standard for Scalable Coherent Interface (SCI), 1992. IEEE 1596-1992.

[60] Intel. A Platform 2015 Workload Model: Recognition, Mining and Synthesis Move the Computers to the Era of Tera. ftp://download.intel.com/technology/computing/archinnov/platform2015/download/RMS.pdf, 2006.

[61] Intel. Multi-core from Intel. http://www.intel.com/multi-core/, 2006.

[62] Intel. Tera-scale Computing. http://www.intel.com/technology/techresearch/terascale/, 2006.

[63] R. Iyer. CQoS: a Framework for Enabling QoS in Shared Caches of CMP Platforms. In Proceedings of the 18th Annual International Conference on Supercomputing, 2004.

[64] A. Jaleel, M. Mattina, and B. Jacob. Last-Level Cache (LLC) Performance of Data-mining Workloads on a CMP–A Case Study of Parallel Bioinformatics Workloads. In Proceedings of the 12th International Symposium on High Performance Computer Architecture, 2006.

[65] J. Jeong and M. Dubois. Cost-Sensitive Cache Replacement Algorithms. In Proceedings of the 9th International Symposium on High-Performance Computer Architecture, 2003.

[66] S. Jiang and X. Zhang. LIRS: an Efficient Low Inter-reference Recency Set Replacement Policy to Improve Buffer Cache Performance. In Proceedings of the 2002 ACM SIGMETRICS international conference on Measurement and modeling of computer systems, 2002.

[67] L. Jin, J. Lee, and S. Cho. A Flexible Data to L2 Cache Mapping Approach for Future Multicore Processors. In ACM Workshop on Memory Systems Performance and Correctness (MSPC), 2006.

[68] L. K. John. More on Finding a Single number to Indicate Overall Performance of a Benchmark Suite. SIGARCH Computer Architecture News, 32(1), 2004.

[69] T. L. Johnson, D. A. Connors, M. C. Merten, and W. mei W. Hwu. Run-Time Cache Bypassing. IEEE Transaction of Computer, 48(12): 1338–1354, 1999.

[70] H. Kannan, F. Guo, L. Zhao, R. Illikkal, R. Iyer, D. Newell, Y. Solihin, and C. Kozyrakis. From Chaos to QoS: Case Studies in CMP Resource Management. In Proceedings of 2006 Workshop on Design, Architecture and Simulation of Chip Multi-Processors (dasCMP 2006), 2006.

[71] S. Kaxiras and J. R. Goodman. Improving CC-NUMA performance using Instruction-based Prediction. In Proceedings of the 5th International Symposium on High Performance Computer Architecture, Jan. 1999.

[72] C. Kim, D. Burger, and S. W. Keckler. An Adaptive, Non-uniform Cache Structure for Wire-Delay Dominated On-chip Caches. In Proceedings of the 10th International Conference on Architectural Support for Programming Languages and Operating Systems, 2002.

[73] S. Kim, D. Chandra, and Y. Solihin. Fair Cache Sharing and Partitioning in a Chip Multiprocessor Architecture. In Proceedings of the 13th International Conference on Parallel Architecture and Compilation Techniques, 2004.

[74] M. Kondo, H. Sasaki, and H. Nakamura. Improving Fairness, Throughput and Energy-Efficiency on a Chip Multiprocessor through DVFS. In Proceedings of 2006 Workshop on Design, Architecture and Simulation of Chip Multi-Processors (dasCMP 2006), 2006.

[75] P. Kongetira, K. Aingaran, and K. Olukotun. Niagara: A 32-Way Multithreaded SPARC Processor. IEEE Micro, 25(2): 21–29, 2005.

[76] C.-C. Kuo, J. B. Carter, R. Kuramkote, and M. R. Swanson. AS-COMA: An Adaptive Hybrid Shared Memory Architecture. In Proceedings of the 1998 International Conference on Parallel Processing, pages 207–216, 1998.

[77] A.-C. Lai and B. Falsafi. Memory Sharing Predictor: The Key to a Speculative Coherent DSM. In Proceedings of the 26th International Symposium on Computer Architecture, May 1999.

[78] A.-C. Lai and B. Falsafi. Selective, Accurate, and Timely Self-Invalidation Using Last-Touch Prediction. In Proceedings of the 27th International Symposium on Computer Architecture, 2000.

[79] J. Laudon and D. Lenoski. The SGI Origin: A ccNUMA Highly Scalable Server. In Proceedings of the 24th Annual International Symposium on Computer Architecture, pages 241–251, 1997.

[80] A. R. Lebeck and D. A. Wood. Dynamic Self-Invalidation: Reducing Coherence Overhead in Shared-Memory Multiprocessors. In Proceedings of the 22nd International Symposium on Computer Architecture, June 1995.

[81] D. Lenoski, J. Laudon, K. Gharachorloo, W.-D. Weber, A. Gupta, J. Hennessy, M. Horowitz, and M. Lam. The Stanford DASH Multiprocessor. IEEE Computer, 25(3): 63–79, Mar. 1992.

[82] F. Li, C. Nicopoulos, T. Richardson, Y. Xie, V. Narayanan, and M. Kandemir. Design and Management of 3D Chip Multiprocessors Using Network-in-Memory. In Proceedings of the 33th Annual International Symposium on Computer Architecture, June 2006.

[83] J. S. Liptay. Structural Aspects of the System/360 Model 85, Part II: The cache. IBM Journal of Research and Development, 7(1), 1968.

[84] C. Liu, A. Sivasubramaniam, and M. Kandemir. Organizing the Last Line of Defense before Hitting the Memory Wall for CMPs. In Proceedings of the 10th IEEE Symposium on High-Performance Computer Architecture, Feb. 2004.

[85] J. L. Lo, L. A. Barroso, S. J. Eggers, K. Gharachorloo, H. M. Levy, and S. S. Parekh. An Analysis of Database Workload Performance on Simultaneous Multithreaded Processors. In Proceedings of the 25th Annual International Symposium on Computer Architecture, pages 39–50, 1998.

[86] K. Luo, J. Gummaraju, and M. Franklin. Balancing Throughput and Fairness in SMT Processors. In Proceedings of 2001 IEEE International Symposium on Performance Analysis of Systems and Software, 2001.

[87] P. Magnusson, M. Christensson, J. Eskilson, D. Forsgren, G. Hållberg, J. Högberg, F. Larsson, A. Moestedt, and B. Werner. Simics: A Full System Simulation Platform. IEEE Computer, 35(2): 50–58, Feb 2002.

[88] M. M. K. Martin, M. D. Hill, and D. A. Wood. Token Coherence: Decoupling Performance and Correctness. In Proceedings of the 30th Annual International Symposium on Computer Architecture, June 2003.

[89] M. M. K. Martin, D. J. Sorin, B.M. Beckmann, M. R. Marty, M. Xu, A. R. Alameldeen, K. E. Moore, M. D. Hill, and D. A. Wood. Multifacet's General Execution-driven Multiprocessor Simulator (GEMS) Toolset. Computer Architecture News, 2005.

[90] R. L. Mattson, J. Gecsei, D. R. Slutz, and I. L. Traiger. Evaluation Techniques for Storage Hierarchies. IBM Systems Journal, 9(2): 78–117, 1970.

[91] N. Megiddo and D. S. Modha. ARC: A Self-Tuning, Low Overhead Replacement Cache. In Proceedings of the 2nd USENIX Conference on File and Storage Technologies, 2003.

[92] A. Moga and M. Dubois. The Effectiveness of SRAM Network Caches in Clustered DSMs. In Proceedings of the 4th International Symposium on High Performance Computer Architecture, 1998.

[93] S. S. Mukherjee and M. D. Hill. Using Prediction to Accelerate Coherence Protocols. In Proceedings of the 25th International Symposium on Computer Architecture, pages 179–190, June 1998.

[94] O. Mutlu, J. Stark, C. Wilkerson, and Y. N. Patt. Runahead Execution: An Effective Alternative to Large Instruction Windows. IEEE Micro, 23(6): 20–25, 2003.

[95] A. K. Nanda, A.-T. Nguyen, M. M. Michael, and D. J. Joseph. High-throughput Coherence Control and Hardware Messaging in Everest. IBM Journal of Research and Development, 45(2), 2001.

[96] B. A. Nayfeh, L. Hammond, and K. Olukotun. Evaluation of Design Alternatives for a Multiprocessor Microprocessor. In Proceedings of the 23rd Annual International Symposium on Computer Architecture, May 1996.

[97] K. J. Nesbit, N. Aggarwal, J. Laudon, and J. E. Smith. Fair Queuing Memory Systems. In Proceedings of the 39th Annual International Symposium on Microarchitecture, 2006.

[98] H. Oi and N. Ranganathan. Utilization of Cache Area in On-Chip Multiprocessor. In Proceedings of the Second International Symposium on High Performance Computing (ISHPC '99), 1999.

[99] M. S. Papamarcos and J. H. Patel. A Low-overhead Coherence Solution for Multiprocessors with Private Cache Memories. In Proceedings of the 11th Annual International Symposium on Computer Architecture, 1984.

[100] A. K. Parekh and R. G. Gallager. A Generalized Processor Sharing Approach to Flow Control in Integrated Services Networks: the Single-node Case. IEEE Transaction of Networks, 1(3): 344–357, 1993.

[101] D. Patterson, T. Anderson, N. Cardwell, R. Fromm, K. Keeton, C. Kozyrakis, R. Thomas, and K. Yelick. A Case for Intelligent RAM. IEEE Micro, 17(2): 34–44, 1997.

[102] D. A. Patterson. Computer Science Education in the 21st Century. Communication of ACM, 49(3): 27–30, 2006.

[103] C. Percival. BSDCan Conference, 2005, http://www.daemonology.net/hyperthreading-considered-harmful.

[104] P. Petoumenos, G. Keramidas, H. Zeffer, S. Kaxiras, and E. Hagersten. STATSHARE: A Statistical Model for Managing Cache Sharing via Decay. In MoBS 2006, June 2006.

[105] M. K. Qureshi, D. N. Lynch, O. Mutlu, and Y. N. Patt. A Case for MLP-Aware Cache Replacement. In Proceedings of the 33th Annual International Symposium on Computer Architecture, June 2006.

[106] M. K. Qureshi and Y. N. Patt. Utility-Based Cache Partitioning: A Low-Overhead, High-Performance, Runtime Mechanism to Partition Shared Caches. In MICRO-39, 2006.

[107] M. K. Qureshi, D. Thompson, and Y. N. Patt. The V-way Cache: Demand Based Associativity via Global Replacement. In Proceedings of the 32nd Annual International Symposium on Computer Architecture, 2005.

[108] N. Rafique, W.-T. Lim, and M. Thottethodi. Architectural Support for Operating System-Driven CMP Cache Management. In Proceedings of the 15th International Conference on Parallel Architecture and Compilation Techniques, 2006.

[109] R. Rajwar and J. R. Goodman. Speculative Lock Elision: Enabling Highly Concurrent Multithreaded Execution. In 34th International Symposium on Microarchitecture, December, 2001.

[110] Tilera Co. Tilera Announces the World's First 100-core Processor with the New TILE-Gx Family http://www.tilera.com/news_&_events/press_release_091026.php, 2009.

[111] A. Saulsbury, T. Wilkinson, J. Carter, and A. Landin. An Argument for Simple COMA. In Proceedings of the 1st IEEE Symposium on High-Performance Computer Architecture, 1995.

[112] T. Sherwood, B. Calder, and J. Emer. Reducing Cache Misses Using Hardware and Software Page Placement. In Proceedings of the 13th international conference on Supercomputing, 1999.

[113] P. Shivakumar and N. P. Jouppi. CACTI 3.0: An Integrated Cache Timing, Power, and Area Model, 2001.

[114] J. E. Smith. Characterizing Computer Performance with a Single Number. Communication of ACM, 31(10), 1988.

[115] A. Snavely and D. M. Tullsen. Symbiotic Jobscheduling for a Simultaneous Multithreaded Processor. In Proceedings of the 9th International Conference on Architectural Support for Programming Languages and Operating Systems, 2000.

[116] A. Snavely, D. M. Tullsen, and G. Voelker. Symbiotic Jobscheduling with Priorities for a Simultaneous Multithreading Processor. In The 2002 ACM SIGMETRICS International Conference on Measurement and Modeling of Computer Systems, 2002.

[117] K. So and R. N. Rechtschaffen. Cache Operations by MRU Change. IEEE Transaction of Computers, 37(6): 700–709, 1988.

[118] V. Soundararajan, M. Heinrich, B. Verghese, K. Gharachorloo, A. Gupta, and J. Hennessy. Flexible Use of Memory for Replication/Migration in Cache-coherent DSM Multiprocessors. In Proceedings of the 25th Annual International Symposium on Computer Architecture, 1998.

[119] E. Speight, H. Shafi, L. Zhang, and R. Rajamony. Adaptive Mechanisms and Policies for Managing Cache Hierarchies in Chip Multiprocessors. In Proceedings of the 32th Annual International Symposium on Computer Architecture, June 2005.

[120] P. Stenström, M. Brorsson, and L. Sandberg. Adaptive Cache Coherence Protocol Optimized for Migratory Sharing. In Proceedings of the 20th Annual International Symposium on Computer Architecture, May 1993.

[121] H. S. Stone, J. Turek, and J. L. Wolf. Optimal Partitioning of Cache Memory. IEEE Transaction of Computers, 41(9): 1054–1068, 1992.

[122] R. Subramanian, Y. Smaragdakis, and G. H. Loh. Adaptive Caches: Effective Shaping of Cache Behavior to Workloads. In MICRO-39, 2006.

[123] G. E. Suh, S. Devadas, and L. Rudolph. A New Memory Monitoring Scheme for Memory-aware Scheduling and Partitioning. In HPCA-8, 2002.

[124] G. E. Suh, L. Rudolph, and S. Devadas. Dynamic Partitioning of Shared Cache Memory. Journal of Supercomputing, 28(1): 7–26, 2004.

[125] D. G. Sullivan and M. I. Seltzer. Isolation with Flexibility: A Resource Management Framework for Central Servers. In USENIX 2000 Annual Technical Conference, June 2000.

[126] M. Takahashi, H. Takano, E. Kaneko, and S. Suzuki. A Shared-bus Control Mechanism and a Cache Coherence Protocol for a High-performance On-chip Multiprocessor. In Proceedings of the 2nd IEEE Symposium on High-Performance Computer Architecture, 1996.

[127] A. S. Tanenbaum and A. S. Woodhull. Operating Systems: Design And Implementation, 2/E. Prentice Hall, 1997.

[128] J. M. Tendler, J. S. Dodson, J. S. F. Jr., H. Le, and B. Sinharoy. IBM Power4 System Microarchitecture. IBM Journal of Research and Development, 46(1): 5–26, 2002.

[129] K. Vardarajan, S. Nandy, V. Sharda, A. Bharadwaj, R. Iyer, S. Makineni, and D. Newell. Molecular Caches: A caching structure for dynamic creation of application-specific Heterogeneous cache regions. In MICRO-39, 2006.

[130] B. Verghese, S. Devine, A. Gupta, and M. Rosenblum. Operating System Support for Improving Data Locality on CC-NUMA Compute Servers. In Proceedings of the Seventh International Conference on Architectural Support for Programming Languages and Operating Systems, 1996.

[131] B. Verghese, A. Gupta, and M. Rosenblum. Performance Isolation: Sharing and Isolation in Sharedmemory Multiprocessors. In Proceedings of the 8th international Conference on Architectural Support for Programming Languages and operating systems, 1998.

[132] C. A. Waldspurger. Memory Resource Management in VMware ESX Server. In Proceedings of the 5th Symposium on Operating Systems Design and Implementation, 2002.

[133] C. A. Waldspurger and W. E. Weihl. Lottery Scheduling: Flexible Proportional-Share Resource Management. In First Symposium on Operating System Design and Implementation, 1994.

[134] H.-S. Wang, X. Zhu, L.-S. Peh, and S. Malik. Orion: a power-performance simulator for interconnection networks. In 35th International Symposium on Microarchitecture, pages 294–305, November 2002.

[135] Z. Wang. Cooperative Software and Hardware Caching for Next Generation Memory Systems. PhD thesis, University of Massachussets, Amherst, 2004.

[136] S. C. Woo, M. Ohara, E. Torrie, J. P. Singh, and A. Gupta. The SPLASH-2 Programs: Characterization and Methodological Considerations. In the 22th Annual International Symposium on Computer Architecture, 1995.

[137] T. Y. Yeh and G. Reinman. Fast and Fair: Data-Stream Quality of Service. In CASES'05, Sep 2005.

[138] N. E. Young. On-line File Caching. In Proceedings of the 9th Annual ACM-SIAM Symposium on Discrete Algorithms, 1998.

[139] M. Zhang and K. Asanovic. Victim Migration: Dynamically Adapting Between Private and Shared CMP Caches. Technical Report MIT-CSAIL-TR-2005-064, MIT CSAIL, 2005.

[140] M. Zhang and K. Asanovic. Victim Replication: Maximizing Capacity while Hiding Wire Delay in Tiled CMPs. In Proceedings of the 32th Annual International Symposium on Computer Architecture, June 2005.

[141] Z. Zhang and J. Torrellas. Reducing Remote Conflict Misses: NUMA with Remote Cache versus COMA. In Proceedings of the 3rd IEEE Symposium on High-Performance Computer Architecture, 1997.

14

Market-Oriented Resource Management and Scheduling: A Taxonomy and Survey

Saurabh Kumar Garg and Rajkumar Buyya
Cloud Computing and Distributed Systems (CLOUDS) Laboratory,
Department of Computer Science and Software Engineering, University of Melbourne,
Victoria, Australia

14.1 Introduction

The shift of grids from providing computer power on a sharing basis to commercial purposes, even though it has not fully unfolded and is still mostly limited to research, has led to various technical advancements and paved a way to make utility grids a reality. Those advancements favour the application of market-based mechanisms for Grid systems by providing various prerequisites on the technical and economic sides. The creation of pervasive grid requires integration view of scalable system architecture, resource management and scheduling, and market models as shown in Figure 14.1.

This chapter summarizes the recent advances toward the vision of utility grids. In Section 14.2, it starts with an overview of grid and utility computing, and then Section 14.3 specifies all the requirements of a utility grid. Section 14.4 presents an abstract model to conceptualize essential infrastructure needed to support this vision. Then, a taxonomy and survey of the current market-oriented and system-oriented schedulers is provided in Sections 14.5 and 14.6, examining the contribution and the out-standing issues of each system in terms of utility grid's requirements. Section 14.7 summarizes the gaps based on the taxonomy and survey of market-oriented systems. Finally, in Section 14.8, we summarize the whole chapter.

14.2 Overview of Utility Grids and Preliminaries

A utility grid imitates a market scenario consisting of two key players (that is, Grid Service Consumers (GSCs) and Grid Service Providers (GSPs)). Each of these players is generally self-interested and wants to maximize their utility (Figure 14.2). Consumers are users who have resource requirements to execute their applications. The resource requirement varies depending on the application model. For instance, parallel applications demand multiple CPUs at the same

Cooperative Networking, First Edition. Edited by Mohammad S. Obaidat and Sudip Misra.
© 2011 John Wiley & Sons, Ltd. Published 2011 by John Wiley & Sons, Ltd.

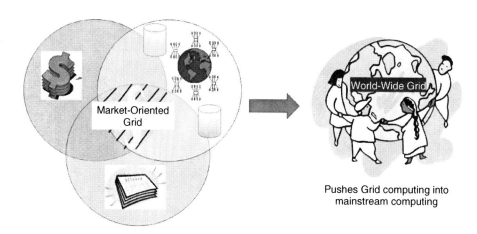

Figure 14.1   A view of market-oriented grid pushing grid into mainstream computing.

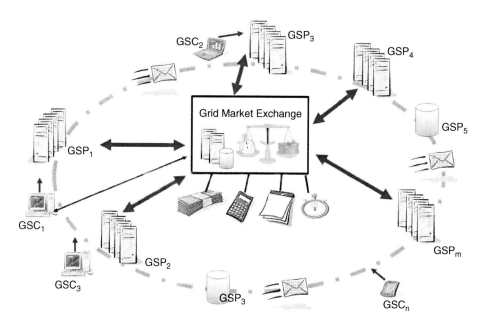

Figure 14.2   A grid market exchange managing self-interested entities (providers and consumers).

time with equal configuration, and similar network bandwidth between resources. The consumers are willing to compensate providers for using their resources in the form of real money or barter [16]. On the other hand, providers are the owners of resources (that is, disk, CPU) which satisfy consumer needs. They can advertise their resources using other agents of the grid such as Grid Market Directories [68]. It is the responsibility of resource providers to ensure the user's application gets executed according to Service Level Agreement (SLA) [15] signed with the consumer.

To ease and control the buying and selling process there are other players in the utility grid, such as grid market place or grid market exchange [43][46], which allow various consumers and providers to publish their requirements and goods (compute power or storage) respectively. The market exchange service provides a shared trading infrastructure designed to support different market-oriented systems. It provides transparent message routing among participants, authenticated messages and logging of messages for auditing. It can coordinate the users and lower the delay in acquiring resources. Moreover, it can help in price control and reduces the chances of the market being monopolized. Personalized Brokers are another kind of middle agent which, based on users' QoS requirements, perform the function of resource monitoring, resource discovery and job submission. These brokers hide all the complexity of grids from users. Each of the utility grid players (that is, consumer (or user agents such as personalized broker), provider, market exchange) has different requirements and goals. These requirements are discussed in detail and summarized in the next section.

14.3   Requirements

In this section, we discuss the main bottlenecks or infrastructural enhancements required for utility grids. In general, consumers and providers need mechanisms and tools that facilitate the description of their requirements and facilitate decision making to achieve their goals such as minimization of cost while meeting QoS requirements.

14.3.1   Consumer Side Requirements

User-centric Brokers: These brokers are user agents which discover and schedule jobs on to resources according to the user priorities and application's QoS requirements such as budget, deadline and number of CPU required [41], [63]. These brokers hide the heterogeneity and complexity of resources available in the grid. On behalf of users, user-centric brokers provide functionalities such as application description, submission and scheduling; resource discovery and matching; and job monitoring. The user broker can also perform negotiation and bidding in an auction conducted by a market exchange or provider for acquiring resources.

Bidding/Valuation Mechanism: In the utility grid, a variety of market models can exist simultaneously such as a commodity and auction market. To participate in both of the market model, users need to know the valuation of their applications in the form of budget which estimates the user's requirements. For example, in an auction market, users need to bid in order to grab a resource, and in such cases, budget or valuation can help brokers to bid on behalf of the users. In summary, consumers need utility models to allow them to specify resource requirements and constraints.

Market-oriented Scheduling Mechanisms: In traditional grids, generally users want to schedule their applications on the resources which can provide the minimum response time, and satisfy other QoS requirements in terms of memory and bandwidth. In utility grids, one additional factor comes into the picture that is, monetary cost, which requires new mechanisms as a user may relax some of their other requirements to save on monetary cost. Thus, one of the objectives of new scheduling mechanisms will be to execute the user application on the cheapest resource which can satisfy the user's QoS requirements. The market-oriented mechanisms can vary depending on the market model, user's objective (such as reduce time or cost), and application model (require multiple types of resources).

Allocation of Multiple Resources: Depending on the application model, the consumer may want to run its application on multiple resources provided by more than one resource provider; for example, scheduling of a large parameter sweep across a number of providers, performing distributed queries across multiple databases, or creating a distributed multi-site work flow. Thus, brokers should have capabilities to schedule applications and obtain resources from multiple resource sites.

Estimation of Resource Usage: In general, due to the heterogeneity of hardware and different input sizes, it is difficult to describe precisely execution time, and the requirement of an application which can vary drastically. In the traditional grid, an important research problem is how to profile the runtime of an application since execution time can affect not only the resource utilization but also cause delays for users. In the utility grid, this requirement becomes more critical as over estimation and under estimation of resource requirements can lead to tangible loss in the form of real money. Currently, the resource providers such as Amazon sell their computer resources in time blocks. In addition, if many users compete for the same resource, the resource availability, depending on individual user's requirement, can vary from minutes to days. Thus, users must estimate their resource needs in advance. Thus, profiling tools and mechanisms are required for efficient resource allocation in terms of utility.

14.3.2   Resource Provider Side Requirements

Resource Management Systems: These systems interact with underline hardware infrastructure and control the allocation of resources and job scheduling. In a market-oriented system, the advance reservation function is required to allocate resources in future, and also to track the availability of resources which can be advertised. Thus, the reservation system should be able to provide the guaranteed resource usage time (based on SLA) and support estimation of the future resource offers. Grid Middleware such as Globus has components such as advance reservation, but to support the market-oriented reservation and scheduling, they should be integrated with a module that supports various market-oriented scheduling mechanisms and models.

Pricing/Valuation Mechanism: In utility grids, a resource provider's main objective is to maximize their profit not just the efficiency of the system, thus mechanisms are required to set the resource price based on the market supply and demand, and the current level of resource utilization. These prices can be static or can vary dynamically based on the resource demand. For example, an academic user may require more resources due to a conference deadline and, thus, is willing to pay more.

Admission Control and Negotiation Protocols: As stated before, in market-oriented systems, all participants are self-interested and want to maximize their utility. Thus, providers need to decide which user application they should accept or negotiate for. Since there may be a chance of reservation cancellation by users, thus the mechanisms such as resource over provisioning may be required. SLA is also needed to be formulated once a user request is accepted for reservation. In addition, depending on how providers want to lease their resources, they may choose different market model for negotiation. For example, the simplest negotiation is required in the commodity models, while the bargaining model requires more intelligent negotiation protocol.

Commoditization of the Resources: Unlike many other markets, commoditization of the resources is one of the major difficult problems, which complicates the reservation and allocation decisions. For instance, for a compute intensive application, it is meaningless to just allocate CPUs without some memory. How much memory should be allocated when hardware infrastructure contain shared memory? Even for storage some small CPU cycles are required. Thus, intelligent resource partitioning techniques are required to overcome the hardware difficulties.

14.3.3   Market Exchange Requirements

An Information and Market Directory is required for advertising participants, available resources and auctions. It should support heterogeneous resources, as well as provide support for different resource specifications. This means that the market ought to offer functionalities for providing, for instance, both storage and computation with different qualities and sizes.

Support for Different Market Models: Multiple market models are needed to be designed and deployed as resource providers and consumers have different goals, objectives, strategies and requirements that vary with time [43]. If there are multiple sellers for the same good, a double auction which aggregates supply and demand generally yields higher efficiency. If there is only one seller (for example, in a differentiated service market for complex services), supporting single-sided auction protocols may be desirable. The negotiation protocol also depends on the user application. For example, in the case applications with soft deadlines, the large scheduling cycle helps in collecting a higher number of bids and offers for auction. This may lead to more efficient allocation than clearing continuously, since the allocation can be based on more resource information and has more degrees of freedom in optimizing efficiency (and/or other objectives). On the contrary, users having urgent resource requirements may prefer an immediate allocation, thus the commodity model will be a better choice for negotiation. Consequently, the market exchange must clearly support multiple negotiation protocols.

Reputation and Monitoring System: In general, it is assumed that after the scheduling mechanism has determined the allocation and resultant pricing, the market participants adhere to the market's decisions and promises. In reality, however, this does not happen due to several reasons such as untruthful behaviour of participants, failure while communicating the decision, and failure of resources. Consequently, there is a need for reputation mechanisms that prevent such circumstances by removing distrustful participants. However, there is a strong need for monitoring systems that can detect any SLA violation during execution. In grids, the reason for a job failure or a corruption of results is hard to detect since it can occur due to several reasons such as intentional misbehaviour of the resource provider, or technical reasons which are neither controlled by the user nor the provider. The monitoring systems should support reputation system for early detection of violations and responsible participant. An important challenge is thus to design such intelligent monitoring systems.

Banking system (Accounting, Billing, Payment mechanism): In the market exchange, an accounting system is necessary to record all the transactions between resource providers and consumers. The accounting system especially records the resource usage and charges the consumer as per the usage agreement between the consumer and provider.

Meta-scheduling/Meta-Brokering: The market exchange provides the services such as meta-scheduling of consumer applications on multiple resource providers in case several consumers require simultaneous access to resources. For instance, a large parameter sweep application requires resources across the number of providers, performing distributed queries across multiple databases, or creating a distributed multi-site workflow. Thus, the meta-scheduling service does two tasks for their clients (that is, resource discovery and efficiently scheduling applications according to client's objectives). It can act as an auctioneer in case a client wants to hold an auction.

Currency Management: For ensuring the fair and efficient sharing of resources, and a successful market, a well-defined currency system is essential. Two kinds of currency models are proposed that is, virtual and real currency. Both of these currency models have advantages and disadvantages based on their managerial requirements. The virtual currency is generally deployed [52] due to its low risk and low stakes in case of mismanagement or abuse. However, virtual currency requires careful initial and ongoing management, and lacks flexibility. For buying and selling resources in a real commercial environment, the real currency is preferred due to several reasons. The most important reason is that the real currency formats (for example, USD, Euro, and so on) are universally recognized and are easily transferable and exchanged, and are managed outside the scope of a grid market exchange, by linked free markets and respective government policy.

Security and Legal System: To avoid spamming, there should be a security system for user registration. All the services of the exchange must be accessed by authorized users. Similarly, there is also a need for legal support that can resolve various conflicts between providers and consumers, such as violations of SLA [42]. Thus, the legal support can be built with the help of some authoritative agency such as a country's government.

14.4 Utility Grid Infrastructural Components

Based on the above requirements, in this section we discuss various infrastructure required for a fully functional utility grid. Figure 14.3 outlines an abstract model for utility grid that identifies essential components. This model can be used to explore how existing grid middleware such as user-centric brokers, meta-schedulers and RMSs can be leveraged and extended to incorporate market-oriented mechanisms.

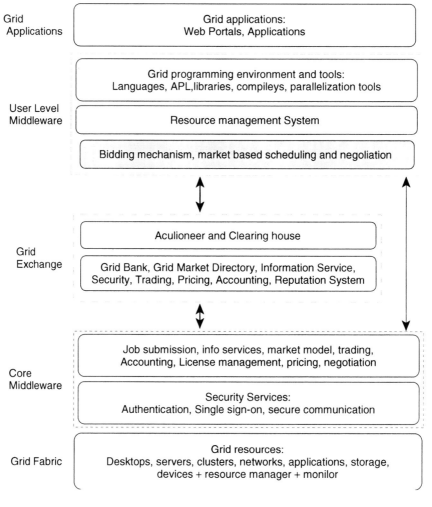

Figure 14.3 Utility grid components.

The utility grid consists of multi-layer middleware for each participant: users (grid application, user level middleware), grid exchange, and providers (core middleware and grid fabric). The architecture of each of the components should be generic enough to accommodate different negotiation models for resource trading. Except for grid exchange and highlighted components, most of the components are also present in traditional grids.

The lowest layer is the **grid fabric** that consists of distributed resources such as computers, networks, storage devices and scientific instruments. These computational resources are leased by providers, thus the resource usage is needed to be monitored periodically to inform the above layers about free resources which can be rented out. The resource managers in this layer have the responsibility of scheduling applications.

The **core middleware** offers the interface for negotiation with grid exchange and user-level middleware. It offers services such as remote process management, co-allocation of resources, storage access, information registration and discovery, security, and aspects of QoS such as resource reservation and trading. These services hide the heterogeneity at the fabric level. The support for accounting, market model and pricing mechanisms is vital for resource providers since such support enables them to participate in the utility grid. The pricing mechanism decides how resource requests are charged. The pricing of resource usage by consumers can depend on several variables such as submission time (peak/offpeak), pricing rates (fixed/changing) or availability of resources (supply/demand). Pricing serves in the utility grid as an efficient and cost-effective medium for resource sharing. The accounting mechanism maintains the actual usage of resources so that the final cost can be computed and charged from the consumers. The market model defines the negotiation protocol that can be used to serve different resource requests depending on their effect on the provider's utility.

The user-level grid middleware and applications also need to be enhanced to satisfy the requirements discussed in the previous section. A new layer is needed in the brokers to give users functionality of automatic bidding and negotiation. This layer also discovers resources based on a user's requirements such as deadline and budget. Automated negotiation capabilities are needed to be added to allow brokers to interact with grid exchange and provider's middleware, and maintain SLAs.

In traditional grids, the users and providers generally interact on a one-to-one basis rather than using third party services. Similar to other markets, in utility grids, since negotiation with a provider is more complex due to the involvement of money and flexible resource reservation, the third party services become essential. Thus, utility grids require **grid exchange middleware** which can act as a buyer or seller on behalf of users and resource providers respectively. They require the capability of auctions and clearing house to match user resource demand to available resources. This middleware needs infrastructures such as a meta-broker which matches of users and providers, Grid Market Directory (GMD) that allows resource advertisements, negotiation protocols, a reputation system, security and price control. Meta-broker is a key component of the market-exchange. It acts as an auctioneer or clearing house, and thus schedules a user's application on the desired resources.

14.5 Taxonomy of Market-Oriented Scheduling

There are several proposed taxonomies for scheduling in distributed and heterogeneous computing [69], [70]. However, none of these taxonomies focus on the market-oriented scheduling mechanism in grids. Here, we present the taxonomy, which emphasizes the practical aspects of market-oriented scheduling in grids and its difficulties which are vital to achieve utility-based grid computing in practice. We can understand works in market-oriented scheduling (as shown in

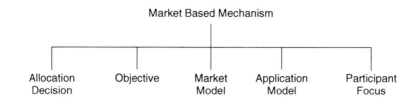

Figure 14.4 Higher level division of research works in market-oriented scheduling.

Figures 14.4 and 14.5) from five major perspectives, namely market model, allocation decision, market objective, application model and participant focus.

14.5.1 Market Model

The market model refers to the mechanism used for trading between consumers and providers. Any particular market model cannot solve all the special requirements of different participants. Having different characteristics each model is the most profitable to its participants depending on the grid situation. For example, when the number of participants in terms of consumers and providers is almost the same, the double auction is a much better choice. Due to the differences in the applicability of auctions, various authors have applied and analyzed the efficiency achieved [43]. Various market models that can be applied to market-oriented grid computing, include the following:

14.5.1.1 Game Theory

If grid participants only interact in the form of an allocation game with different payoffs as a result of specific actions employing various strategies, a game theoretical setting can be assumed. This approach is generally applied to ease the congestion of a common resource or network which can lead to reduction in the overall utility of the system. There are two types of solution approaches in this context:

- To avoid excessive use of the resources one can use a game with self-interested economic agents (non-cooperative games).
- To achieve a load balancing effect the unselfish distribution of tasks over resources can be achieved by using cooperative game agents.

The cooperative and non-cooperative games for resource sharing and allocation often employ the 'Nash bargaining' approach, where bargainers negotiate for a fair 'contract point' within a feasible solution set [Nash 1950]. The application of game theory-based approaches is not very common for resource allocation in market-oriented grid computing. Feldman et al. [23] indicated in their analysis that the price-anticipating allocation scheme can result in higher efficiency and fairness at equilibrium. In their approach, resource price is estimated by total bids placed on a machine. Kwok et al. [37] proposed a Nash equilibrium-oriented allocation system with hierarchical organized game. They used a reputation index instead of virtual budget or monetary unit for job acceptance decisions.

14.5.1.2 Proportional Share

The proportional share introduced and implemented in real cluster based systems such as Tycoon [38] is to decrease the response time of jobs and to allocate them fairly. Neumann et al. [58] proposed a similar approach such as proportional share where shares of resources are distributed using a discriminatory pay-as-bid mechanism to increase the efficiency of allocation and for the maximization of resource provider profit. This model makes an inherent assumption that resources are divisible, which is generally not the case when a single CPU is needed to be allocated; a situation which is quite usual in cooperative problem-solving environments such as clusters (in single administrative domain).

In the proportional share based market model, the percentage of resource share allocated to an application is proportional to the bid value in comparison to other users' bids. The users are allocated credits or tokens, which can be used for gaining access to resources. The value of each credit depends on the resource demand and the value that other users place on the resource at the time of usage. One major drawback of proportional share is that users do not get any QoS guarantee.

14.5.1.3 Commodity Market Model

In this model, resource providers specify their resource price and charge users according to the amount of resources they consume. The resource allocation mechanisms consist of finding prices and allocations such that each economic participant maximizes their utility. One of the first evaluation works in grids on commodity market was presented by Wolski et al. [64] who analyzed and compared commodity markets with other auction models. Many commercial providers such as Amazon [4] are using commodity market models with fixed and dynamic pricing.

The determination of the equilibrium price is crucial since it acts as a great tool in resource allocation decisions. The prices depend on various factors such as investment and management cost of resource provider, current demand and supply, and future markets [53], [2]. According to the prices, users can adopt any strategy to decrease their spending while getting satisfactory QoS level [49], [56]. Various systems have been designed to automate this process; such as Dornemann et al. [17] designed and implemented a workflow system based on a Business Process Execution Language (BPEL) to support on-demand resource provisioning. Ernemann et al. [20] presented an economic scheduling system for grid environments. HA-JES (Kang et al. [33]) presented an algorithm to increase revenue for providers whose resources are underutilized, by ensuring high availability to users. To determine the equilibrium price, Stuer et al. [59] presented a strategy for commodity resource pricing in dynamic computational grids. They proposed some refinements to the application of Smale's method for finding price equilibria in such a grid market for price stability, allocative efficiency and fairness.

14.5.1.4 Contract-Net

In the contract net protocol, the user advertises its demand and invites resource owners to submit bids [55][67][32]. Resource owners check these advertisements with respect to their requirements. In this case, the advertisement is favourable and the resource owners respond with bids. The user consolidates all bids, compares them and selects the most favourable bids. The bidding process has only two outcomes: the bid is accepted or rejected in its entirety.

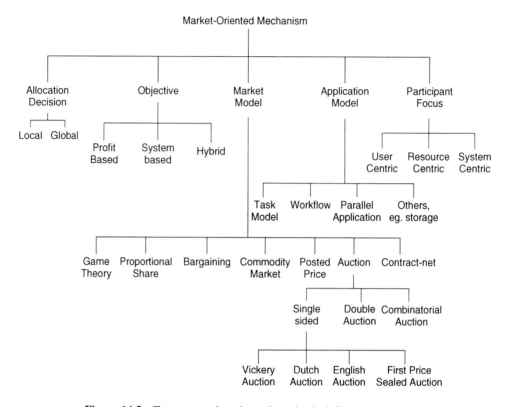

Figure 14.5 Taxonomy of market oriented scheduling mechanisms.

14.5.1.5 Bargaining

Bargaining models are employed in bi-lateral negotiations between providers and consumers, and do not rely on third parties to mediate the negotiation. During the negotiation, each player applies concessions until mutual agreement is reached by alternating offers [62]. Li and Yahyapour [39] proposed a concurrent bilateral negotiation model for grid resource management. The bargaining problem in grid resource management is difficult because while attempting to optimize utility, negotiation agents need to: (i) negotiate for simultaneous access to multiple resources, (ii) consider the (market) dynamics of a computational grid, and (iii) be highly successful in acquiring resources to reduce delay overhead in waiting for resources.

14.5.1.6 Posted Price

This is similar to the commodity market. In this model, providers may also make special offers such as discounts for new clients; differentiate prices across peak and off-peak hours. Prices do not vary relative to the current supply and demand but are fixed over a period of time.

14.5.1.7 Auction

An auction is a process of trading resources by offering them up for bidding, and selling the items to the highest bidder. In economic terms, it is also a method to determine the value of a resource whose price is unknown. An auction is organized by an auctioneer, who distributes grid resources from the providers to the users. Thus, the mechanism consists of determining the winner and setting the price. The auctions can be divided into three types based on participants and commodity exchanged: a) Single-sided auction, b) Double-sided auction, and c) Combinatorial auctions.

- **Single-sided Auction:** Single-sided auctions are mechanisms, where only buyers or sellers can submit bids or asks. Even though the single-sided auction is the most widely applied market model, it often leads to inefficient allocation [51]. The most prominent single sided auctions are Vickrey Auction, Dutch Auction, First Price Sealed Bid (FPSB), and English Auction.

  a. **English Auction:** In the English auction, the auctioneer begins the auction with a reserve price (lowest acceptable price) [16]. Auction continues in rounds with increasing bid prices, until there is no price increase. The item is then sold to the highest bidder.

  b. **Dutch Auction:** In the Dutch auction the auctioneer begins with a high asking price which is lowered until some participant is willing to accept the auctioneer's price or a predetermined minimum price is reached. That participant pays the last announced price. This type of auction is convenient when it is important to auction resources quickly, since a sale never requires more than one bid.

  c. **Vickrey Auction:** A Vickrey auction is a sealed-bid auction, where bidders submit sealed bids. The highest bidder wins, paying the price of the second highest bid. This gives bidders incentives to bid their true value. When multiple identical units are auctioned, one obvious generalization is to have all bidders pay the amount of the highest non-winning bid.

  d. **First Price Sealed Bid (FPSB) Auction:** In this type of auction, all bidders simultaneously submit bids so that no bidder knows the bid of any other participant [16]. The highest bidder pays the price they submitted. In this case, the bid strategy is a function of one's private value and the prior belief of other bidders' valuations. The best strategy is bid less than its true valuation and it might still win the bid, but it all depends on what the others bid.

- **Double Sided Auction:** In double auction, both providers and users submit bids which are then ranked highest to lowest to generate demand and supply profiles. From the profiles, the maximum quantity exchanged can be determined by matching selling offers (starting with the lowest price and moving up) with demand bids (starting with the highest price and moving down). This format allows users to make offers and providers to accept those offers at any particular moment. In double auction, the winner determination depends on different aspects such as aggregation, resource divisibility and if goods are homogeneous or are heterogeneous. Aggregation can come from the supplier side or from the buyer side. If no aggregation is allowed then each bid can be exactly matched to one ask. Divisible goods can be allocated partially. In the case that the bidder wants the entire good or nothing then its bid is considered indivisible. Kant et al. [34] proposed and compared various types of double auctions to investigate its efficiency for resource allocation in grid. Tan et al. [61] proposed stable continuous double auction to overcome high volatility.

- **Combinatorial Auctions:** The grid users may require a combination of multiple resources such as CPUs, memory and bandwidth. Combinatorial auction allows users and providers to trade a bundle of multiple resources. It is advantageous to users as they do not need to participate in multiple negotiations with providers for each resource required. Moreover, in some cases it also leads to cost benefits. In combinatorial auction, users express their preferences as bundles of resources that need to be matched. The providers submit their tasks and the auctioneer solves the optimization problem of allocation. The only drawback of combinatorial auction is the NP-hardness [42] of the matching problem which makes it inapplicable for large scale settings. Various variants of combinatorial auction are proposed in the literature to allocate computational resources among grid users [35], [50].

14.5.2 Allocation Decision

In grids, the resource allocation to users can be done at two points. It can be initiated either by an individual provider (local) or a middleman such as meta-scheduler or auctioneer (global). In 'local', the trading decisions are based on the information provided by one resource provider. Generally, in this case, users approach the resource provider directly to buy or bid for a resource bundle advertised. For instance, to buy computer resources from Amazon [4], users can directly negotiate with the Amazon service [4]. Most of the single-sided auctions fall into this category.

In 'global', the trading decisions are based on the global information of multiple providers. Users utilize the services of an auctioneer in the market exchange to get the required resources. Thus, the auctioneer makes the decision on behalf of the users to buy resources from providers. The two-sided auctions fall into this category. The local decision point is more scalable but can lead to contention. While the global decision point is more optimized and coordinates demand fairly.

14.5.3 Participant Focus

The two major parties in grid computing, namely, resource consumers who submit various applications, and resources providers who share their resources, usually have different motivations when they join the grid. The participant focus identifies the market side for which market oriented systems or mechanisms are explicitly designed to achieve benefit.

Application-Centric

• In the application centric, mechanisms are designed such that application can be executed on the resources that meet user requirement within budget or deadline.

Resource-Centric

• Similarly, a resource-centric mechanism focuses mainly on resource providers by fulfilling their desired utility goal in terms of resource utilization and profit.

System Centric

• In the market scenario, there may be middlemen such as meta-brokers or meta-schedulers who act like an exchange, and coordinate and negotiate the resource allocations between users and producers. They try to maximize the utility for both users and providers. Thus, the resource allocation decision involves multiple users and providers.

In utility grids, mechanisms are required that can cater for the needs of both sides of the market. For instance, they should be able to satisfy end-users' demand for resources while giving enough incentive to resource providers to join the market. Moreover, the specific requirements of participants should not be neglected. It is also possible for market-oriented resource management systems (RMS) to have multiple participants focus, such as in double auctions.

14.5.4 Application Type

Market-oriented resource allocation mechanisms need to take into account job attributes to ensure that different job types with distinct requirements can be fulfilled successfully. The application model affects not only the scheduling mechanism but also other aspects of the utility grid such as resource offerings by providers, negotiation with providers, and formation of SLAs and their monitoring. For applications consisting of independent tasks, all the tasks can be distributed across multiple providers and thus optimization of the user's utility is easier. For the parallel application model, all tasks are needed to be mapped on a single resource site. In the workflow type of application, there is a precedence order existing in tasks; that is, a task cannot start until all its parent tasks are done. This will require coordination between multiple providers, and the scheduling problem needs to be fault tolerant since single failure may result in a large utility loss.

14.5.5 Allocation Objective

The market-oriented mechanisms can be used to achieve different objectives both in utility grids and traditional grids. The allocation objective of mechanism can be profit-oriented, system-oriented, or a hybrid of both of them. The objectives of various participants decide the trading relationship between them. The profit-based objective in terms of monetary gains, in general, encourages competition between participants. Thus, each participant tries to maximize their own utility without considering other consumers.

The objective of market-oriented scheduling mechanism could be to achieve optimization of system metrics such as utilization, fairness, load balancing and response time. This application of market-oriented mechanism, categorized in taxonomy as 'system based objective', is quite common. For example, OurGrid [6] uses a resource exchange mechanism termed network of favours which is used to share resources among distributed users. Bellagio [8] is another system deployed on PlanetLab for increasing system utilization on non peak time. The objective of market-oriented scheduling mechanisms can be of hybrid type. For example, a user may simultaneously want to minimize the response time and the cost of application execution. A provider may accept less profitable application to increase its utilization rather than waiting for more profitable jobs.

14.6   Survey of Grid Resource Management Systems

Grid RMSs chosen for the survey can be classified into two broad categories: market-oriented or system-oriented (presented in the following sections). Since this survey focuses on utility computing, market-oriented RMSs are surveyed to understand current technological advances and identify outstanding issues that are yet to be explored so that more practical RMSs can be implemented in future. On the other hand, surveying system-oriented RMSs allow analyzing and examining the applicability, and suitability of these systems for supporting utility grids in practice. This in turn helps us to identify possible strengths of these systems that may be leveraged for utility computing environments. In traditional grids, a user accesses the grid services either through User Brokers (UB) or Local Resource Managers (LRM). These systems schedule the jobs using system-centric approaches which optimize the metrics such as response time and utilization. We call such schedulers 'System-Oriented Schedulers' to differentiate them from the schedulers in market-oriented grids (or utility grids).

14.6.1 Survey of Market-Oriented Systems

Table 14.1 shows a summary listing of the existing market-oriented RMSs and brokers that have been proposed by researchers for various computing platforms. In the market-oriented grid, there are also three entities which participate in scheduling. One is the user broker that provides

Table 14.1 Market-Oriented Scheduling Systems

Name	Architecture	Economic Model	Mechanism	Traded Commodity	Pricing	Target Platform	Application Model	User Role
Tycoon (RMS)	Distributed, centralized	Proportional Share	Proportional Share	CPU Cycles	Pricing based on bids	Clusters, Grids	Task allocation	Discrete bid
Spawn (RMS)	Decentralized	Auction	Vickery Auction	CPU Time	Second price	Cluster	Task allocation	Discrete bid
Bellagio (RMS)	Centralized	Combinatorial Auction	Vickery Auction	CPUs and Storage	Second price	P2P	Not Specified	Bidding
Sharp (RMS)	Centralized	Commodity Market	Leases	CPU Time	Fixed	Grid	Lease allocation	Lease request to Sharp
Shirako (I)	Centralized	Not specified seems to be commodity	Negotiation, leasing generic	Virtual Machine and Storage	NA	Virtual Machines	Not specified	Lease request to broker
OCEAN (I)	Distributed	Bargaining, Tendering, Contract-Net, Continuous Double Auction	Discovering potential seller and Negotiation	CPU Cycles	Fixed	Any distributed resource	Not Specified	Discover and Negotiation.
CatNets (I)	Decentralized	Bargaining	Negotiation	Complex Services and Basic Services	Through negotiation. Dynamic pricing depend on available servers	Grid and Service-Oriented Computing	Not Specified	Bidding type
Nimrod-G (UB)	Centralized	Commodity Market, Spot Market, and Contract-Net for price establishment	Deadline and Budget Constrained Algorithms	CPU Cycles and Storage	Fixed pricing	World Wide Grid (resources Grid enabled using Globus middleware)	Independent multiple tasks & data parallel applications	Time and Budget

SORMA (I)	Centralized	Combinatorial Auction	Greedy Mechanism	NS	K-Pricing, based on auction	Commercial Providers	Not Specified (Simulation are based on independent tasks)	bidding
GridEcon (I)	Centralized	Commodity Market	NS	Resources managed by commercial service providers	Fixed	Commercial Resource Providers	Not Specified	Price specified by resource
Gridbus (UB)	NA	Commodity Market	Time and Budget based Algorithm	Compute and Storagee	NA	Commercial Providers	Bag of Task and workflow	Budget and time
Java Market (I)	Centralized	Commodity Market	Cost-based Greedy	Compute	Fixed	WWW	Java program	Bidding done by resources
Mariposa (RMS)	Centralized	Commodity Market	Cost minimization	Storage	Pricing based on load and historical info	Distributed database	Data	Budget and queries
GRIA (RMS)	P2P	Contract-based	NS	Compute	Through negotiation between providers.	Grid	NS	NA
PeerMart (RMS)	P2P	Auction	Double Auction	NS	Mean Price based on matched ask and bid	P2P	NS	Bids
G-Commerce (I)	Centralized	Commodity Market, Auction	NA	NA	Dynamic Pricing	Simulates hypothetical consumers and produces	NA	Bids

access to users on multiple resources. The resource providers are other entities who also have resource brokers that facilitate the admission control, pricing and negotiation with user brokers. The negotiation between users and resource providers can be on a one-to-one basis or through a third entity, that is, market exchange. The market exchange also provides other services such as resource discovery, banking, buying and selling computer services.

14.6.1.1 Tycoon

Tycoon [38] is a market-based distributed resource allocation system based on Proportional Share scheduling algorithm. The user request with the highest bid is allocated the processor time slice. The bid is computed as the pricing rate that the user pays for the required processor time. Tycoon allocates the resources to the self-interested users in environments where service hosts are unreliable because of frequent changes in the availability. Tycoon distinguishes itself from other systems by separating the allocation mechanism (which provides incentives) from the agent strategy (which interprets preferences). This simplifies the system and allows specialization of agent strategies for different applications while providing incentives for applications to use resources efficiently and resource providers to provide valuable resources. A host self-manages its local selection of applications, thus maintaining decentralized resource management. Hosts are heterogeneous since they are installed in various administrative domains and owned by different owners. Tycoon's distributed market allows the system to be fault tolerant and to allocate resources with low latency.

14.6.1.2 Spawn

Spawn [36] uses sealed-bid second-price auctions for market-oriented resource allocation in a network of heterogeneous computer nodes. Users place bids to purchase CPU resources for executing hierarchy-based concurrent programs in auctions held privately by each computer node and are not aware of other users' bids. The concurrent applications are then represented using a tree structure where a hierarchy of tasks expand or shrink in size depending on the resource cost. This mechanism limits the ability of customers to express fine-grained preferences for services.

14.6.1.3 Bellagio

The Bellagio [8] is a resource management system that allocates resources using combinatorial auction in order maximize aggregate end-user utility. Users identify their resources of interest via a SWORD [45] based resource discovery mechanism and register their preference to a centralized auctioneer for said resources over time and space as a combinatorial auction bids using a bidding language, which support XOR bids [44]. The bids are formulated in the virtual currency. The auction employed in Bellagio is periodic. Unlike other work that focuses on the contention for a single resource (CPU cycles), they are motivated by scenarios where users express interest in 'slices' of heterogeneous goods (for example, disk space, memory, bandwidth). Bellagio employs Share [13] for resource allocation in order to support a combinatorial auction.

14.6.1.4 SHARP

SHARP [26] is not exactly a complete resource management system, but it is an architecture to enable secure distributed resource management, resource control and sharing across sites and trust domains. The real management and enforcement of allocations are created by resource provider middleware which process the tickets and leases issued by SHARP. SHARP stands for Secure Highly Available Resource Peering and is based around timed claims that expire after a specified period, following a classical lease model. The resource claims are split into two phases. In the first phase, a user agent obtains a 'ticket', representing a soft claim that represents a probabilistic claim on a specific resource for a period of time. In the second phase, the ticket must be converted into a concrete reservation by contracting the resources site authority and requesting a 'lease'. These two phases allow SHARP system to become oversubscribed by issuing more tickets than it can support. SHARP also presents a very strong security model to exchange claims between agents, either user agents or third party brokers, and that then achieves identification, non-repudiation, encryption and prevents man-in-the-middle attacks.

14.6.1.5   Shirako

Shirako [31] is a generic and extensible system that is motivated by SHARP for on-demand leasing of shared networked resources across clusters. Shirako framework consists of distributed brokers which provision the resources advertised by provider sites to the guest applications. Thus, it enables users to lease groups of resources from multiple providers over multiple physical sites through broker service. Site authorities compute and advertise the amount of free resource by issuing resource tickets to the selected brokers. When a broker approves a request, it issues a ticket that is redeemable for a lease at a site authority. The ticket specifies the resource type, resource units granted and the validation period. Shirako allows 'flexible' resource allocation through leases which can be re-negotiated and extended via mutual agreement. A request can be defined as 'elastic' to specify that a user will accept fewer resources if its full allocation is not available. Requests can be 'deferrable' if a user will accept a later start time than what is specified in the lease if that time is unavailable. The function of broker is to prioritize the request and match to appropriate resource type and quantity. Provider side is represented by site authorities that run Cluster on Demand [12] to configure the resources allocated at remote sites.

14.6.1.6   OCEAN

OCEAN (Open Computation Exchange and Arbitration Network) [46] is a market based system for matching user applications with resources in the high performance computing environments, such as Cluster and Grid computing. It consists of all major components required to build utility grid, such as user node which submit trade proposals, computational resource and underlying market mechanism. Ocean first discovers potential sellers by announcing a buyer's trade proposal using optimized P2P search protocol. Then, the user node can negotiate with sellers based on the rules dynamically defined in a XML format. The ability to define negotiation rules is a remarkable characteristic of OCEAN that allows the adaptation of the economic model to diverse applications. The two possible negotiation allowed by OCEAN are 'yes/no' and automated bargain.

14.6.1.7   CATNET

CATNET Project [21] proposed a Catallaxy based market place where trading is divided into two layers, the application and service layer. The notion of a Catallaxy based market for grids was proposed by Austrian economist F.A. von Hayek. In this market, prices evolve from the actions of economically self-interested participants who try to maximize their own gain whilst having limited information available to them. In the application layer, complex services are mapped to basic services. The service layer maps service requests to actual resources provided by local resource managers. There are two markets which operate simultaneously – one for buying resources by service providers from resource providers, and another for buying services by clients from service providers. Thus, the client is not aware of the details of the resource provider, and vice versa. The prices are fixed in two markets by bilateral bargaining. CATNETS offers very interesting features but lacks comprehensive support (for example, monitoring, multi platform deployment).

In both layers, the participants have varying objectives which change dynamically and unpredictably over time. In the application/service layer, a complex service is a proxy which negotiates the access to bundles of basic service capabilities for execution on behalf of the application. Basic services provide an interface to access computational resources Agents representing the complex services, basic services and resources participate in a peer-to-peer trading network, on which requests are disseminated and when an appropriate provider is found, agents engage in a bilateral bargaining [22].

14.6.1.8 GridBus Broker

Gridbus Broker [63] is a single user resource broker that supports access to both computational and data grids. It transparently interacts with multiple types of computational resources which are exposed by various local grid middleware's such as Globus, Alchemi, Unicore and Amazon EC2, and scheduling systems such as PBS and Condor. By default, it implements two scheduling strategies that take into account budget and deadline of applications. Additionally, the design of the broker allows the integration of custom scheduling algorithms. Job-monitoring and status-reporting features are provided. Gridbus broker supports two application models, that is, parametricsweep and workflow.

14.6.1.9 Nimord/G

Nimrod/G is an automated and specialized resource management system, which allows execution of parameter sweep applications on grid to scientists and other types of users. Nimrod/G follows mainly the commodity market model and provides four budget and deadline based algorithms [10] for computationally-intensive applications. Each resource provider is compensated for sharing their resources by the users. The users can vary their QoS requirement based on their urgency and execution expense. Nimrod/G consists of a Task Farming Engine (TFE) for managing an execution, a Scheduler that talks to various information services and decides resource allocations, and a Dispatcher that creates Agents and sends them to remote nodes for execution. It is widely used by the scientific community for their computation-intensive simulations in the areas of bio-informatics, operations research, ecological modelling, and Business Process Simulation.

14.6.1.10 SORMA

Based on market engineering principles, the SORMA project [43] proposed an Open Grid Market which is built above the existing RMSs. It consists of self-interested resource brokers and user-agents. The users submit their bids for resources to the Open Grid Market using an autonomous bidding agent. On the other side of the market, the resource side bidding agents publish automatically available resources based on their predefined policies. The Open Grid Market matches requesting and offering bids and executes them against each other using Combinatorial Auction. The matches (that is, allocations) are formulated in SLAs (that is, contracts). The grid middleware is responsible for the resource provisioning and the payment system (such as PayPal) for the monetary transfer of funds. The open infrastructure of Open Grid Market allows various resource providers with different virtualization platforms or managers to easily plug in the market.

14.6.1.11 GridEcon

The GridEcon Project [3] proposed a market exchange technology that allows many (small and medium) providers to offer their resources for sale. To support buying and selling of resources, GridEcon market offers various services that makes exchange of commodity convenient, secure, and safe. The GridEcon market also proposed to design a series of value-added services on top of the market exchange (for example, insurance against resource failures, capacity planning, resource quality assurance, stable price offering), ensuring quality of the traded goods for grid users. Currently, GridEcon supports the only commodity market model where commercial resource providers can advertise their spare resources. The fixed pricing is used to allow users to sell and buy resources. The GridEcon market delegates the real resource management to commercial service providers.

14.6.1.12 Java Market

Java Market [5] is one of the oldest market-oriented systems developed by John Hopkins University It is an Internet-wide meta-computing system that brings together people who have worked to execute and people who have spare computing resources. One can sell CPU cycles by pointing Java-enabled browser to portal and allow execution of applets in a QoS-based computational market. The goal of Java Market is to make it possible to transfer jobs to any participating machine. In addition, in Java Market, resource provider receives payments or awards which are function of execution time of job and amount of work done.

14.6.1.13 Mariposa

Mariposa [57] is a distributed database system developed at the University of California. It supports query processing and storage management based on budget. Users submit queries with time-dependent budget to brokers who then select servers for executing the queries based on two protocols. One protocol is expensive as it solicits bids from all servers, requiring many communication messages. The expensive protocol adopts a greedy algorithm that aims to minimize cost to schedule sub-queries so as to select the cheapest server for the user. The other protocol is cheap since it selects specific server based on historical information. In Mariposa, bids on queries are based on local and selfish optimization of each user.

14.6.1.14 GRIA

GRIA (Grid Resources for Industrial Applications) [60] is a web-service based grid middleware for Business-to-Business (B2B) service procurement and operation. It aims at the development of business models and processes that make it feasible and cost-effective to offer and use computational services securely in an open grid market exchange. It also helps the industries to achieve better utilization and manage demand peaks on resources. GRIA software is based on and uses web service standard specifications and tools such as Apache AXIS. GRIA aiming to make Grid Middleware reliable for industrial application, thus, provides various software packages for performance estimation and QoS, workflow enforcement, cluster management, security and inter-operability semantics. Thus, each service provider using GRIA middleware has an account service and a resource allocation service, as well as services to store and transfer data files, and execute jobs which process these data files. The service provider's interaction is based on B2B model for accounting and QoS agreement.

14.6.1.15 PeerMart

PeerMart [28] is a Peer-to-Peer market based framework which allows completely decentralized trading of services between peers. It includes the capability of dynamic pricing and efficient price dissemination and services discovery over a P2P network. Using PeerMart, peers can bid prices for services, which enable them to govern the desired service performance. PeerMart implements an economically efficient distributed double auction mechanism where each peer is responsible for matching several services. PeerMart uses the overlay network infrastructure to map the services onto particular sets of peers following a fully distributed and redundant approach for high reliability and scalability to the number of participating peers. Its main limitation is the tight integration of auctions model in the framework, making it inflexible with respect to the market model.

14.6.1.16 G-Commerce

G-Commerce [65] provides a framework for trading computer resources (CPU and hard disk) in commodity markets and Vickrey auctions. By comparing the results of both market strategies in terms of price stability, market equilibrium, consumer efficiency and producer efficiency, the G-commerce project concludes that commodity market is a better choice for controlling grid resources than the existing auction strategies. It is argued and shown in simulations that this model achieves better price predictability than auctions. Thus, G-commerce is a grid resource allocation system based on the commodity market model where providers decide the selling price after considering long-term profit and past performance. The simulated auctions are winner-takes-it-all auctions and not proportional share, leading to reduced fairness and starvation. Furthermore, the auctions are only performed locally and separately on all hosts, leading to poor efficiency.

14.6.2   System-Oriented Schedulers

For over a decade various technologies have enabled applications to be deployed on the grids, including grid middleware such as Globus [24], Legion [11] and gLite [7]; schedulers such as Application Level Schedulers (AppLeS) [9]; and resource brokers including Gridbus Resource Broker[63], Nimrod/G [1], Condor-G [25], and GridWay [30]. These meta-schedulers or resource management systems interact with Local Schedulers or Grid Middlewares of various resource sites. The Local Scheduler supported such as Load Sharing Facility (LSF) [14], Open Portable Batch System (Open PBS [29] and Grid Engine (SGE/N1GE) [27]. In following section, we will discuss some of the Scheduling systems in detail and compare them using Table 14.2.

14.6.2.1 Community Scheduler Framework (CSF)

The Community Scheduler Framework (CSF) [54] is an open source tool set for implementing a grid meta-scheduler using the Globus Toolkit Services. The grid meta-scheduler provides an environment for dispatching jobs to various resource managers. CSF was developed by Platform Computing in cooperation with the Jilin University, China. The CSF provides plug-in for various heterogeneous schedulers such as Platform Load Sharing Facility (LSF), Open Portable Batch

Table 14.2 System-Oriented Scheduling Systems

Name	Allocation Mechanism	Scheduling Type	Scheduling Objective	Architecture	Application Type	Advance Reservation	Target Platform
CSF	Round Robin & reservation Based	Online	NA	Centralized	Task Model	Yes	LSF. PBS. SGE, Globus
CCS	FCFS	Online/ Interactive	Minimizing Response Time	Decentralized	Parallel Application	Yes	NA
GridWay	Greedy/Adaptive scheduling	Online	Minimize Response Time	Centralized	Parallel application. Parametric Sweep	No	Globus
Maob (Silver)	Fairshare, job prioritization	Online	Load balancing and response time minimization	Centralized	Task Model	Yes	Globus. Maui Scheduler
Condor/G	Matchmaking	Online	NA	Centralized	NS	No	Globus. Unicore. NorduGrid
Grubber/ Di-Grubber	FCFS	Online	NS	Decentralized	Task Model	Yes	Globus
eNanos	Job Selection policies based on arrival time and deadline; Resource Selection based on EST	Batch/periodic	Minimizing response time,	Centralized	Task Model	No	Globus
APST	Divisible Load Scheduling-based algorithms	Single application	Optimize response time	Centralized	Parametric-Sweep	Yes/No	Globus

System (Open PBS) and Grid Engine (SGE/N1GE), however CSF is designed for the best compliance with Platform LSF. CSF implements by default two basic scheduling algorithms, that is, Round-robin and reservation based algorithm. For the later algorithm, CSF facilitates advance reservation of resources for its users. Thus, users can make resource reservation using Resource Manager Tool of CSF, in order to guarantee the resource availability at a specified time. In addition, it also provides submission and monitoring tools to its users.

14.6.2.2   Computing Centre Software (CCS)

CCS [47] is vendor-independent resource management software that manages geographically distributed High Performance Computers. It is analogous to Globus and consists of three main components the CCS, which is a vendor-independent Local Resource Management Schedulers (LRMSs) for local HPC systems; the Resource and Service Description (RSD), used by the CCS to specify, and map hardware and software components of computing environments; and the Service Coordination Layer (SCL), which coordinates the resource usage across multiple computing sites. CCS schedules and maps interactive and parallel jobs using an enhanced first-come-first-served (FCFS) scheduler with backfilling. Deadline scheduling is another feature of CCS that gives the flexibility to improve the system utilisation by scheduling batch jobs at the earliest convenient and at the latest possible time. It also supports jobs with reservation requirements. At the meta-computing level, the Centre Resource Manager (CRM) is a layer above the CCS islands that exposes CSS scheduling features. When a user submits an application, the CRM maps the user request to the static and dynamic information regarding available resources through Centre Information Server (CIS). Centre Information Server (CIS) is a passive component that contains information about resources and their status. Once the CRM finds resources, it interacts with selected CCS islands for resource allocations. If not all resources are available, the CRM either re-schedules the request or rejects it.

14.6.2.3   GridWay

GridWay [30] is a meta-scheduler framework developed by a team working for Distributed Architecture Group from Universidad Complutense in Madrid, Spain. GridWay provides a transparent job execution management and resource brokering to the end user in a 'submit and forget' fashion. GridWay uses the Globus GRAM to interface with remote resources and, thus it can support all remote platforms and resource managers (for example fork, PBS and LSF) compatible with Globus. GridWay offers only simple scheduling capabilities even though custom scheduling algorithms are also supported. By default, GridWay follows the 'greedy approach', implemented by the round-robin algorithm. The collective scheduling of many jobs is not supported by the meta-scheduler. GridWay also provides sophisticated resource discovery, scheduling, constant monitoring and self-adaptive job migration to increase performance. Thus, an application is able to decide about resource selection as it operates, that is, it can modify its requirements and request a migration. GridWay also enables the deployment of virtual machines in a Globus Grid.

14.6.2.4   Moab (Silver) Grid Scheduler

Moab Grid Scheduler is a grid metascheduler developed by Cluster Resources Inc. Moab allows combining the resources from multiple high performance computing systems while providing a common user interface to all resources. It supports intelligent load balancing and advanced allocation allowing a job to be run over multiple machines in a homogeneous way or in a heterogeneous way resulting in better overall utilisation and better time. Moab supports all major scheduling systems and even optionally relies on Globus Toolkit Grid middleware for security and user account management purposes. It manages the resources on any system where Moab Workload Manager (a part of Moab Cluster Suite) is installed. Moab Workload Manager is a policy engine that allows sites to control the allocation of available resources to jobs. The meta-scheduler supports fine-grained grid level fairness policies. Using these policies, the system manager may configure complex throttling rules, fairshare, a hierarchical prioritization, and cooperation with allocation managers. Moab also has support for advanced reservations. This feature enables scheduling techniques such as backfilling; deadline based scheduling, QoS support and grid scheduling. One of the most interesting features going to be added in Moab is support for resource selection based on utility function where job completion time, resource cost, and other parameters are taken into account. These features allow easy transition of Moab meta-scheduler to market-oriented grid.

14.6.2.5   Condor–G

Condor-G [25] is a fault tolerant job submission system that can access various computing resources which employs software from Globus and Condor [40] to allocate resources to users in multiple domains. Condor-G is not a real broker but a job manager, thus it does not support scheduling policies but it provides framework to implement scheduling architecture about it. Condor-G can cooperate with the following middleware: Globus Toolkit (2.4.x - 4.0.x), Unicore and NorduGrid, and it can submit jobs to Condor, PBS and Grid Engine (SGE/N1GE) scheduling systems. Condor's Classified Advertisement language (ClassAd) MatchMaking tool allows users to specify which resource to allocate. The mechanism allows both jobs and machines to describe attributes about themselves, their requirements and preferences, and matches result in a logical-to physical binding. The GlideIn mechanism is also provided in Condor-G that starts a daemon processes which can advertise resource availability which is used by Condor-G to match locally queued jobs to resources advertised. The command-line interface is provided to perform basic job management such as submitting a job, indicating executable input and output files and arguments, querying a job status or cancelling a job. The most striking capability of Condor-G is its failure management which can deal with crashes at various levels.

14.6.2.6   GRUBER/DI-GRUBER

To avoid bottleneck of a central broker, DI-Gruber [19] is implemented as a completely distributed resource broker. It has been developed as an extension of the SLA based GRUBER broker deployed on the Open Science Grid. The GRUBER system [18] consists of four main components. The engine implements several algorithms necessary for determining optimized resource assignments. The site monitor acts as a data provider that publishes the status of Grid resources. The site selector provides the information about sites which is used for selecting a resource provider for execution of new tasks. It communicates with the engine to select the resource provider. The queue manager resides on submitting hosts, deciding which jobs can be executed at what time. The GRUBER can be utilized as the queue manager that controls the start time of jobs and enforces Virtual Organization (VO) policies, or as a site recommender when the queue manager is not available.

14.6.2.7   eNanos Resource Broker

eNANOS [48] is a general purpose OGSI-compliant resource broker developed by the Barcelona Supercomputing Center. It abstracts grid resource use and provides an API-based model of Grid access. The eNanos Grid resource broker is implemented on top of Globus Toolkit (GT) and supports both GT2 and GT3. It focuses on resource discovery and management, failure handling, and dynamic policy management for job scheduling and resource selection. The eNanos grid resource broker provides dynamic policy management and multi-criteria user requirements which are described in an XML document. These multi-criteria descriptions are used for resource filtering and ranking. The job scheduling in eNanos broker is divided into three phases, first is to select job to be schedule, second to select resource for selected job, and finally using meta-policy which consists of selection of the best job and the best resource. For job scheduling, several polices are implemented such as FIFOjobPolicy (First In First Out), REALTIMEjobPolicy (minimizes REALTIME = deadline time-estimated time of job finalization), EDFjobPolicy (Earlest Deadline First). Similarly for resource selection RANKresPolicy (resource selection based in the greatest rank obtained from the resource filtering process), ESTresPolicy (Earliest Starting Time, based in the estimated waiting time for a job in a local queue). Jobs are queued up in local system, and periodically scheduled by the resource broker.

14.6.2.8 AppLeS Parameter Sweep Template (APST)

APST [9] is an application level scheduler that provides an environment for scheduling and deploying large-scale parameter sweep applications (PSAs) on grid platforms. APST supports scheduling and job submission on different grid middleware and schedulers that take into account PSAs with data requirements. The APST scheduler allocates resources based on several parameters including predictions of resource performance, expected network bandwidths and historical data. The scheduler takes help of tools such as DataManger and ComputeManager to deploy and monitor data transfers and computation respectively which in turn get information from sources such as Network Weather Service (NWS) [66] and the Globus Monitoring and Discovery Service (MDS) [36]. AppLeS interacts directly with resource managers, perform all application management tasks, including, for example, file staging, and can enact collations of applications. APST is compatible with different low-level grid middleware through the use of Actuators and also allows for different scheduling algorithms to be implemented.

14.7 Discussion and Gap Analysis

After understanding the basic features of various market-oriented and system-oriented schedulers, based on presented taxonomy and requirements of utility grids, we can identify several outstanding issues that are yet to be explored to adopt a grid for creating a utility computing environment.

14.7.1 Scheduling Mechanisms

The market oriented scheduling mechanisms vary based on a market model used for trading resources. For example, if auction is the main market model then strategies for bidding and auction selection are required to maximize the chances of winning the auction. While in a commodity model, aggregation of resources from different provider is required to maximize user's utility. The challenges which are needed to be tackled more deeply can be categorized as following:

14.7.1.1   Support for Multiple QoS Parameters

In utility grids, other than traditional QoS requirements of users such as response time, additional QoS issues are needed to be addressed. For example, for HPC application, one has to minimize the execution time, thus, the resource capability and availability becomes essential which may be contradictory to the budget constraint. Many factors, such as deadline, resource reliability and security, need to be considered with monetary cost while making a scheduling decision on utility grids. Similarly, the resource management system should support QoS based scheduling and monitoring to deliver the quality service.

14.7.1.2   Support for Different Application Type

The current market-oriented scheduling mechanisms mainly support simpler applications/job models such as parameter-sweep. But, in reality, more advanced job models that comprise parallel job processing type and multiple-task data intensive applications, such as message-passing and workflow applications are also required by users. Thus, advanced algorithms, which require concurrent negotiation with multiple resource providers, are needed to be designed.

14.7.1.3   Support for Market-Oriented Meta-Scheduling Mechanisms

Currently most economics based approaches to grid scheduling are studied using an auctions perspective [51] (Schnizler et al., 2008). However, auctions based scheduling may not be always suitable, particularly when users want immediate access to resources or they are part of the same community. As an example of a user community, we consider the financial institution Morgan Stanley that has various branches across the world. Each branch has computational needs and QoS (Quality of Service) constraints that can be satisfied by grid resources. In this scenario, it is more appealing for the company to schedule various applications in a coordinated manner. Furthermore, another goal is to minimize the cost of using resources to all users across the community. Thus, mechanisms are required for user selection, and then resource allocation for utility maximization across all users. To tackle the problem of coordination and utility maximization among concurrent users, this thesis proposes market-oriented meta-scheduling mechanisms in the three scenarios.

14.7.2   Market Based Systems

In the previous sections, we discussed major systems which support market based mechanisms to buy and sell resources, and execute applications. Some of the most important outstanding issues from user, provider and market exchange perspective are presented as follows:

14.7.2.1   User Level Middleware

User level Middleware such as Gridbus broker [63] and GridWay [30] are designed only to participate in commodity market model. Moreover, they are not designed to trade in market exchange for leasing resources. Thus, these infrastructure supports is needed to provide flexibility for user to trade resources in any market. Moreover, automatic bidding support is required to participate in auctions used by systems such as Tycoon [38].

14.7.2.2   Market Exchange

As discussed previously, users and providers can also start negotiation using market exchange's services. The market exchange needed to match multiple users to multiple providers. The market exchange systems, such as Catnet [21], Bellagio [8], GridEcon [3] and SORMA [43], have restrictive price setting and negotiation policies. In a real market exchange, the choice of negotiation and pricing protocols are decided by participants in the system. This flexibility is critical because the choice of negotiation protocol (auction, commodity market, and one-to-one) and pricing (fixed, variable) can affect the participants utility enormously depending on the current demand and supply. As the number of consumers and providers grows, scalability of the market exchange will be become an issue. Thus, some of the components such as job submission and monitoring which are already well supported by user brokers and meta-schedulers can be delegated to each user. It makes the system more decentralized in the sense that, market exchange mainly acts as the middleman for matching users demand to providers supply, and other responsibilities during job submission and execution will be delegated to user and provider's brokers (or resource management systems). A reputation system would also complement the market exchanges by removing the unreliable and malicious users from the market. In addition to that, market exchanges are needed to be flexible enough to provide the participants to use market protocol of their choice. It will require co-existence of multiple negotiations between consumers and providers.

14.7.2.3   Core Middleware (Resource Level Middleware)

Similar to user level middleware, the existing resource middleware needed to be extended to participate in market exchange. In addition to that, these systems support simple job models, and thus more advanced job models such as parallel applications and workflows needed to be considered. In addition to that, SLA monitoring is required to ensure that user's QoS satisfaction.

14.8   Summary

In this chapter, a taxonomy and survey of market based resource allocation approaches and systems are presented. This chapter also provides an overview of the key requirements and components that are needed to be added in the grid middleware to support utility grids. The taxonomy categorized the market based resource allocation approaches from five different angles: (i) allocation decisions, (ii) mechanism's objective, (iii) market model, (iv) application model, (v) participant focus. This taxonomy is then used to classify various Grid middleware, which helps to examine current state-of-the-art in utility grids, thus identifying gaps which are needed to be filled in. In the utility grid, several projects are working to solve issues from both user and provider perspectives. The rapid emergence of utility computing infrastructures such as Amazon's Elastic Cloud, combined with industrial and academic High Performance Computing (HPC) demand, has increased the development of open marketplaces. However, significant work is required to fully benefit from the capabilities of utility grids. Specifically, the need for coordination between different participants while maximising their utilities has not been addressed. The lack of coordination can lead to underutilization of resources and starvation of low income users. Thus, market-based meta-scheduling mechanisms are required to tackle such challenges, particularly in situations where a scheduling decision can affect more than just the immediate users and providers. The market-exchange infrastructure will also be helpful in enabling such coordination and utility maximisation of utility grid participants.

References

[1] D. Abramson, R. Buyya and J. Giddy. A computational economy for Grid Computing and its Implementation in the Nimrod-G resource broker. *Future Generation Computing System*, 18(8): 1061–1074, 2002.

[2] D. Allenotor and R. Thulasiram. Grid Resources Pricing: A Novel Financial Option based Quality of Service-Profit Quasi-Static Equilibrium Model. In Proceedings of the 2008 9th IEEE/ACM International Conference on Grid Computing, IEEE Computer Society, 2008.

[3] J. Altmann, C. Courcoubetis, J. Darlington and J. Cohen. In GridEcon-The Economic-Enhanced Next-Generation Internet, In Proceedings of the 4th International Worksop on Grid Economics and Business Models, Rennes, France, 2007.

[4] Amazon. Amazon. elastic compute cloud (ec2). http://www.amazon.com/ec2/2009.

[5] Y. Amir, B. Awerbuch and R. Borgstrom. The Java market: Transforming the Internet into a metacomputer. Department of Computer Science, Technical Report CNDS-98-1, Johns Hopkins University, 1998.

[6] N. Andrade, W. Cirne, F. Brasileiro and P. Roisenberg. OurGrid: An approach to easily assemble grids with equitable resource sharing. *Lecture Notes in Computer Science*, 2862: 61–86, 2003.

[7] P. Andreetto, S. Andreozzi, G. Avellino, S. Beco, A. Cavallini, M. Cecchi, V. Ciaschini, A. Dorise, F. Giacomini, A. Gianelle, et al. The gLite workload management system. In Journal of Physics: Conference Series, volume 119, page 062007. Institute of Physics Publishing, 2008.

[8] A. AuYoung, B. Chun, A. Snoeren and A. Vahdat. Resource allocation in federated distributed computing infrastructures. In Proceedings of the 1st Workshop on Operating System and Architectural Support for the On-demand IT InfraStructure, NV, USA, 2004.

[9] F. Berman and R. Wolski. The AppLeS Project: A Status Report. In Proceedings of the 8th NEC Research Symposium, Berlin, Germany, May 1997.

[10] R. Buyya, M. Murshed, D. Abramson and S. Venugopal. Scheduling Parameter Sweep Applications on Global Grids: A Deadline and Budget Constrained Cost-Time Optimisation Algorithm. *International Journal of Software: Practice and Experience (SPE)*, 35(5): 491–512, 2005.

[11] S. Chapin, J. Karpovich and A. Grimshaw. The Legion resource management system. In Proceedings of the 5th Workshop on Job Scheduling Strategies for Parallel Processing, San Juan, Puerto Rico, Apr. 1999. IEEE CS Press, Los Alamitos, CA, USA.

[12] J. Chase, D. Irwin, L. Grit, J. Moore and S. Sprenkle. Dynamic virtual clusters in a grid site manager. In Proceedings of the Twelfth International Symposium on High Performance Distributed Computing, Seattle, Washington, USA, 2003.

[13] B. Chun, C. Ng, J. Albrecht, D. Parkes and A. Vahdat. Computational resource exchanges for distributed resource allocation. Technical report, http://citeseer. ist.psu.edu/706369.html, 2004.

[14] P. Computing. LSF-Load Sharing Facility, 2003.

[15] K. Czajkowski, I. Foster, C. Kesselman, V. Sander and S. Tuecke. SNAP: A protocol for negotiating service level agreements and coordinating resource management in distributed systems. *Lecture Notes in Computer Science*, 2537: 153–183, 2002.

[16] E. David, R. Azoulay-Schwartz, and S. Kraus. Protocols and strategies for automated multi-attribute auctions. In Proceedings of the first international joint conference on Autonomous agents and multiagent systems: part 1, pages 77–85. ACM New York, NY, USA, 2002.

[17] T. Dornemann, E. Juhnke and B. Freisleben. On-Demand resource provisioning for BPEL workflows using Amazon's Elastic Compute Cloud. In Proceedings of the 2009 9th IEEE/ACM International Symposium on Cluster Computing and the Grid, Shanghai, China, 2009.

[18] C. Dumitrescu and I. Foster. Gruber: a grid resource usage sla broker, In Proceedings of 2005 Euro-Par Parallel Processing conference, Monte de Caparica, Portugal, 2005.

[19] C. Dumitrescu, I. Raicu and I. Foster. Di-gruber: A distributed approach to grid resource brokering. In Proceedings of the 2005 ACM/IEEE conference on Supercomputing (SC'05), Seattle, WA, USA, November 2005.

[20] C. Ernemann, V. Hamscher and R. Yahyapour. Economic scheduling in Grid Computing. In Proceeding of the 7th international workshop on Job scheduling strategies for parallel processing, Cambridge, MA, USA, June 16, 2001.

[21] T. Eymann, M. Reinicke, O. Ardaiz, P. Artigas, F. Freitag and L. Navarro. Decentralized resource allocation in application layer networks. In Proceedings of the 3rd International Symposium on Cluster Computing and the Grid, Tokyo, Japan, 2003.

[22] T. Eymann, M. Reinicke, F. Freitag, L. Navarro, O′. Arda′iz and P. Artigas. A Hayekian self-organization approach to service allocation in computing systems. *Advanced Engineering Informatics*, 19(3): 223–233, 2005.

[23] M. Feldman, K. Lai and L. Zhang. A price-anticipating resource allocation mechanism for distributed shared clusters. In Proceedings of the 6th ACM conference on Electronic commerce, Vancouver, Canada, 2005.

[24] I. Foster and C. Kesselman. Globus: A metacomputing infrastructure toolkit. *International Journal of High Performance Computing Applications*, 11(2): 115, 1997.

[25] J. Frey, T. Tannenbaum, M. Livny, I. Foster and S. Tuecke. Condor-G: A Computation Management Agent for Multi-Institutional Grids. *Cluster Computing*, 5(3): 237–246, July 2002.

[26] Y. Fu, J. Chase, B. Chun, S. Schwab and A. Vahdat. SHARP: An architecture for secure resource peering. In Proceedings of the nineteenth ACM symposium on Operating systems principles, Bolton Landing, New York, USA, 2003.

[27] W. Gentzsch et al. Sun grid engine: Towards creating a compute power grid. In Proceedings of the 1st International Symposium on Cluster Computing and the Grid, page 35, 2001.

[28] D. Hausheer and B. Stiller. Peermart: the technology for a distributed auctionbased market for peer-to-peer services. In Proceedings of the It IEEE International Conference on Communications, Brisbane, Australia, 2005.

[29] R. Henderson. Job scheduling under the portable batch system, In Proceedings of 1995 Worksop on Job Scheduling Strategies for Parallel Processing, Santa Barbara, CA, USA, 1995.

[30] E. Huedo, R. Montero, I. Llorente, D. Thain, M. Livny, R. van Nieuwpoort, J. Maassen, T. Kielmann, H. Bal, G. Kola, et al. The GridWay framework for adaptive scheduling and execution on grids. *Software-Practice and Experience*, 6(8), 2005.

[31] D. Irwin, J. Chase, L. Grit, A. Yumerefendi, D. Becker and K. Yocum. Sharing networked resources with brokered leases. In Proceedings of the USENIX Technical Conference, Boston, MA, USA, 2006.

[32] L. Kale, S. Kumar, M. Potnuru, J. DeSouza and S. Bandhakavi. Faucets: efficient resource allocation on the computational grid. In Proceedings of the 33rd International Conference on Parallel Processing, Quebec, Canada, 2004.

[33] W. Kang, H. Huang and A. Grimshaw. A highly available job execution service in computational service market. In Proceedings of the 8th IEEE/ACM International Conference on Grid Computing, Austin, Texas, USA, 2007.

[34] U. Kant and D. Grosu. Double auction protocols for resource allocation in grids. In Proceedings of the International Conference on Information Technology: Coding and Computing (ITCC05), pages 366–371, 2005.

[35] T. Kelly. Generalized knapsack solvers for multi-unit combinatorial auctions: Analysis and application to computational resource allocation. Technical Report HPL- 2004-21, HP Labs, Palo Alto, CA, USA, 2004.

[36] H. Keung, J. Dyson, S. Jarvis, and G. Nudd. Performance evaluation of a grid resource monitoring and discovery service. *IEE Proceedings - Software*, 150(4): 243–251, 2003.

[37] Y. Kwok, S. Song and K. Hwang. Selfish grid computing: Game-Theoretic modeling and NAS performance results. In Proceedings of the Fifth IEEE International Symposium on Cluster Computing and the Grid (CCGrid'05), Cardiff, UK, 2005.

[38] K. Lai, L. Rasmusson, E. Adar, L. Zhang and B. Huberman. Tycoon: An implementation of a distributed, market-based resource allocation system. *Multiagent and Grid Systems*, 1(3): 169–182, 2005.

[39] J. Li and R. Yahyapour. Learning-based negotiation strategies for grid scheduling. In Proceedings of the International Symposium on Cluster Computing and the Grid (CCGRID2006), Singapore, 2006.

[40] M. Litzkow, M. Livny and M. W. Mutka. Condor - a hunter of idle workstations. In Proceedings of the 8th Int'l Conference of Distributed Computing Systems, San Jose, CA, June 1988.

[41] R. Montero, E. Huedo and I. M. Llorente. Grid Scheduling Infrastructures with the GridWay Metascheduler, 2006.

[42] M. Narumanchi and J. Vidal. Algorithms for distributed winner determination in combinatorial auctions. *Lecture Notes in Computer Science*, 3937: 43, 2006.

[43] D. Neumann, J. Stoesser, A. Anandasivam, and N. Borissov. Sorma-building an open grid market for grid resource allocation. *Lecture Notes in Computer Science*, 4685: 194, 2007.

[44] N. Nisan. Bidding and allocation in combinatorial auctions. In EC '00: Proceedings of the 2nd ACM conference on Electronic commerce, pages 1–12, New York, NY, USA, 2000. ACM.

[45] D. Oppenheimer, J. Albrecht, D. Patterson and A. Vahdat. Scalable wide-area resource discovery. In USENIX WORLDS, volume 4. Citeseer, 2005.

[46] P. Padala, C. Harrison, N. Pelfort, E. Jansen, M. Frank and C. Chokkareddy. OCEAN: The open computation exchange and arbitration network, a market approach to meta computing. In Proceedings of the 2nd International Symposium on Parallel and Distributed Computing, Ljubljana, Slovenia, 2003.

[47] F. Ramme, T. Romke and K. Kremer. A distributed computing center software for the efficient use of parallel computer systems. In Proceedings of the 1994 International Conference on High Performance Computing and Networking, Munich, Germany, 1994.

[48] I. Rodero, F. Guim, J. Corbalan and J. Labarta. eNANOS: Coordinated Scheduling in Grid Environments. In Proceedings of the 2005 International Conference on Parallel Computing (ParCo), Malaga, Spain, 2005.

[49] T. Sandholm, K. Lai, J. Ortiz and J. Odeberg. Market-Based Resource Allocation using Price Prediction in a High Performance Computing Grid for Scientific Applications. In Proceedings of the 15th IEEE International Symposium on High Performance Distributed Computing, Paris, France, 2006.

[50] B. Schnizler. MACE: a multi-attribute combinatorial exchange Negotiation, Auctions, and Market Engineering, 2: 84–100, 2006.

[51] B. Schnizler. Resource allocation in the Grid. A market engineering approach, Ph.D.thesis. Studies on eOrganisation and Market Engineering, 2007.

[52] Shneidman, C. Ng, D. Parkes, A. AuYoung, A. Snoeren, A. Vahdat and B. Chun. Why markets could (but dont currently) solve resource allocation problems in systems. In 10th USENIX Workshop on Hot Topics in Operating System, Santa Fe, NM, USA, 2005.

[53] G. Singh, C. Kesselman and E. Deelman. Adaptive pricing for resource reservations in shared environments. In Proceedings of the 8th IEEE/ACM International Conference on Grid Computing, Texas. USA, 2007.

[54] C. Smith. Open source metascheduling for virtual organizations with the community scheduler framework (CSF). Technical report, 2003.

[55] R. Smith. The Contract Net Protocol High-Level Communication and Control in a Distributed Porblem Solver, *IEEE Transactions On Computer*, 4: 1104–1113, 1980.

[56] O. Sonmez and A. Gursoy. A novel economic-based scheduling heuristic for computational grids. *International Journal of High Performance Computing Applications*, 21(1): 21, 2007.

[57] M. Stonebraker, R. Devine, M. Kornacker, W. Litwin, A. Pfeffer, A. Sah and C. Staelin. An economic paradigm for query processing and data migration in Mariposa. In 3rd International Conference on Parallel and Distributed Information Systems, Austin, Texas, USA, 1994.

[58] J. Stosser, P. Bodenbenner, S. See and D. Neumann. A Discriminatory Pay-as- Bid Mechanism for Efficient Scheduling in the Sun N1 Grid Engine. In Proceedings of the 41st Annual Hawaii International Conference on System Sciences, Hawaii, 2008.

[59] G. Stuer, K. Vanmechelen and J. Broeckhove. A commodity market algorithm for pricing substitutable Grid resources. *Future Generation Computer Systems*, 23(5): 688–701, 2007.

[60] M. Surridge, S. Taylor, D. De Roure and E. Zaluska. Experiences with GRIA: industrial applications on a Web Services Grid. In Proceedings of the First International Conference on e-Science and Grid Computing, Melbourne, Australia, 2005.

[61] Z. Tan and J. Gurd. Market-based grid resource allocation using a stable continuous double auction. In Proceedings of the 8th IEEE/ACM International Conference on Grid Computing, Austin, Texas, USA, 2007.

[62] S. Venugopal, X. Chu and R. Buyya. A negotiation mechanism for advance resource reservation using the alternate offers protocol. In Proceedings of the 16th International Workshop on Quality of Service (IWQoS 2008), Netherlands, 2008.

[63] S. Venugopal, K. Nadiminti, H. Gibbins and R. Buyya. Designing a resource broker for heterogeneous grids. *Software-Practice and Experience*, 38(8): 793–826, 2008.

[64] R. Wolski, J. Plank, J. Brevik and T. Bryan. Analyzing market-based resource allocation strategies for the computational grid. *International Journal of High Performance Computing Applications*, 15(3): 258, 2001.

[65] R. Wolski, J. Plank, J. Brevik and T. Bryan. G-commerce: Market formulations controlling resource allocation on the computational grid. In International Parallel and Distributed Processing Symposium (IPDPS), pages 46–66. Citeseer, 2001.

[66] R. Wolski, N. Spring and J. Hayes. The Network Weather Service: A Distributed Resource Performance Forecasting Service for Metacomputing. *Journal of Future Generation Computing Systems*, 15: 757–768, 1999.

[67] L. Xiao, Y. Zhu, L. Ni and Z. Xu. GridIS: An Incentive-Based Grid Scheduling. In 19th IEEE International Parallel and Distributed Processing Symposium, 2005. Proceedings, pages 65b–65b, 2005.

[68] J. Yu, S. Venugopal and R. Buyya. A market-oriented grid directory service for publication and discovery of grid service providers and their services. *The Journal of Supercomputing*, 36(1): 17–31, 2006.

[69] C. Yeo and R. Buyya. A Taxonomy of Market-based Resource Management Systems for Utility-Driven Cluster Computing. *Software Practice and Experience*, 36(13): 1381, 2006.

[70] J. Altmann, M. Ion, A. Adel, and B. Mohammed. A Taxonomy of Grid Business Models. In Proceedings of the 4th International Workshop on Grid Economics and Business Models, Rennes, France, 2007.

Glossary

3G Third Generation.

3GPP Third Generation Partnership Project.

ACS Ambient Control Space.

ACK Acknowledgment.

ADCR Adaptive Distributed Cross layer Routing Algorithm.

Additive White Gaussian Noise (AWGN) AWGN is a channel model in which signals are distorted by a linear addition of wideband or white noise with a constant spectral density and a Gaussian distribution of amplitude.

AFN Associated Feedback Mechanism.

Alert Clustering Group security alerts into clusters (or threads) based on some similarity measure, such as IP addresses or port numbers.

Alert Correlation The combination of fragmented information contained in the alert sequences and interpreting the whole flow of alerts.

Amplify-and-forward (AF) A relaying technique in which the relay amplifies the signal and then forwards the amplified the signal to the destination.

AN Ambient Network.

Analog Network Coding (ANC) ANC is effectively a PNC but exploits signal interference at intermediate nodes to increase the throughput in the network while relaxing the synchronization between mixed signals under severe conditions.

Angle of Arrival measurement, or AoA, is a method for determining the direction of propagation of a radio-frequency wave incident on an antenna array.

ANHASA Ambient Networks Heterogeneous Access Selection Architecture.

ANI Ambient Network Interface.

AP Access Point.

ARI Ambient Resource Interface.

ARQ Automatic Repeat request.

Cooperative Networking, First Edition. Edited by Mohammad S. Obaidat and Sudip Misra.
© 2011 John Wiley & Sons, Ltd. Published 2011 by John Wiley & Sons, Ltd.

Artificial Neural Network (ANN) usually called 'neural network' (NN), is a mathematical model or computational model that is inspired by the structure and/or functional aspects of biological neural networks. It consists of an interconnected group of artificial neurons and processes information using a connectionist approach to computation.

AS Active Set.

ASI Ambient Service Interface.

AT Audit Trail.

BC Boundary Controller.

BER Bit Error Rate.

Best-relay Selection When there are multiple relays, the best relay only (e.g., the one that maximizes the signal-to-noise ratio) is selected for relaying.

Binary Phase Shift Keying (BPSK) BPSK is a 2-state phase modulation scheme.

Botnet A group of computers which are compromised and controlled by a hacker.

Broker An entity that creates and maintains relationships with multiple resource providers. It acts as a liaison between cloud services customers and cloud service providers, selecting the best provider for each customer and monitoring the services.

BS Base Station.

Bundle Protocol data unit of the DTN bundle protocol. Represents aggregates of entire blocks of application-program data and metadata.

Bursty Amplify and Forward (BAF) In BAF the source transmits bursts using only a fraction of the available time slot at high SNR but ensures being observed by the relay.

CAIA Cooperative Ad hoc Intrusion analysis.

Calibration It is a comparison between measurements – one of known magnitude or correctness made or set with one device and another measurement made in as similar a way as possible with a second device.

CARM Coordinated Access Resource Management.

CCA Cooperative Cache Agent.

CCA Cooperative Collision Avoidance.

CCD Cooperative Cache Daemon.

CCSL Cooperative Cache Supporting Library.

CCW Cooperative Collision Warning.

CDP Cell Density Packet.

CEDAR Core Extraction Distributed Ad hoc Routing.

Channel State Information (CSI) CSI in wireless communications refers to known channel properties of a communication link.

CITRA Cooperative Intrusion Trace back and Response Architecture.

CM Cognitive Manager.

Code Division Multiple Access (CDMA) CDMA is a medium-access method in which different users are allocated different code sequences.

Coded Cooperation Coded cooperation strategy integrates relay cooperation with channel coding.

Coherence Engine Directory responsible of granting cache coherence on chip.

Combining Techniques These techniques combine multiple received signals at the receiver to create a robust signal.

Complete Complementary codes (CC) CC codes are pairs of sequences with the useful property that their cross-correlation out-of-phase auto-correlation coefficients sum to zero. Binary complementary sequences were first introduced by Marcel J. E. Golay in 1949. In 1961–1962 Golay gave several methods for constructing sequences of length 2^N.

Compress and Forward (CF) CF is a relaying strategy that exploits redundancies present within the correlated signals received by the relay and the destination to compress the received signal with certain distortion rates.

Computer Virus A computer program that can insert/copy itself into one or more files without the permission or knowledge of the user and perform some (possibly null) actions.

Cooperation Cooperation is the act of network nodes using their storage, bandwidth, and energy resources to mutually enhance the overall network performance. In a cooperative environment, network nodes collaborate with each other, storing and distributing bundles not only in their own interest, but also in the interest of other nodes.

Cooperative Beamforming In this technique, cooperating nodes forwards the signal to the receiver after multiplying it by certain weights to optimize some function (e.g., maximize the signal-to-noise ratio).

Cooperative Cache Partitioning (CCP) Heuristic-based hybrid cache allocation scheme that integrates MTP with CC. This technique uses a heuristic-based partitioning algorithm to determine MTP partitions and a weight-based integration policy to decide when to use MTP or CC's LRU-based capacity sharing policy.

Cooperative Caching (CC) Cache organization which uses a Centralized Coherence Engine (CCE) to grant coherence on-chip and private caches. This organization is also able to share cache capacity through spilling.

Cooperative Diversity A technique that achieves diversity (availability of multiple copies of the received signal at the receiver) using the cooperation relaying.

Cooperative Network Coding (CNC) CNC exploits both advantages of cooperative diversity and network coding.

Cooperative Relaying A technique that uses intermediate nodes (between the source and destination) as relays.

CPU Central Processing Unit.

CQoS Cooperative Quality of Service.

CR Cognitive Radio.

CRRM Common Radio Resource Management.

CS Candidate Set.

CSMA Carrier Sense Multiple Access.

CSP Cluster-based Self-organizing.

Cyclic Redundancy Check (CRC) CRC is a fixed length binary sequence that is calculated for every transmitted and received block of data in order to decide whether the block is received correctly or with errors.

Data Aggregation is the process of joining multiple packets together into a single transmission unit, in order to reduce the overhead associated with each transmission.

Data Fusion is generally defined as the use of techniques that combine data from multiple sources and gather that information in order to achieve inferences, which will be more efficient and potentially more accurate than if they were achieved by means of a single source.

DC Discovery Coordinator.

Decode-and-forward (DF) A relaying technique in which the relay detects the signal first, then regenerates it and finally forwards it to the destination.

Delay-tolerant networking (DTN) Network research topic focused on the design, implementation, evaluation, and application of architectures and protocols that intend to enable data communication among heterogeneous networks operating on different transmission media.

Decode and Forward (DF) DF is a relaying strategy where upon receiving a source signal each relay regenerates a new signal by demodulating and decoding the received signal then re-encoding and modulating it before forwarding to the destination.

Distributed Cooperative Caching (DCC) Distributed version of the Cooperative Caching. It uses several Distributed Coherence Engines (DCEs) to grant coherence. Coherence engines have a different organization which requires less energy but also a replacement mechanism.

Diversity The availability of multiple copies of the received signal at the receiver.

Diversity Gain is the increase in SNR due to some diversity scheme (Frequency, time, polarization, space, multi-user, cooperative diversity).

Diversity–Multiplexing Tradeoff (DMT) DMT is the tradeoff between diversity and multiplexing gain in a MIMO system.

DoA Denial of Service.

DoS Denial of Service.

DS Detected Set.

DSDV Destination Sequence Distance Vector.

DSR Dynamic Source Routing.

Duty Cycle is the fraction of time that a system is in an 'active' state.

Dynamic Decode and Forward (DDF) DDF is a relaying strategy in which the relay listens to the source and as soon as it is able to decode the received data correctly, it starts encoding and relaying to the destination.

Elastic Cooperative Caching (ElasticCC) Distributed cache organization which uses Private/Shared caches and DCEs. Through independent repartitioning units adapts cache assignation to each node based on application requirements.

Equal Gain Combining (EGC) A combining technique in which the received multiple copies are added after weighing them with the same weight.

Ergodic Capacity or 'average capacity' is defined as the probabilistic average of the Shannon capacity for AWGN channel over all instantaneous channel fading realization states.

FE Functional Entity.

Frequency Division Multiple Access (FDMA) FDMA is a channel access method, used in multiple-access protocols, that gives users an individual allocation of one or several frequency bands, or channels in a multiple access.

Full-duplex system, or sometimes double-duplex system, allows communication in both directions, and, unlike half-duplex, allows this to happen simultaneously.

GBN Go Back N.

GLL Generic Link Layer.

GLL_AL Generic Link Layer Abstraction Layer.

GLL_C Generic Link Layer Control.

Grid Computing The use of pools of resources onto which applications or services may be dynamically provisioned and re-provisioned to meet the goals of one or more enterprises, whilst improving both efficiency and agility.

Grid node/site is a machine/location that is capable of receiving and executing work that is distributed in a grid.

Grid Resource It is the physical machine/site where job will be executed.

GVI Geo-localized Virtual Infrastructure.

GyTAR: improved Greedy Traffic Aware Routing protocol.

Half-duplex system allows communication in both directions, but only one direction at a time (not simultaneously).

Ham mails Legitimate emails.

HOLM HandOver and Locator Management.

HRS Hybrid Reputation System.

HWN Hybrid Wireless Network.

I2V Infrastructure to Vehicles.

Idle listening is listening to receive possible traffic that is not sent.

IDS Intrusion Detection System.

IFTIS Infrastructure-Free Traffic Information System.

Incremental Relaying A relaying technique in which the relay does not forward the signal to the destination unless the direct signal is not adequate for acceptable detection.

IP Multicast In the IP Multicast model, multicast service is implemented at the IP layer. It retains the IP interface, and introduces the concept of open and dynamic groups, which greatly inspires later proposals. Given that the network topology is best-known in the network layer, multicast routing in this layer is also the most efficient.

IP Internet Protocol.

ISO International Standardization Organization.

Job a metadata object that specifies processes that create output.

JRRM Joint Radio Resource Management.

Localization is a process used to determine the location of one position relative to other defined reference positions.

LTE Long Term Evolution.

MAC Medium Access Control.

Malware Code designed to exploit or damage a computer, server, or network.

MANET Moble Ad hoc NETwork.

Market Exchange An environment where agents (or obejcts) compete for resources. They may have certain capabilities, notably a budget, that limit how much of a service they can afford. Marketplace supply and demand forces control emergent behaviour.

Mary Quadrature Phase Shift Keying (M-QAM) M-QAM is a M-state Quadrature Amplitude and phase modulation scheme.

Master-slave is a model of communication where one device or process has unidirectional control over one or more other devices. In some systems a master is elected from a group of eligible devices, with the other devices acting in the role of slaves.

Max-flow Min-cut Formula states that in a flow network, the maximum amount of flow passing from a source node to a sink node is equal to the minimum capacity that needs to be removed from the network so that no flow can pass between them.

Maximum Ratio Combining (MRC) A combining technique in which the received multiple copies are added after weighing them with different weights (the weight of each received signal is equal to the conjugate of the signal).

Mesh-based Approach Mesh-based overlay designs do not construct and maintain an explicit structure for delivering data. The underlying argument is that, rather than constantly repair a structure in a highly dynamic peer-to-peer environment, the availability of data can be used to guide the data flow. To handle this, most of such approaches adopt pull-based techniques. More explicitly, nodes maintain a set of partners, and periodically exchange data availability information with the partners. A node may then retrieve unavailable data from one or more partners, or supply available data to partners.

Middleware is computer software that connects software components or some people and their applications. The software consists of a set of services that allows multiple processes running on one or more machines to interact.

MRA Multi-Radio Access.

MRRM Multi-Radio Resource Management.

mTreebone is a cooperative mesh-tree design that leverages both tree-based and mesh-based approaches. The key idea is to identify a set of stable hosts to construct a tree-based backbone, called treebone, with most of the data being pushed over this backbone. These stable hosts, together with others, are further organized through an auxiliary mesh overlay, which facilitates the treebone to accommodate host dynamics and fully exploit the available bandwidth between overlay hosts.

Multi-agent System (MAS) is a system composed of multiple interacting intelligent entities called agents.

Multi-Carrier Code Division Multiple Access (MC-CDMA) MC-CDMA is an OFDM modulation system operating on a code division multiple access system.

Multiple-Input and Multiple-Output (MIMO) MIMO is the use of multiple antennas at both the transmitter and receiver to improve communication performance.

Multiple-Input Single-Output (MISO) MISO is the use of multiple antennas at the transmitter.

Multiple Time-Sharing Partitioning (MTP) Cache partitioning policy which adds a time-sharing aspect on top of multiple spatial partitioning to solve conflicts between competing threads and optimize throughput and fairness over the long time.

Mutual Information of two random variables is a quantity that measures the mutual dependence of the two variables.

NAK Negative Acknowledgement.

NCSW Node Cooperative Stop and Wait ARQ.

Network Coding (NC) is a technique where nodes, instead of simply forwarding the packets of information they receive, they mix the packets then forward them to the destination nodes.

Network Spread Coding (NSC) NSC is an additional approach to PNC that brings features of NC and spread spectrum technique together for exploiting advantages of better bandwidth efficiency offered by NC with higher robustness of spread spectrum against interference and noise due to use of linearly independent complete complementary (CC) sequences.

NLoS None Line of Sight.

Non-orthogonal Amplify and Forward (NAF) NAF is a relaying strategy in which sources are allowed to transmit all the time.

NRS Negative Reputation System.

NSC-Decode Spread and Forward (NSC-DSF) NSC-DSF is NSC using decode spread and forward the relay nodes.

NSC-Forward (NSC-F) NSC-F is NSC using simple forward at the relay nodes.

ODAM Optimized Dissemination of Alarm Messages.

OFDM Orthogonal Frequency Division Multiplexing.

Opportunistic Network is a network that falls into the general category of delay-tolerant networks (DTNs), characterized by opportunistic contacts (i.e., communication opportunities between network nodes happen unexpectedly).

Orthogonal Frequency Division Multiple Access (OFDMA) OFDMA is a multi-user version of the Orthogonal Frequency Division Multiplexing (OFDM) digital modulation scheme, in which multiple-access is achieved by assigning subsets of subcarriers to individual users.

OSI Open Systems Interconnection.

Outage Capacity is defined as the maximum data rate that can be sent over a channel for a given outage probability.

Outage event occurs when signals are transmitted at higher rates than the mutual information of the fading channel.

Out-of-band Signaling Separation of control information from the data bundles; control information is sent separately from data bundles via different transmission links.

Packet-level NC is NC performed at the network layer.

PANA Protocol for carrying Authentication and Network Access.

Partial Decode and Forward (PDF) PDF is a relaying strategy in which the relay is not forced to decode the entire received message; instead the source message is split and transmitted to the relay in parts.

PDF Probability Density Function.

Peer-to-Peer (P2P) computing or networking is a distributed application architecture that partitions tasks or work loads between peers.

Peer-to-Peer Media Streaming pushes functionality to the users actually participating in the multicast group. Instead of being handled by a single entity, administration, maintenance, responsibility for the operation of such a peer-to-peer system are distributed among the users, and the research focuses on simultaneous media content broadcast using the application end-point architecture. While the proposals for peer-to-peer media streaming differ on a wide-range of dimensions, they can be broadly classified into two categories based on the overlay structure used for data dissemination, namely, tree-based and mesh-based approaches.

PER Packet Error Rate.

Physical layer Network Coding (PNC) PNC is NC at the physical-layer in order to make use of the broadcasting nature of the wireless channel.

PRCSMA Persistent Relay Carrier Sensing Multiple Access Protocol.

Proxy Caching Proxy caching is preciously used in web service. In the context of media streaming, the proxy caching technique is used to exploit the temporal locality of client requests for streaming media content and deploys a group of proxies to cooperatively utilize caching space, balance loads and improve the overall performance of streaming media content.

PRS Positive Reputation System.

PSD Power Spectral Density.

Quadrature Phase Shift Keying (QPSK) QPSK is a 4-state phase modulation scheme.

Quality of service (QoS) is the ability to provide different priority to different applications, users, or data flows, or to guarantee a certain level of performance to a data flow.

Radio Signal Strength (RSS) refers to the magnitude of the electric field at a reference point that is a significant distance from the transmitting antenna.

RAT Radio Access Technology.

Rayleigh Fading Channels Rayleigh fading channel models assume that the magnitude of a signal that has passed through a communications channel will vary randomly, or fade, according to a Rayleigh distribution – the radial component of the sum of two uncorrelated Gaussian random variables.

Relaying This refers to the delivery of the transmitted signal to the destination through intermediate nodes (relays).

Replica Cache block in a valid coherence state with multiple on-chip copies.

Resource Hog A network node that, on average, attempts to send more of its own data and possibly forward less peer data than a typical well-behaved node.

Resource Management The process of controlling and assigning the radio resource (channel) in wireless communication systems.

RN Relay Network.

Scheduler A grid-based system that is responsible for routing a job to proper physical node or selected processor. The scheduling software identifies a processor on which to run a specific grid job that has been submitted by a user. Schedulers, along with load balancers, provide a function to route jobs to available resources based on SLAs, capabilities, and availability.

Selection Combining (SC) A combining technique in which one signal only (the best one in terms of the signal-to-noise ratio) is selected from the multiple received signals.

Self-organizing Map (SOM) or self-organizing feature map (SOFM) is a type of artificial neural network that is trained using unsupervised learning to produce a low-dimensional (typically two-dimensional), discretized representation of the input space of the training samples, called a map.

Service Level Agreement (SLA) A contractual agreement by which a service provider defines the level of service, responsibilities, priorities, and guarantees regarding availability, performance, and other aspects of the service.

Service-oriented Architecture A collection of services which communicate with each other. Services can be internal or external to the enterprise utilizing the service-oriented architecture.

Shannon Capacity is defined as the channel capacity as the maximum data rate that can be sent over the channel while ensuring a reliable communication.

SIFS Short Inter-Frame Space.

Single-point-of-failure The scenario that compromising one node can lead to the whole network dysfunctional.

Singlet Cache block in a valid coherence state which is the only on-chip copy.

Signal to Noise Ratio (SNR) SNR is the ratio of signal power to the noise power corrupting the signal.

Space-time Processing The processing of signals by the transmitter and/or the receiver in time and space (space here refers to different antennas at different locations).

Spatial Multiplexing Gain is achieved by transmitting different signals over spatially independent fading channels.

Spilling Sharing mechanism, also known as N-Chance forwarding technique, which allows sharing cache capacity among private caches. When a singlet is replaced in a cache it is forwarded to another one N times before being evicted from the chip.

Spyware A malware that is installed surreptitiously on a personal computer to collect information about the user, including their browsing habits without their informed consent.

SR Selective Repeat.

STBC Space Time Block Coding.

Store-carry-and-forward Bundle delivery paradigm that is used to overcome partitioning in a DTN-based network.

Superposition Transmission This refers to the transmission of more than one signal at the same time by adding them with different weights.

SW Stop and Wait.

Symbol Error Rate (SER) SER is the number of symbol errors due to the transmission channel per second.

TBP Ticket Based Probing.

TCP Transmission Control Protocol.

Time Difference of Arrival (TDOA) is a method to locate a receiver by measuring the time difference of a signal transmitted from three or more synchronised transmitters.

Time Division Multiple Access (TDMA) TDMA is a channel access method for shared medium networks. It allows several users to share the same frequency channel by dividing the signal into different time slots.

Time Synchronization is timekeeping which requires the coordination of events to operate a system in unison.

Tree-based Approach In tree-based approach, peers are organized into structures (typically trees) for delivering data, with each data packet being disseminated using the same structure. Nodes on the structure have well-defined relationships, for example, 'parent-child' relationships in trees. Such approaches are typically push-based, that is, when a node receives a data packet, it also forwards copies of the packet to each of its children.

TRG TRIggering.

Trojan A program with an overt (documented or known) effect and a covert (undocumented or unexpected) effect.

UDP User Datagram Protocol.

UMTS Universal Mobile Telecommunications System.

Utility Computing The technology, tools and processes that collectively deliver, monitor and manage IT as a service to users.

V2I Vehicle to Infrastructure.

V2V Vehicle to Vehicle.

VANET Vehicular Ad hoc Networks.

Vehicular-Delay Tolerant Networking (VDTN) A network architecture for vehicular communications based on the delay-tolerant network architecture, which integrates the concepts of end-to-end, asynchronous, and variable-length bundle oriented communication, Internet protocol over VDTN, and out-of-band signaling.

Virtualization Abstraction of key properties of a physical device and presentation of those properties to users of the device. In IT, this includes servers, network and disk and tape storage. It is most commonly implemented using a software layer between the physical device, or devices, and the application that wants to use it.

VS Validated Set.

Walsh Spreading Sequences are orthogonal sequences that are assigned to different users in CDMA system.

Wireless Ad Hoc Network is a decentralized wireless network. The network is ad hoc because it does not rely on a preexisting infrastructure, such as routers in wired networks or access points in managed (infrastructure) wireless networks. Instead, each node participates in routing by forwarding data for other nodes, and so the determination of which nodes forward data is made dynamically based on the network connectivity.

WLAN Wireless Local Area Network.

Worm Computer viruses which actively search for victims.

WSN Wireless Sensor Network.

Wyner Zive Coding (WZC) is a compression technique that gives the information-theoretic rate-distortion bounds for memoryless coding of directly observed data with side information at the decoder.

ZRP Zone Routing Protocol.

Index

Note: Figures and Tables are indicated by *italic page numbers*, terms in glossary by **emboldened numbers**

Cooperative Networking, First Edition. Edited by Mohammad S. Obaidat and Sudip Misra.
© 2011 John Wiley & Sons, Ltd. Published 2011 by John Wiley & Sons, Ltd.

WAdSNs (Wireless Ad Hoc Sensor Networks)
    (*continued*)
  middleware 46–8
  multi-agent systems in 48–50
  routing in 41–3
  SLC techniques 36–41
Walsh spreading sequences 198, **317**
wildlife tracking networks 103, 112
WiMax 7, 118
wireless ad hoc network(s) 154, **317** *see also*
    WAdSNs
WLANs (Wireless Local Area Networks) 7,
    118, 154
  authentication and access control in 68
Worminator 137, *138*, 142

worm(s) 134, **317**
  detection of *138*, 140–2
WRANs (Wireless Regional Area Networks)
    126
WSNs (Wireless Sensor Networks) 12, 192
  characteristics 13
  environment limitations 16
  hardware limitations 16
  network limitations 16
  security issues 15–19
WZC (Wyner–Ziv Coding) 197, **317**

zombies 18
ZRP (Zone Routing Protocol) 9

CPSIA information can be obtained at www.ICGtesting.com
Printed in the USA
267299BV00002B/1-84/P